T0260215

Ecological Models and Data in R

Ecological Models and Data in R

Benjamin M. Bolker

PRINCETON UNIVERSITY PRESS

PRINCETON AND OXFORD

In the United Kingdom: Princeton University Press,
3 Market Place, Woodstock, Oxfordshire OX20 1SY

Library of Congress Cataloging-in-Publication Data

Bolker, Benjamin M., 1967–
    Ecological models and data in R / Benjamin M. Bolker.
        p. cm.
    Includes bibliographical references and index.
    ISBN 978-0-691-12522-0 (alk. paper)
    1. Ecology—Statistical methods.   2. Ecology—Mathematical models.
3. Ecology—Simulation methods.   4. R (Computer program language)   I. Title.
    QH541.15.S72 B65 2008
    577.01'5118—dc22   2007060388

British Library Cataloging-in-Publication Data is available

This book has been composed in Sabon

pup.princeton.edu

10 9 8 7 6 5 4 3 2 1

Contents

Acknowledgments

Lots of people have helped me start and finish this book. I would like to thank:

The R community, for building such a useful tool.

Various research institutions that have provided me room to work over the years: the Department of Zoology at the University of Florida, the Mathematical Biosciences Institute, and the NERC Centre for Population Biology at Silwood Park.

All the students and colleagues who have brought me such interesting and challenging problems over the years, especially those who took the time to find errors or make suggestions: Jorge Ahumada, Chad Brassil, David Buck, Lin Cassidy, Lew Coggins, Rick Condit, Mick Crawley, Ian Dworkin, Ian Fiske, Shane Geange, Gregor Gorjanc, Tom Hobbs, Rico Holdo, Holly Kindsvater, Aaron MacNeil, Julien Martin, Mike McCoy, Jeremy Mendoza, Jess Metcalf, Toshinori Okuyama, Alfredo Rios, Stuart Sandin, Nat Seavy, Darren Shaw, Hank Stevens, Adrian Stier, Don Strong, Maria Uriarte, Denis Valle, Will Wilson, and several anonymous reviewers. I apologize if I forgot your name, or if your suggestions slipped through the cracks or were just too hard to implement.

For finding errors in the first printing: Ian Carroll, Jeanne DeNoyer, Aaron Ellison, Jack Finn, Nicole Gottdanker, Barney Luttbeg, Duncan Menge, Mike Meredith, Fernando Miguez, John Poulsen, Martin Schlather, Gavin Simpson, Fernando Taboada, Francois Teste, Eric Walters, and Will White.

Researchers who generously contributed their original data sets as examples: James D. Thomson, Russ Schmitt and Sally Holbrook, Mike Dodd and Jonathan Silvertown, James Vonesh, and Jackie Wilson.

My Ph.D. and postdoctoral advisers, Bryan Grenfell and Steve Pacala, for introducing me to mathematical and statistical ecology.

My parents, Joan and Ethan Bolker, for thorough and thoughtful comments on the manuscript and for establishing the expectation that writing books is just what people do.

And last but not least Aidan and especially Tara for their love, patience, and support.

Essentially, all models are wrong
but some are useful.

—*George Box*

Ecological Models and Data in R

1 Introduction and Background

This chapter gives a broad overview of the philosophy and techniques of ecological modeling. A small data set on seed removal illustrates the three most common frameworks for statistical modeling in ecology: frequentist, likelihood-based, and Bayesian. The chapter also reviews what you should know to get the most out of the book, discusses the R language, and spells out a step-by-step process for building models of ecological systems.

If you're impatient with philosophical discussion, you can read Section 1.4 and the R supplement at the end of the chapter and move on to Chapter 2.

1.1 Introduction

This book is about combining models with data to answer ecological questions. Pursuing this worthwhile goal will lead to topics ranging from basic statistics, to the cutting edge of modern statistics, to the nuts and bolts of computer programming, to the philosophy of science. Remember as we go along not to miss the ecological forest for the statistical trees; all of these complexities are in the service of answering ecological questions, and the most important thing is to keep your common sense about you and your focus on the biological questions you set out to answer. "Does this make sense?" and "What does this answer really mean?" are the two questions you should ask constantly. If you cannot answer them, back up to the last point you understood.

If you want to combine models with data, you need to use statistical tools. Ecological statistics has gotten much more complicated in the last few decades. Research papers in ecology now routinely refer to likelihood, Markov chain Monte Carlo, and other arcana. This new complexity arises from the explosion of cheap computing power, which allows us to run complicated tests quickly and easily—or at least more easily than before. But there is still a lot to know about how these tests work, which is what this book is about. The good news is that we can now develop statistical methods that directly answer our ecological questions, adapting statistics to the data rather than vice versa. Instead of asking "What is the probability of observing at least this much variability among the arcsine-square-root-transformed counts of seeds in different treatments?" we can ask "Is the number of seeds removed consistent with standard foraging theory, and what are the attack rates and handling times of predators? Do the attack rates or handling times increase with mean seed size? With the time that the seeds have been available? Is there evidence for variability among seeds?" By customizing statistical tests we can squeeze more information, and more relevant information, from expensive data. Building your own statistical tests is not easy, but it is really no harder than using any of the other tools ecologists have picked up in their ongoing effort to extract meaning from the natural world (stable isotope techniques, radiotelemetry, microsatellite population genetics, geographic information systems, otolith analysis, flow cytometry, mist netting ... you can probably identify several more from your own field). Custom statistical techniques are just another set of tools in the modern ecologist's toolbox; the information this book presents should show you how to use them on your own data, to answer your own questions.

For example, Sandin and Pacala (2005) combined population counts through time with remote underwater video monitoring to analyze how the density of reef fishes in the Caribbean affected their risk of predation. The classic approach to this problem would be to test for a significant correlation between density and mortality rate, or between density and predator activity. A positive correlation between prey population density and the number of observed predator visits or attacks would suggest that prey aggregations attract predators. If predator attacks on the prey population are proportional to population density, then the predation rate per prey *individual* will be independent of density; predator attacks would need to accelerate with increasing population density in order for predators to regulate the prey population. One could test for positive correlations between prey density and per capita mortality to see whether this is so.

However, correlation analysis assumes the data are bivariate normally distributed, while linear regression assumes a linear relationship between a predictor variable and a normally distributed response variable. Although one can sometimes transform data to satisfy these assumptions, or simply ignore minor violations, Sandin and Pacala took a more powerful approach: they built explicit models to describe how the absolute and per capita predator visits or mortality depended on prey population density. For example, the absolute mortality probability would be $r_0 + r_1 n$ and the per capita mortality probability would be $(r_0 + r_1 n)/n$ if predator visits are proportional to prey density. They also used realistic binomial and Poisson probability distributions to describe the variation in the data, rather than assuming normality (a particularly awkward assumption when there are lots of zeros in the data). By doing so, they were able to choose among a variety of possible models and conclude that predators induce *inverse* density dependence in this system (i.e., that smaller prey populations experience higher per capita mortality, because predators are present at relatively constant numbers independent of prey density). Because they fitted models rather than running classical statistical tests on transformed data, they were also able to estimate meaningful parameter values, such as the increase in predator visits per hour for every additional prey individual present. These values are more useful than p (significance) values, or than regression slopes from transformed data, because they express statistical information in ecological terms.

1.2 What This Book Is Not About

1.2.1 What You Should Already Know

To get the most out of the material presented here you should already have a good grasp of basic statistics, be comfortable with computers (e.g., have used Microsoft Excel to deal with data), and have some rusty calculus. But attitude and aptitude are more important than previous classroom experience. Getting into this material requires some hard work at the outset, but it will become easier as you brush up on basic concepts.*

STATISTICS

I assume that you've had the equivalent of a one-semester undergraduate statistics course. The phrases *hypothesis test, analysis of variance, linear regression, normal distribution* (maybe even *Central Limit Theorem*) should be familiar to you, even if you don't remember all of the details. The basics of experimental design—the meaning of and need for randomization, control, independence, and replication in setting up experiments, the idea of statistical power, and the concept of pseudoreplication (Hurlbert, 1984; Hargrove and Pickering, 1992; Heffner et al., 1996; Oksanen, 2001)—are essential tools for any working ecologist, but you can learn them from a good introductory statistics class or textbook such as Gotelli and Ellison (2004) or Quinn and Keough (2002).†

Further reading: If you need to review statistics, try Crawley (2002), Dalgaard (2003), or Gotelli and Ellison (2004). Gonick and Smith's 1993 *Cartoon Guide to Statistics* gives a gentle introduction to some basic concepts, but you will need to go beyond what they cover. Sokal and Rohlf (1995), Zar (1999), and Crawley (2005, 2007) cover a broader range of classical statistics. For experimental design, try Underwood (1996), Scheiner and Gurevitch (2001), or Quinn and Keough (2002) (the latter two discuss statistical analysis as well).

COMPUTERS

This book will teach you how to use computers to understand data. You will be writing a few lines of R code at a time rather than full-blown computer programs, but you will have to go beyond pointing and clicking. You need to be comfortable with computers, and with using spreadsheets like Excel to manipulate data. Familiarity with a mainstream statistics package like SPSS or SAS will be useful, although you

---

* After teaching with Hilborn and Mangel's excellent book *The Ecological Detective* (1997) I wanted to write a book that included enough nitty-gritty detail for students to tackle their own problems. If this book feels too hard for you, consider starting with *The Ecological Detective*—but consider reading *ED* in any case.

† Ideally, you would think about how you will analyze your data before you go into the field to collect it. This rarely happens. Fortunately, if your observations are adequately randomized, controlled, independent, and replicated, you will be able to do *something* with your data. If they aren't, no fancy statistical techniques can help you.

should definitely use R to work through this book instead of falling back on a familiar software package. (If you have used R already, you'll have a big head start.) You needn't have done any programming.

MATH

Having "rusty" calculus means knowing what a derivative and an integral are. While it would be handy to remember a few of the formulas for derivatives, a feeling for the meanings of logarithms, exponentials, derivatives, and integrals is more important than the formulas (you'll find the formulas in the appendix). In working through this book you will have to *use* algebra, as much as calculus, in a routine way to solve equations and answer questions. Most of the people who have taken my classes were very rusty when they started.

Further reading: Adler (2004) gives a very applied review of basic calculus, differential equations, and probability, while Neuhauser (2003) covers calculus in a more rigorous and traditional way, but still with a biological slant.

ECOLOGY

I have assumed you know some basic ecological concepts, since they are the foundation of ecological data analysis. You should be familiar, for example, with exponential and logistic growth from population ecology; functional responses from predator-prey ecology; and competitive exclusion from community ecology.

Further reading: For a short introduction to ecological theory, try Hastings (1997) or Vandermeer and Goldberg (2004) (the latter is more general). Gotelli (2001) is more detailed. Begon et al. (1996) gives an extremely thorough introduction to general ecology, including some basic ecological models. Case (1999) provides an illustrated treatment of theory, while Roughgarden (1997) integrates ecological theory with programming examples in MATLAB. Mangel (2006) and Otto and Day (2007), two new books, both give basic introductions to the "theoretical biologist's toolbox."

1.2.2 Other Kinds of Models

Ecologists sometimes want to "learn how to model" without knowing clearly what questions they hope the models will answer, and without knowing what kind of models might be useful. This is a bit like saying "I want to learn to do experiments" or "I want to learn molecular biology": Do you want to analyze microsatellites? Use RNA inactivation to knock out gene function? Sequence genomes? What people usually mean by "I want to learn how to model" is "I have heard that modeling is a powerful tool and I think it could tell me something about my system, but I'm not really sure what it can do."

Ecological modeling has many facets. This book covers only one: statistical modeling, with a bias toward mechanistic descriptions of ecological patterns. The next section briefly reviews a much broader range of modeling frameworks and gives some starting points in the modeling literature in case you want to learn more about other kinds of ecological models.

1.3 Frameworks for Modeling

This book is primarily about how to combine models with data and how to use them to discover the answers to theoretical or applied questions. To help fit statistical models into the larger picture, Table 1.1 presents a broad range of dichotomies that cover some of the kinds and uses of ecological models. The discussion of these dichotomies starts to draw in some of the statistical, mathematical, and ecological concepts I suggested you should know. However, if a few are unfamiliar, don't worry—the next few chapters will review the most important concepts. Part of the challenge of learning the material in this book is a chicken-and-egg problem: to know why certain technical details are important, you need to know the big picture, but the big picture itself involves knowing some of those technical details. Iterating, or cycling, is the best way to handle this problem. Most of the material introduced in this chapter will be covered in more detail in later chapters. If you don't completely get it this time around, hang on and see if it makes more sense the second time.

1.3.1 Scope and Approach

The first set of dichotomies in the table subdivides models into two categories, one (theoretical/strategic) that aims for general insight into the workings of ecological processes and one (applied/tactical) that aims to describe and predict how a particular system functions, often with the goal of forecasting or managing its behavior. Theoretical models are often mathematically difficult and ecologically oversimplified, which is the price of generality. Paradoxically, although theoretical models are defined in terms of precise numbers of individuals, because of their simplicity they are usually used only for qualitative predictions. Applied models are often mathematically simpler (although they can require complex computer code) but tend to capture more of the ecological complexity and quirkiness needed to make detailed predictions about a particular place and time. Because of this complexity their predictions are often less general.

The dichotomy of mathematical versus statistical modeling says more about the culture of modeling and how different disciplines go about thinking about models than about how we should actually model ecological systems. A mathematician is more likely to produce a deterministic, dynamic process model without thinking very much about noise and uncertainty (e.g., the ordinary differential equations that make up the Lotka-Volterra predator-prey model). A statistician, on the other hand, is more likely to produce a stochastic but static model that treats noise and uncertainty carefully but focuses more on static patterns than on the dynamic processes that produce them (e.g., linear regression).*

---

* Of course, both mathematicians and statisticians are capable of more sophisticated models than the simple examples given here.

TABLE 1.1
Modeling dichotomies

Scope and approach	
abstract	concrete
strategic	tactical
general	specific
theoretical	applied
qualitative	quantitative
descriptive	predictive
mathematical	statistical
mechanistic	phenomenological
pattern	process

Technical details	
analytical	computational
dynamic	static
continuous	discrete
population-based	individual-based
Eulerian	Lagrangian
deterministic	stochastic

Sophistication	
simple	complex
crude	sophisticated

Each column contrasts a different qualitative style of modeling. The loose association of descriptors in each column gets looser as you work downward.

The important difference between phenomenological (pattern) and mechanistic (process) models will be with us throughout the book. Phenomenological models concentrate on observed patterns in the data, using functions and distributions that are the right shape and/or sufficiently flexible to match them; mechanistic models are more concerned with the underlying processes, using functions and distributions based on theoretical expectations. As usual, shades of gray abound; the same function could be classified as either phenomenological or mechanistic depending on why it was chosen. For example, you could use the function $f(x) = ax/(b + x)$ (a Holling type II functional response) as a mechanistic model in a predator-prey context because you expected predators to attack prey at a constant rate and be constrained by handling time, or as a phenomenological model of population growth simply because you wanted a function that started at zero, was initially linear, and leveled off as it approached an asymptote (see Chapter 3). All other things being equal, mechanistic models are more powerful since they tell you about the underlying processes driving patterns. They are more likely to work correctly when extrapolating beyond the observed conditions. Finally, by making more assumptions, they allow you to extract more information from your data—with the risk of making the *wrong* assumptions.[*]

Examples of theoretical models include the Lotka-Volterra or Nicholson-Bailey predator-prey equations (Hastings, 1997); classical metapopulation models for single (Hanski, 1999) and multiple (Levins and Culver, 1971; Tilman, 1994) species; simple food web models (May, 1973; Cohen et al. 1990); and theoretical ecosystem models (Agren and Bosatta, 1996). Applied models include forestry and biogeochemical cycling models (Blanco et al. 2005), fisheries stock-recruitment models (Quinn and Deriso, 1999), and population viability analysis (Morris and Doak, 2002; Miller and Lacy, 2005).

Further reading: Books on ecological modeling overlap with those on ecological theory listed on p. 4. Other good sources include Nisbet and Gurney (1982; a well-written but challenging classic), Gurney and Nisbet (1998; a lighter version), Haefner (1996; broader, including physiological and ecosystem perspectives), Renshaw (1991; good coverage of stochastic models), Wilson (2000; simulation modeling in C), and Ellner and Guckenheimer (2006; dynamics of biological systems in general).

1.3.2 Technical Details

Another set of dichotomies characterizes models according to the methods used to analyze them or according to the decisions they embody about how to represent individuals, time, and space.

An analytical model is made up of equations solved with algebra and calculus. A computational model consists of a computer program which you run for a range of parameter values to see how it behaves.

Most mathematical models and a few statistical models are dynamic; the response variables at a particular time (the state of the system) feed back to affect the response variables in the future. Integrating dynamical and statistical models is challenging (see Chapter 11). Most statistical models are static; the relationship between predictor and response variables is fixed.

One can specify how models represent the passage of time or the structure of space (both can be continuous or discrete); whether they track continuous population densities (or biomass or carbon densities) or discrete individuals; whether they consider individuals within a species to be equivalent or divide them by age, size, genotype, or past experience; and whether they track the properties of individuals

---

[*] For an alternative, classic approach to the tradeoffs between different kinds of models, see Levins (1966) (criticized by Orzack and Sober (1993); Levins's (1993) defense invokes the fluidity of model-building in ecology).

(individual-based or Eulerian) or the number of individuals within different categories (population-based or Lagrangian).

Deterministic models represent only the average, expected behavior of a system in the absence of random variation, while stochastic models incorporate noise or randomness in some way. A purely deterministic model allows only for qualitative comparisons with real systems; since the model will never match the data *exactly*, how can you tell if it matches closely enough? For example, a deterministic food web model might predict that introducing pike to a lake would cause a trophic cascade, decreasing the density of phytoplankton (because pike prey on sunfish, which eat zooplankton, which in turn consume phytoplankton); it might even predict the expected magnitude of the change. To test this prediction with real data, however, you would need some kind of statistical model to estimate the magnitude of the average change in several lakes (and the uncertainty), and to distinguish between observed changes due to pike introduction and those due to other causes (measurement error, seasonal variation, weather, nutrient dynamics, population cycles, etc.).

Most ecological models incorporate stochasticity crudely, by simply assuming that there is some kind of (perhaps normally distributed) variation, arising from a combination of unknown factors, and estimating the magnitude of that variation from the variation observed in the field. We will go beyond this approach, specifying different sources of variability and something about their expected distributions. More sophisticated models of variability enjoy some of the advantages of mechanistic models: models that make explicit assumptions about the underlying causes of variability can both provide more information about the ecological processes at work and get more out of your data.

There are essentially three kinds of random variability:

- *Measurement error* is the variability imposed by our imperfect observation of the world; it is always present, except perhaps when we are counting a small number of easily detected organisms. It is usually modeled by the standard approach of adding normally distributed variability around a mean value.
- *Demographic stochasticity* is the innate variability in outcomes due to random processes even among otherwise identical units. In experimental trials where you flip a coin 20 times you might get 10 heads, or 9, or 11, even though you're flipping the same coin the same way each time. Likewise, the number of tadpoles out of an initial cohort of 20 eaten by predators in a set amount of time will vary between experiments. Even if we controlled everything about the environment and genotype of the predators and prey, we would still see different numbers dying in each run of the experiment.
- *Environmental stochasticity* is variability imposed from "outside" the ecological system, such as climatic, seasonal, or topographic variation. We usually reserve environmental stochasticity for unpredictable variability, as opposed to predictable changes (such as seasonal or latitudinal changes in temperature) which we can incorporate into our models in a deterministic way.

The latter two categories, demographic and environmental stochasticity, make up *process variability*,* which, unlike measurement error, affects the future dynamics of the ecological system. (Suppose we expect to find three individuals on an isolated

---

* Process variability is also called *process noise* or *process error* (Chapter 10).

island. If we make a measurement error and measure zero instead of three, we may go back at some time in the future and still find them. If an unexpected predator eats all three individuals (process variability), and no immigrants arrive, any future observations will find no individuals.) The conceptual distinction between process and measurement error is most important in dynamic models, where the process error has a chance to feed back on the dynamics.

The distinctions between stochastic and deterministic effects, and between demographic and environmental variability, are really a matter of definition. Until you get down to the quantum level, any "random" variability can in principle be explained and predicted. What determines whether a tossed coin will land heads-up? Its starting orientation and the number of times it turns in the air, which depends on how hard you toss it (Keller, 1986). What determines exactly which and how many seedlings of a cohort die? The amount of energy with which their mother provisions the seeds, their individual light and nutrient environments, and encounters with pathogens and herbivores. Variation that drives mortality in seedlings—e.g., variation in available carbohydrates among individuals because of small-scale variation in light availability—might be treated as a random variable by a forester at the same time that it is treated as a deterministic function of light availability by a physiological ecologist measuring the same plants. Climatic variation is random to an ecologist (at least on short time scales) but might be deterministic, although chaotically unpredictable, to a meteorologist. Similarly, the distinction between demographic variation, internal to the system, and environmental variation, external to the system, varies according to the focus of a study. Is the variation in the number of trees that die every year an internal property of the variability in the population or does it depend on an external climatic variable that is modeled as random noise?

1.3.3 Sophistication

I want to make one final distinction, between simple and complex models and between crude and sophisticated ones. One could quantify simplicity versus complexity by the length of the description of the analysis or by the number of lines of computer script or code required to implement a model. Crudity and sophistication are harder to recognize; they represent the conceptual depth, or the amount of *hidden* complexity, involved in a model or statistical approach. For example, a computer model that picks random numbers to determine when individuals give birth and die and keeps track of the total population size, for particular values of the birth and death rates and starting population size, is simple and crude. Even simpler, but far more sophisticated, is the mathematical theory of random walks (Okubo, 1980) which describes the same system but—at the cost of challenging mathematics— predicts its behavior for *any* birth and death rates and any starting population sizes. A statistical model that searches at random for the line that minimizes the sum of squared deviations of the data is crude and simple; the theory of linear models, which involves more mathematics, does the same thing in a more powerful and general way. Computer programs, too, can be either crude or sophisticated. One can pick numbers from a binomial distribution by virtually flipping the right number of coins and seeing how many come up heads, or by using numerical methods that arrive at the same result far more efficiently. A simple R command like `rbinom`, which picks random binomial deviates, hides a lot of complexity.

The value of sophistication is generality, simplicity, and power; its costs are opacity and conceptual and mathematical difficulty. In this book, I will take advantage of many of R's sophisticated tools for optimization and random number generation (since in this context it's more important to have these tools available than to learn the details of how they work), but I will avoid many of its sophisticated statistical tools, so that you can learn from the ground up how statistical models really work and make your models work the way you want them to rather than being constrained by existing frameworks. Having reinvented the wheel, however, we'll briefly revisit some standard statistical frameworks like generalized linear models and see how they can solve some problems more efficiently.

1.4 Frameworks for Statistical Inference

This section will explore three different ways of drawing statistical conclusions from data—frequentist, Bayesian, and likelihood-based. While the differences among these frameworks are sometimes controversial, most modern statisticians know them all and use whatever tools they need to get the job done; this book will teach you the details of those tools, and the distinctions among them.

To illustrate the ideas I'll draw on a seed predation data set from Duncan and Duncan (2000) that quantifies how many times seeds of two different species disappeared (presumably taken by seed predators, although we can't be sure) from observation stations in Kibale National Park, Uganda. The two species (actually the smallest- and largest-seeded species of a set of eight species) are *Polyscias fulva* (pol: seed mass < 0.01 g) and *Pseudospondias microcarpa* (psd: seed mass ≈ 50 g).

1.4.1 Classical Frequentist

Classical statistics, which are part of the broader *frequentist* paradigm, are the kind of statistics typically presented in introductory statistics classes. For a specific experimental procedure (such as drawing cards or flipping coins), you calculate the probability of a particular outcome, which is defined as *the long-run average frequency of that outcome in a sequence of repeated experiments*. Next you calculate a *p-value*, defined as the probability of that outcome *or any more extreme outcome* given a specified null hypothesis. If this so-called *tail probability* is small, then you reject the null hypothesis; otherwise, you fail to reject it. But you don't accept the null hypothesis if the tail probability is large; you just fail to reject it.

The frequentist approach to statistics (due to Fisher, Neyman, and Pearson) is useful and very widely used, but it has some serious drawbacks—which are repeatedly pointed out by proponents of other statistical frameworks (Berger and Berry, 1988). It relies on the probability of a series of outcomes that didn't happen (the tail probabilities), and which depend on the way the experiment is defined; its definition of probability depends on a series of hypothetical repeated experiments that are often impossible in any practical sense; and it tempts us to construct straw-man

TABLE 1.2
Seed removal data

	pol	psd
Any taken (*t*)	26	25
None taken	184	706
Total (*N*)	210	731

null hypotheses and make convoluted arguments about why we have failed to reject them. Probably the most criticized aspect of frequentist statistics is their reliance on *p*-values, which when misused (as frequently occurs) are poor tools for scientific inference. To abuse *p*-values seems to be human nature; we act as though alternative hypotheses (which are usually what we're really interested in) are "true" if we can reject the null hypothesis with $p < 0.05$ and "false" if we can't. In fact, when the null hypothesis is true we still find $p < 0.05$ one time in twenty (we falsely reject the null hypothesis 5% of the time, by definition). If $p > 0.05$, the null hypothesis could still be false but we have insufficient data to reject it. We could also reject the null hypothesis in cases where we have lots of data, even though the results are biologically insignificant—that is, if the estimated effect size is ecologically irrelevant (e.g., a 0.01% increase in plant growth rate with a 30°C increase in temperature). More fundamentally, if we use a so-called *point null hypothesis* (such as "the slope of the relationship between plant productivity and temperature is zero"), common sense tells us that the null hypothesis *must* be false, because it can't be exactly zero—which makes the *p*-value into a statement about whether we have enough data to detect a nonzero slope, rather than about whether the slope is actually different from zero. Working statisticians will tell you that it is better to focus on estimating the values of biologically meaningful parameters and finding their confidence limits rather than worrying too much about whether *p* is greater or less than 0.05 (Yoccoz, 1991; Johnson, 1999; Osenberg et al. 2002)—although Stephens et al. (2005) remind us that hypothesis testing can still be useful.

Looking at the seed data, we have a 2×2 table (Table 1.2). If t_i is the number of times that species *i* seeds disappear and N_i is the total number of observations of species *i*, then the observed proportions of the time that seeds disappeared for each species are (pol) $t_1/N_1 = 0.124$ and (psd) $t_2/N_2 = 0.034$. The overall proportion taken (which is not the average of the two proportions since the total numbers of observations for each species differ) is $(t_1 + t_2)/(N_1 + N_2) = 0.054$. The ratio of the predation probabilities (proportion for pol/proportion for psd) is $0.124/0.034 = 3.62$. The ecological question we want to answer is "Is there differential predation on the seeds on these two species?" (Given the sample sizes and the size of the observed difference, what do you think? Do you think the answer is likely to be statistically significant? How about biologically significant? What assumptions or preconceptions does your answer depend on?)

A frequentist would translate this biological question into statistics as "What is the probability that I would observe a result this extreme, or more extreme, given the sampling procedure?" More specifically, "What proportion of possible outcomes would result in observed ratios of proportions greater than 3.62 *or*

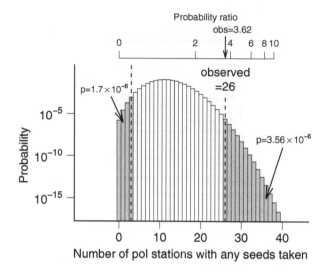

Figure 1.1 Classical frequentist analysis. Fisher's exact test calculates the probability of a given number of pol stations having seeds taken under the null hypothesis that both species have the same predation probability. The total probability that as many or more pol stations had seeds taken, *or that the difference was more extreme in the other direction,* is the two-tailed frequentist *p*-value ($3.56 \times 10^{-6} + 1.70 \times 10^{-6} = 5.26 \times 10^{-6}$). The top axis shows the equivalent in seed predation probability ratios. (*Note*: I put the *y*-axis on a log scale because the tails of the curve are otherwise too small to see, even though this change means that the area under the curve no longer represents the total probability.)

smaller than $1/3.62 = 0.276$?" (Figure 1.1). Fisher's exact test (`fisher.test` in R) calculates this probability, as a one-tailed test (proportion of outcomes with ratios greater than 3.62) or a two-tailed test (proportion with ratios greater than 3.62 or less than its reciprocal, 0.276); the two-tailed answer in this case is 5.26×10^{-6}. According to Fisher's original interpretation, this number represents the strength of evidence against the null hypothesis, or (loosely speaking) for the alternative hypothesis—that there is a difference in seed predation rates. According to the Neyman-Pearson decision rule, if we had set our acceptance cutoff at $\alpha = 0.05$, we could conclude that there was a *statistically significant* difference in predation rates.

We needn't fixate on *p*-values: the R command for Fisher's test, `fisher.test`, also tells us the 95% confidence limits for the difference between rates.* In terms of probability ratios, this example gives (2.073, 6.057), which as expected does not include 1. Do you think a range of a 107% to a 506% increase in seed predation probability[†] is significant?

---

* R expresses the difference in predation rates in terms of the *odds ratio*—if there are t_1 seeds taken and $N_1 - t_1$ seeds not taken for species 1, then the odds of a seed being taken are $t_1/(N_1 - t_1)$ and the odds ratio between the species is $(t_1/(N_1 - t_1))/(t_2/(N_2 - t_2))$. The odds ratio and its logarithm (the *logit* or log-odds ratio) have nice statistical properties.

[†] These values are the confidence limits on the probability ratios, minus 1, converted into percentages: for example, a probability ratio of 1.1 would represent a 10% increase in predation.

1.4.2 Likelihood

Most of the book will focus on frequentist statistics, but not the standard version that you may be used to. Most modern statistics uses an approach called *maximum likelihood estimation*, or approximations to it. For a particular statistical model, maximum likelihood finds the set of parameters (e.g., seed removal rates) *that makes the observed data* (e.g., the particular outcomes of predation trials) *most likely to have occurred*. Based on a model for both the deterministic and stochastic aspects of the data, we can compute the *likelihood* (the probability of the observed outcome) given a particular choice of parameters. We then find the set of parameters that makes the likelihood as large as possible, and take the resulting *maximum likelihood estimates* (MLEs) as our best guess at the parameters. So far we haven't assumed any particular definition of probability of the parameters. We could decide on confidence limits by choosing a likelihood-based cutoff, for example, by saying that any parameters that make the probability of the observed outcomes at least one-tenth as likely as the maximum likelihood are "reasonable." For mathematical convenience, we often work with the logarithm of the likelihood (the *log-likelihood*) instead of the likelihood; the parameters that give the maximum log-likelihood also give the maximum likelihood. On the log scale, statisticians have suggested a cutoff of 2 log-likelihood units (Edwards, 1992), meaning that we consider any parameter reasonable that makes the observed data at least $e^{-2} \approx 1/7.4 = 14\%$ as likely as the maximum likelihood.

However, most modelers add a frequentist interpretation to likelihoods, using a mathematical proof that says that, across the hypothetical repeated trials of the frequentist approach, the distribution of the negative logarithm of the likelihood itself follows a χ^2 (chi-squared) distribution.[*] This fact means that we can set a cutoff for differences in log-likelihoods based on the 95th percentile of the χ^2 distribution, which corresponds to 1.92 log-likelihood units, or parameters that lower the likelihood by a factor of $e^{1.92} = 6.82$. The theory says that the estimated value of the parameter will fall farther away than that from the true value only 5% of the time in a long series of repeated experiments. This rule is called the *Likelihood Ratio Test* (LRT).[†] We will see that it lets us both estimate confidence limits for parameters and choose between competing models.

Bayesians (discussed below) also use the likelihood—it is part of the recipe for computing the posterior distribution—but they take it as a measure of the information we can gain from the data, without saying anything about what the distribution of the likelihood would be in repeated trials.

How would one apply maximum likelihood estimation to the seed predation example? Lumping all the data from both species together at first, and assuming that (1) all observations are independent of each other and (2) the probability of at least one seed being taken is the same for all observations, it follows that the number of times at least one seed is removed is *binomially* distributed (we'll get to the formulas in Chapter 4). Now we want to know how the probability of observing the data (the likelihood $\mathcal{L}$) depends on the probability p_s that at least one seed was

---

[*] This result holds in the *asymptotic* case where we have lots of data, which happens less than we would like—but we often gloss over the fact of limited data and use it anyway.

[†] The difference between log-likelihoods is equivalent to the ratio of likelihoods.

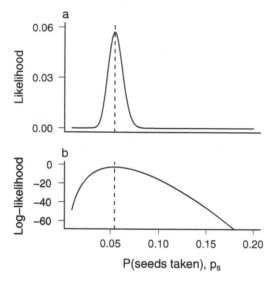

Figure 1.2 Likelihood and log-likelihood curves for removal probability p. Both curves have their maxima at the same point ($p_s = 0.054$). Log-likelihoods are based on natural ($\log_e$ or ln) logarithms.

taken from a particular station by a predator,* and what value of p_s maximizes the likelihood. The likelihood $\mathcal{L}$ is the probability that seeds were taken in 51 out of the total of 941 observations. This probability varies as a function of p_s (Figure 1.2): for $p_s = 0.05$, $\mathcal{L} = 0.048$, while for $p = 0.04$, $\mathcal{L}$ is only 6.16×10^{-3}. As it turns out, the MLE for the probability that seeds were taken in any one trial (p_s) is exactly what we'd expect—51/941, or 0.054—and the likelihood is $\mathcal{L} = 0.057$. (This likelihood is small, but it just means that the probability of any *particular* outcome—seeds being taken in 51 trials rather than 50 or 52—is small.)

To answer the questions that really concern us about the different predation probabilities for different species, we need to allow different probabilities for each species, and see how much better we can do (how much higher the likelihood is) with this more complex model. Now we take the separate values for each species (26 out of 210 and 25 out of 731) and, with a different per-observation probability for each species, compute the likelihoods of each species' data and multiply them (see Chapter 4 for basic probability calculations) or add the log-likelihoods. If we define the model in terms of the probability for psd and the ratio of the probabilities, we can plot a *likelihood profile* for the maximum likelihood we can get for a given value of the ratio (Figure 1.3).

The conclusions from this frequentist, maximum-likelihood analysis are essentially identical to those of the classical frequentist (Fisher's exact test) analyses. The maximum-likelihood estimate equals the observed ratio of the probabilities,

---

* One of the most confusing things about maximum likelihood estimation is that there are so many different probabilities floating around. The likelihood $\mathcal{L}$ is the probability of observing the complete data set (i.e., Prob(seeds were taken 51 times out of 941 observations)); p_s is the probability that seeds were taken in any given trial; and the (one-tailed) frequentist p-value is the probability, given a particular value of p_s, that seeds were taken 51 *or more* times out of 941 observations.

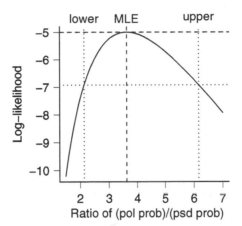

Figure 1.3 Likelihood curve for the ratio of the predation probabilities, showing the maximum likelihood estimate and 95% confidence limits. The null hypothesis value (ratio equal to 1) is just below the lower limit of the horizontal axis.

3.62; the confidence limits are (2.13, 6.16), which do not include 1; and the LRT-based p-value for rejecting the null hypothesis that the probabilities are the same is 3.83×10^{-6}.

Likelihood and classical frequentist analysis share the same philosophical underpinnings. Likelihood analysis is really a particular flavor of frequentist analysis, one that focuses on writing down a likelihood model and then testing for significant differences in the likelihood ratio rather than applying frequentist statistics directly to the observed outcomes. Classical analyses are usually easier because they are built into common statistics packages, and they may make fewer assumptions than likelihood analyses (e.g., Fisher's test is exact while the LRT is valid only for large data sets), but likelihood analyses are often better matched with ecological questions.

1.4.3 Bayesian

Frequentist statistics assumes that there is a "true" state of the world (e.g., the ratio of the species' predation probabilities) which gives rise to a distribution of possible experimental outcomes. The Bayesian framework says instead that the experimental outcome—what we actually saw happen—is the truth, while the parameter values or hypotheses have probability distributions. The Bayesian framework solves many of the conceptual problems of frequentist statistics: answers depend on what we actually saw and not on a range of hypothetical outcomes, and we can legitimately make statements about the probability of different hypotheses or parameter values.

The major fly in the ointment of Bayesian statistics is that in order to make it work we have to specify our *prior beliefs* about the probability of different hypotheses, and these prior beliefs actually affect our answers! One hard-core frequentist ecologist says "Bayesianism means never having to say you're wrong" (Dennis, 1996). It is indeed possible to cheat in Bayesian statistics by setting unreasonably strong priors.* The standard solution to the problem of subjectivity is to assume you are

---

* But if you really want to cheat with statistics you can do it in any framework!

completely ignorant before the experiment (setting a *flat prior*, or "letting the data speak for themselves"), although for technical reasons this isn't always possible. For better or worse, Bayesian statistics operates in the same way as we typically do science: we downweight observations that are too inconsistent with our current beliefs, while using those in line with our current beliefs to strengthen and sharpen those beliefs (statisticians are divided on whether this is good or bad).

The big advantages of Bayesian statistics, besides ease of interpretation, come (1) when we actually have data from prior observations we want to incorporate; (2) in complex models with missing data and several layers of variability; (3) when we are trying to make management decisions based on our data (the Bayesian framework makes it easier to incorporate the effect of unlikely but catastrophic scenarios in decision making). The only big disadvantage (besides the problem of priors) is that problems of small to medium complexity are actually harder with Bayesian approaches than with frequentist approaches—at least in part because most statistical software is geared toward classical statistics.

How would Bayesians answer our question about predation rates? First of all, they would say (without looking at the data) that the answer is "yes"—the true difference between predation rates is certainly not zero. (This discrepancy reflects the difference in perspective between frequentists, who believe that the true value is a fixed number and uncertainty lies in what you observe [or might have observed], and Bayesians, who believe that observations are fixed numbers and the true values are uncertain.) Then they might define a parameter, the ratio of the two proportions, and ask questions about the *posterior distribution* of that parameter—our best estimate of the probability distribution given the observed data and some prior knowledge of its distribution (see Chapter 4). What is the mode (most probable value) of that distribution? What is its expected value, or mean? What is the *credible interval*, which is the interval with equal probability cutoffs below and above the mean within which 95% of the probability falls?

The Bayesian answers, in a nutshell: when using a flat prior distribution, the posterior mode is 3.48 (near the observed proportion of 3.62). The posterior mean is 3.87, slightly larger than the posterior mode since the posterior probability density is slightly asymmetric—the density is skewed to the right (Figure 1.4).* The 95% credible interval, from 2.01 to 6.01, doesn't include 1, so Bayesians would say that there was good evidence against the hypothesis: even more strongly, they could say that the posterior probability that the predation ratio is greater than 1 is 0.998 (the probability that it is less than 1 is 0.002).

If the details of Bayesian statistics aren't perfectly clear at this point, don't worry. We'll explore Bayes' Rule and revisit Bayesian statistics in future chapters.

In this example all three statistical frameworks gave very similar answers, but they don't always. Ecological statisticians are still hotly debating which framework is best, or whether there is a single best framework. While it is important to be clear on the differences among the approaches, and knowing what question each is trying to answer, statisticians commonly move back and forth among them. My own approach is eclectic, agreeing with the advice of Crome (1997) and Stephens et al. (2005) to

---

* While Figure 1.1 showed the probability of each possible discrete outcome (number of seeds taken), Figure 1.4 shows a posterior probability *density* of a continuous parameter, i.e. the relative probability that the parameter lies in a particular range. Chapter 4 will explain this distinction more carefully.

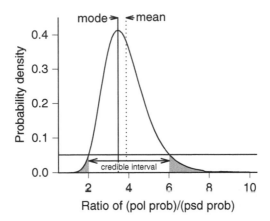

Figure 1.4 Bayesian analysis of seed predation. We calculate the probability density of the ratio of proportions of seeds taken being equal to some particular value, based on our prior (flat, assuming perfect ignorance—and in this case *improper* because it doesn't integrate to 1 [Chapter 4]) and on the data. The most probable value is the mode; the expected value is the mean. The shaded areas contain 5% of the area under the curve and cut off at the same height (probability density); the range between them is therefore the 95% credible interval.

try to understand the strengths and weaknesses of several different approaches and use each one as appropriate.

We will revisit these frameworks in more detail later. Chapter 4 will cover Bayes' Rule, which underpins Bayesian statistics; Chapters 6 and 7 will return to a much more detailed look at the practical details of maximum likelihood and Bayesian analysis. (Textbooks like Dalgaard (2003) cover classical frequentist approaches very well.)

1.5 Frameworks for Computing

To construct your own models, you will need to learn some of the basics of statistical computing. There are many computer languages and modeling tools with built-in statistical libraries (MATLAB, Mathematica) and several statistics packages with serious programming capabilities (SAS, IDL). We will use a system called R, that is both a statistics package and a computing language.

1.5.1 What Is R?

R's developers call it a "language and environment for statistical computing and graphics." This awkward phrase gets at the idea that R is more than just a statistics package. R is closest in spirit to other higher-level modeling languages like MATLAB or MathCAD. It is a dialect of the S computing language, which was written at Bell Labs in the 1980s as a research tool in statistical computing. MathSoft, Inc. (now Insightful Corporation), bought the rights to S and developed it into S-PLUS, a commercial package with a graphical front end. In the 1990s two New Zealand statisticians, Ross Ihaka and Robert Gentleman, rewrote S from scratch, again as a research project. The rewritten (and free) version became immensely popular and is now maintained by an international "core team" of about a dozen well-respected statisticians and computer scientists.

1.5.2 Why Use R?

R is an extremely powerful tool. It is a full-fledged modern computer language with sophisticated data structures; it supports a wide range of computations and statistical procedures; it can produce graphics ranging from exploratory plots to customized publication-quality graphics.

R is free in the sense that you can download it from the Internet, make as many copies as you want, and give them away.* While I don't begrudge spending money on software for research, it is certainly convenient not to have to pay—or to deal with licensing paperwork. This cheapness is vital, rather than convenient, for teachers, independent researchers, people in less-developed countries, and students who are frustrated with limited student versions (or pirated versions) of commercial software.

More important, R is also free in the sense that you can inspect any of the code and change it in any way that you want.† This form of freedom is probably abstract to you at this point—you probably won't need to modify R in the course of your modeling career—but it is a part of the same basic philosophy of the free exchange of information that underlies scientific and academic research in general.

R is the choice of many academic and industrial statisticians, who work to improve it and to write extension packages. If a statistical method has made it into print, the odds are good that there's an R package somewhere that implements it.

R runs well on many computer platforms, including the "big three" (Microsoft Windows, Mac OS X, and Linux). There are only tiny, mostly cosmetic differences in the way that R runs on different machines. You can nearly always move data files and code between operating systems and get the same answers.

R is rapidly gaining popularity. The odds are good that someone in your organization is using R, and there are many resources on the Internet including a very active mailing list. A growing number of introductory books use R (Dalgaard, 2003; Verzani, 2005; Crawley, 2005). There are also books of examples (Maindonald and Braun, 2003; Heiberger and Holland, 2004; Everitt and Hothorn, 2006), more advanced and encyclopedic books covering a range of statistical approaches (Venables and Ripley, 2002; Crawley, 2002), and books on specific topics such as regression analysis (Fox, 2002; Faraway, 2004), mixed-effect models (Pinheiro and Bates, 2000), phylogenetics (Paradis, 2006), and generalized additive models (Wood, 2006) that are geared toward R and S-PLUS users.

1.5.3 Why Not Use R?

R is more difficult than mainstream statistics packages like SYSTAT or SPSS, because it does much more. It would be hard to squeeze all of R's capabilities into a

---

* In programming circles, this freedom is called "gratis" or "free as in beer."
† "Libre" or "free as in speech."

simple graphical user interface (GUI) with menus to guide you through the process of analyzing your data. R's creators haven't even tried very hard to write a GUI, because they have a do-it-yourself philosophy that emphasizes knowing procedures rather than letting the program try to tell you what to do next. John Fox has written a simple GUI for R (called Rcmdr), and the commercial version of R, S-PLUS, does have a graphical user interface—if you can afford it. However, for most of what we will be doing in this book a GUI would not be very useful.

While R comes with a lot of documentation, it's mostly good for reminding you of the syntax of a command rather than for finding out how to do something. Unlike SAS, for which you can buy voluminous manuals that tell you the details of various statistical procedures and how to run them in SAS, R typically assumes that you have a general knowledge of the procedure you want to use and can figure out how to make it work in R by reading the online documentation or a separately published book (including this one).

R is slower than so-called lower-level languages like C and FORTRAN because it is an *interpreted* language that processes strings of commands typed in at the command line or stored in a text file, rather than a *compiled* language that first translates commands into machine code. However, computers are so fast these days that there's speed to burn. For most problems you will encounter the limiting factor will be how fast and easily you can write (and debug) the code, not how long the computer takes to process it. Interpreted languages make writing and debugging faster.

R is memory-hungry. Unlike SAS, which was developed with a metaphor of punch cards being processed one at a time, R tries to operate on the whole data set at once. If you are lucky enough to have a gigantic data set, with hundreds of thousands of observations or more, you will need to find ways (such as using R's capability to connect directly to database software) to do your analysis in chunks rather than loading it all into memory at once.

Unlike some other software such as Maple or Mathematica, R can't do *symbolic* calculation. For example, it can't tell you that the integral of x^2 is $x^3/3 + C$, although it can compute some simple derivatives (using the deriv or D function).

No commercial organization supports R—which may not matter as much as you think. The largest software company in the world supports Microsoft Excel, but Excel's statistical procedures are notoriously unreliable (McCullough and Wilson, 2005). On the other hand, the community of researchers who build and use R are among the best in the world, and R compares well with commercial software (Keeling and Pavur, 2007). While every piece of software has bugs, the core components of R have been used so extensively by so many people that the chances of your finding a bug in R are about the same as the chances of finding a bug in a commercial software package like SAS or SPSS—and if you do find one and report it, it will probably be fixed within a few days.

It is certainly possible to do the kinds of modeling presented in this book with other computing platforms—particularly MATLAB (with appropriate toolboxes), Mathematica, SAS (using the macro language), Excel in combination with Visual Basic, and lower-level languages such as Delphi, Java, C, or FORTRAN. However, I have found R's combination of flexibility, power, and cost make it the best—although not the only—option for statistical modeling in ecology.

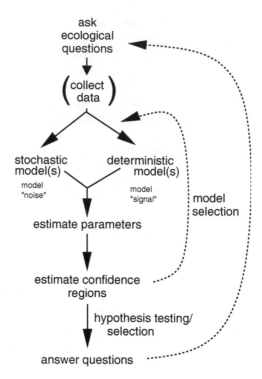

Figure 1.5 Flow of the modeling process.

1.6 Outline of the Modeling Process

After all these caveats and admonitions and before jumping into the nitty-gritty details of modeling particular data, we need an outline or road map of the modeling process (Figure 1.5).

1. **Identify the ecological question.** You have to know what you want to find out before you can start trying to model. You should know what your question is both at a general, conceptual level ("Does disease select against cannibalism in tiger salamander populations?") and at a specific level ("What is the percentage difference in probability of becoming a cannibal for tiger salamander individuals taken from populations *A* and *B*?"). Practice switching back and forth between these two levels. Being either too vague ("I want to explore the population genetics of cannibalism") or too specific ("What is the difference in the intercepts of these two linear regressions?") can impede your progress. Ultimately, knowing how to ask good questions is one of the fundamental skills for any ecologist, or indeed any scientist, and (unfortunately) no recipe can tell you how to do it. Even though I can't teach you to ask good questions, I included it in the list because it is the

first and most important step of any analysis and motivates all the other steps.*

2. **Choose deterministic model(s).** Next, you need to choose a particular mathematical description of the pattern you are trying to describe. The *deterministic* part is the average, or expected pattern in the absence of any kind of randomness or measurement error. It's tempting to call this an "ecological" model, since traditional ecological models are described in deterministic terms, but ecological models can be either deterministic or stochastic.

   The deterministic model can be phenomenological (as simple as "predator density is a linear function of prey density, or $P = a + bV$"); mechanistic (e.g., a type II functional response for predation rate); or even a complex individual-based simulation model. Chapter 3 will remind you of, or introduce you to, a broad range of mathematical models that are useful building blocks for a deterministic model, and provide general tools for getting acquainted with the mathematical properties of deterministic models.

3. **Choose stochastic model(s).** To estimate the parameters of a model, you need to know not just the expected pattern but also something about the variation around the expected pattern. Typically, you describe the stochastic model by specifying a reasonable *probability distribution* for the variation. For example, we often assume that variation that comes from measurement error is normally distributed, while variation in the number of plants found in a quadrat of a specific size is Poisson distributed. Ecologists tend to be less familiar with stochastic building blocks (e.g., the negative binomial or Gamma distributions) than with deterministic building blocks (e.g., linear or Michaelis-Menten functions). The former are frequently covered in the first week of introductory statistics courses and then forgotten as you learn standard statistical methods. Chapter 4 will (re)introduce some basics of probability as well as a wide range of probability distributions useful in building stochastic models.

4. **Fit parameters.** Once you have defined your model, you can estimate both the deterministic parameters (slope, attack rate, handling time, etc.) and stochastic parameters (e.g., the variance or parameters controlling the variance). This step is a purely technical exercise in figuring out how to get the computer to fit the model to the data. Unlike the previous steps, it provides no particular insight into the basic ecological questions. The fitting step does require ecological insight both as input (for most fitting procedures, you must start with some order-of-magnitude idea of reasonable parameter values) and output (the fitted parameters are essentially the answers to your ecological question). Chapters 6 and 7 will go into great detail about the practical aspects of fitting: the basic methods, how to make them work in R, and troubleshooting tips.

5. **Estimate confidence intervals/test hypotheses/select models.** You need to know more than just the best-fit parameters of the model (the *point estimates*, in statistical jargon). Without some measurement of uncertainty, such estimates

---

* In an ideal world, you would identify ecological questions before you designed your experiments and gathered data (!), but in this book I will assume you've already got data (either your own or someone else's) to work with and think about.

are meaningless. By quantifying the uncertainty in the fit of a model, you can estimate confidence limits for the parameters. You can also test ecological hypotheses, from both an ecological and a statistical point of view (e.g., can we tell the difference statistically between the handling times on two different prey types? are these differences large enough to make any practical difference in the population dynamics?). You also need to quantify uncertainty in order to choose the best out of a set of competing models, or to decide how to weight the predictions of different models. All of these procedures—estimating confidence limits, testing the differences between parameters in two models or between a parameter and a null-hypothesis value such as zero, and testing whether one model is significantly better than another—are closely related aspects of the modeling process that we will discuss in Chapter 6.

6. **Put the results together to answer questions/return to step #1.** Modeling is an iterative process. You may have answered your questions with a single pass through steps 1–5, but it is far more likely that estimating parameters and confidence limits will force you to redefine your models (changing their form or complexity or the ecological covariates they take into account) or even to redefine your original ecological questions. You may need to ask different questions, or collect another set of data, to further understand how your system works. Like the first step, this final step is a bit more free-form and general, but there are tools (the Likelihood Ratio test, model selection) that will help (Chapter 6).

I use this approach for modeling ecological systems every day. It answers ecological questions and, more important, it shapes the way I think about data and about those ecological questions. A growing number of studies in ecology use simple but realistic statistical models that do not fit easily into classical statistical frameworks (Butler and Burns, 1993; Ribbens et al., 1994; Pascual and Kareiva, 1996; Ferrari and Sugita, 1996; Damgaard, 1999; Strong et al., 1999; Ricketts, 2001; Lytle, 2002; Dalling et al., 2002; Ovaskainen, 2004; Tracey et al., 2005; Fujiwara et al., 2005; Sandin and Pacala, 2005; Agrawal and Fishbein, 2006; Canham and Uriarte, 2006; Horne and Garton, 2006; Ness et al., 2006; Sack et al., 2006; Wintle and Bardos, 2006). Like any tool, these tools also bias my thinking ("if you have a hammer, everything looks like a nail") and the kinds of questions I like to think about. They are most useful for ecological systems where you want to test among a well-defined set of plausible mechanisms, and where you have measured a few potentially important predictor and response variables. They work less well for generalized "fishing expeditions" where you have measured lots of variables and want to try to sort them out.

1.7 R Supplement

Each chapter ends with a set of notes on R, providing more details of the commands and ideas introduced in the chapter or examples worked in more detail. For this largely conceptual chapter, the notes are about how to get R and how to get it working on your computer.

1.7.1 Installing R; Prebasics

- *Download* R. If R is already installed on your computer, skip this step. If not, here's how to get it from the Web.* Go to the R project home page (http://www.r-project.org) or to CRAN, the repository for R materials (http://cran.r-project.org), and navigate to the binary (precompiled) distributions. Find the latest version for your operating system, download it, and follow the instructions to install it. The installation file is moderately large (the Windows installer for R version 2.5.0 was 28.5 megabytes) but should download easily over a fast connection. It should be fine to accept all the defaults in the installation process.

  R should work well on any reasonably modern computer. Version 2.5.0 requires MacOS 10.2 (or higher) or Windows 98 (or higher), or just about any version of Linux; it can also be compiled on other versions of Unix. MacOS version 10.4.4 or higher and Windows XP or higher are recommended. I developed and ran all the code in the book with R 2.5.0 on a dual-core PC laptop running at 1.66 GHz under Ubuntu Linux 7.04.

  After you have played with R a bit, you may want to take a moment to install extra packages (see below).
- *Start* R. If you are using Windows or MacOS there is probably an R icon on your desktop—click on it. Or use the menus your operating system provides to find R. If you are on a Unix system, you can probably just type R on the command line.
- *Play with* R *a little bit.* When you start R, you will see a *command prompt*—a > that waits for you to type something and hit ENTER. When you type in an expression, R evaluates it and prints the answer:

```
> 2 * 8
```

```
[1] 16
```

```
> sqrt(25)
```

```
[1] 5
```

(The number [1] before the answer says that the answer is the first element in a vector; don't worry about this now.)

If you use an equals sign to assign a value to a *variable*, then R will silently do what you asked. To see the value of the variable, type its name at the command prompt:

```
> x = sqrt(36)
> x
```

```
[1] 6
```

A variable name can be any sequence of alphanumeric characters, as well as "_" or "." (but no spaces), that starts with a letter. Variable names are case-sensitive, so x and X are different variables.

---

* These instructions are accurate at press time—but all software, and stuff from the Web in particular, is subject to change. So details may vary.

For more information, read the *Introduction to* R that comes with your copy of R (look in the documentation section of the menus), get one of the introductory documents from the R Web site, dip into an introductory book (Dalgaard, 2003; Crawley, 2005), or get Lab 1 from `http://press .princeton.edu/titles/8709.html`.

- *Stopping* R. To stop R, type q() (with the empty parentheses) at the command prompt, or choose "Quit" from the appropriate menu. You can say "no" when R asks if you want to save the workspace.

  To stop a long computation without stopping R, type ESCAPE or click on the stop sign on the toolbar (in the R console in Windows or MacOS) or type Control-C (in Unix or MacOS if using the command-line version).

- *The help system.* If you type help.start(), R will open a Web browser with help information. If you type ?cmd, R will open a help page with information on a particular command (e.g., ?sqrt to get information on the square-root command). example(cmd) will run any examples that are included in the help page for command cmd. If you type help.search("topic") (with quotes), R will list information related to topic available in the base system or in any extra installed packages; use ?topic to see the information, perhaps using library(pkg) to load the appropriate package first. help(package="pkg") will list all the help pages for a loaded package. If you type RSiteSearch("topic"), R will search an online database for information on topic. Try out one or more of these aspects of the help system.

- *Install extra packages.* R has many extra packages. You may be able to install new packages from a menu within R. You can always type

```
> install.packages("plotrix")
```

(this installs the plotrix package). You can install more than one package at a time:

```
> install.packages(c("ellipse", "plotrix"))
```

(c stands for "combine" and is the command for combining multiple things into a single object.) If the machine on which you use R is not connected to the Internet, you can download the packages to some other medium (such as a flash drive or CD) and install them later, using the menu or

```
> install.packages("plotrix", repos = NULL)
```

Installing packages may fail if you do not have permission to write to the folder (directory) where R is installed on your computer—which may happen if you are working on a public computer. In this case, R will ask you if it's OK to install the packages in a different location. Say yes, and ignore any warnings about R being unable to update the help index.

Finding information about functions in R packages is a bit tricky. By default, help (or ?) only search for packages that have been loaded with library. The help.search function will tell you about the existence of functions in packages that are installed but not loaded (use help.search("topic", agrep=FALSE) to turn off the sometimes irritating "fuzzy" matching behavior), but to see the help information you have to load the package or specify the package with `help(function, package="pkg")`. `RSiteSearch` (or the R Site Search Sidebar for Firefox, `http://addictedtor.free.fr./rsitesearch/`) are the only ways to find information about functions from packages you have not installed.

Here are all the packages used in this book that are not included with R by default:

```
adapt        bbmle      chron     coda            ellipse     emdbook
gplots       gtools     gdata     MCMCpack        odesolve    plotrix
R2WinBUGS    reshape    rgl       scatterplot3d
```

If you install the emdbook package first (`install.packages ("emdbook")`), load it (`library (emdbook)`), and then run the command `get.emdbook .packages()` (you do need the empty parentheses), it will install these packages for you automatically.

(`R2WinBUGS` is an exception to R's normally seamless cross-platform operation: it depends on a Windows program called WinBUGS. WinBUGS will also run on Linux, and MacOS on Intel hardware, with the help of a program called WINE: see Chapter 6.)

Installing these packages now will save time.

1.7.2 R *Interfaces*

While R works perfectly well out of the box, some interfaces can make your R experience easier. Editors such as Tinn-R (Windows), Kate (Linux), or Emacs/ESS will color R commands and quoted material, allow you to submit lines or blocks of R code to an R session, and give hints about function arguments; the standard MacOS interface has all of these features built in. Graphical interfaces such as JGR (cross-platform) or SciViews (Windows) include similar editors and have extra functions such as a workspace browser for looking at all the variables you have defined. (All of these interfaces, which are designed to facilitate R programming, are in a different category from Rcmdr, which tries to simplify basic statistics in R.) If you are using Windows or Linux I strongly recommend that, once you have tried R a little bit, you download at least an R-aware editor and possibly a GUI to make your life easier. Links to all of these systems can be found at `http://www.r-project.org/ GUI/`.

1.7.3 *Sample Session*

Start R. Then:

Start the Web interface to the help system:

```
> help.start()
```

Seed the pseudo-random-number generator, using an arbitrary integer, to make results match if you start a new session (it's fine to skip this step, but the particular values you get from the random-number commands will be different every time—you won't get exactly the results shown below):

```
> set.seed(101)
```

Create the variable frogs (representing the density of adult frogs in each of 20 populations) from scratch by entering 20 numbers with the c command. Create a second variable tadpoles (the density of tadpoles in each population) by generating 20 normally distributed random numbers, each with twice the mean of the corresponding frogs population and a standard deviation of 0.5:

```
> frogs = c(1.1, 1.3, 1.7, 1.8, 1.9, 2.1, 2.3, 2.4,
+      2.5, 2.8, 3.1, 3.3, 3.6, 3.7, 3.9, 4.1, 4.5,
+      4.8, 5.1, 5.3)
> tadpoles = rnorm(n = 20, mean = 2 * frogs, sd = 0.5)
```

The + at the beginning of the second line is a *continuation character*. If you hit ENTER and R recognizes that your command is unfinished, it will print a + to tell you that you can continue on the next line. Sometimes the continuation character means that you forgot to close parentheses or quotes. To discard what you've done so far and start again, type ESCAPE (on Windows or MacOS) or Control-C (on Linux) or click on the stop sign on the menu.

You can name the *arguments* (n, mean, sd above) in an R function, but R can also recognize the order: tadpoles = rnorm(20,2*frogs,0.5) will give the same answer. In general, however, it's clearer and safer to name arguments.

Notice that R doesn't tell you what's in these variables unless you ask it. Entering a variable name by itself tells R to print the value of the variable:

```
> tadpoles

 [1] 2.036982 2.876231 3.062528 3.707180 3.955385 4.786983
 [7] 4.909395 4.743633 5.458514 5.488370 6.463224 6.202578
[13] 7.913878 6.666590 7.681658 8.103331 8.575123 9.629233
[19] 9.791165 9.574846
```

(The numbers at the beginning of the line are indices.) This rule of printing a variable that is entered on a line by itself also explains why typing q rather than q() prints out R code rather than quitting R. R interprets q() as "run the function q without any arguments"; it interprets q as "print the contents of variable q."

Plot tadpoles against frogs (frogs on the *x* axis, tadpoles on the *y* axis) and add a straight line with intercept 0 and slope 2 to the plot (the result should appear in a new window, looking like Figure 1.6):

```
> plot(frogs, tadpoles)
> abline(a = 0, b = 2)
```

Try calculating the (natural) logarithm of tadpoles and plot it instead:

```
> log_tadpoles = log(tadpoles)
> plot(frogs, log_tadpoles)
```

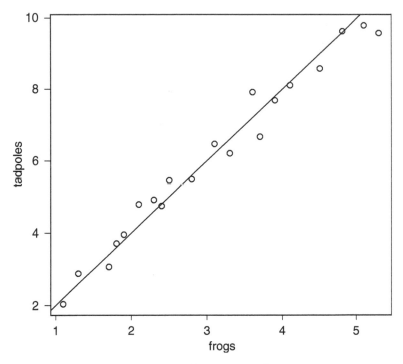

Figure 1.6 Plotting example.

You can get the same plot by typing `plot(frogs,log(tadpoles))` or a similar plot that adjusts the axes rather than the values with `plot(frogs,tadpoles,log="y")`. Use `log10(tadpoles)` to get the logarithm base 10.

Set up a variable n with integers ranging from 1 to 20 (the length of the `frogs` variable) and plot `frogs` against it:

```
> n = 1:length(frogs)
> plot(n, frogs)
```

(You'd get almost the same plot by typing `plot(frogs)`.)

R's default plotting character is an open circle. Open symbols are generally better than closed symbols for plotting because it is easier to see where they overlap, but you could include `pch=16` in the `plot` command if you wanted filled circles instead. Figure 1.7 shows several more ways to adjust the appearance of lines and points in R.

Calculate the mean, standard deviation, and a set of summary statistics for tadpoles:

```
> mean(tadpoles)
```

```
[1] 6.081341
```

```
> sd(tadpoles)
```

```
[1] 2.370449
```

```
> summary(tadpoles)
```

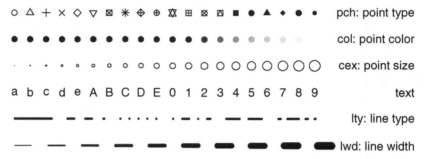

Figure 1.7 Some of R's graphics parameters. Color specification, col, also applies in many other contexts: all colors are set to gray scales here. See ?par for (many more) details on graphics parameters, and one or more of ?rgb, ?palette, or apropos("color") for more on colors.

Min.	1st Qu.	Median	Mean	3rd Qu.	Max.
2.037	4.547	5.845	6.081	7.961	9.791

"1st Qu." and "3rd Qu." represent the first and third quartiles of the data. The summary statistics are displayed to only three significant digits, which can occasionally cause confusion.

Calculate the correlation between frogs and tadpoles:

```
> cor(frogs, tadpoles)
```

```
[1] 0.9870993
```

Test the statistical significance of the correlation:

```
> cor.test(frogs, tadpoles)

	Pearson's product-moment correlation

data:  frogs and tadpoles
t = 26.1566, df = 18, p-value = 8.882e-16
alternative hypothesis: true correlation is not equal to 0
95 percent confidence interval:
 0.9669568 0.9949946
sample estimates:
     cor
0.9870993
```

The p-value here is extraordinarily low because we made up the data with very little noise: you should consider reporting it simply as $p < 0.001$. cor.test does a Pearson correlation test by default, but you can choose other tests; see ?cor.test.

Look for more information on correlations:

```
> help.search("correlation")
```

Now move on to Chapter 2 to see how to deal with real data.

2 Exploratory Data Analysis and Graphics

This chapter covers both the practical details and the broader philosophy of (1) reading data into R and (2) doing exploratory data analysis, in particular graphical analysis. To get the most out of the chapter you should already have some basic knowledge of R's syntax and commands (see the R supplement of the previous chapter).

2.1 Introduction

One of the basic tensions in all data analysis and modeling is how much you have all your questions framed before you begin to look at your data. In the classical statistical framework, you're supposed to lay out all your hypotheses before you start, run your experiments, come back to your office and test those (and only those) hypotheses. Allowing your data to suggest new statistical tests raises the risk of "fishing expeditions" or "data-dredging"—indiscriminately scanning your data for patterns.* Data-dredging is a serious problem. Humans are notoriously good at detecting apparent patterns even when they don't exist. Strictly speaking, interesting patterns that you find in your data after the fact should not be treated statistically, only used as input for the next round of observations and experiments.† Most statisticians are leery of procedures like stepwise regression that search for the best predictors or combinations of predictors from among a large range of options, even though some have elaborate safeguards to avoid overestimating the significance of observed patterns (Whittingham et al., 2006). The worst aspect of such techniques is that in order to use them you must be conservative and discard real patterns, patterns that you originally had in mind, because you are screening your data indiscriminately (Nakagawa, 2004).

---

* "Bible codes," where people find hidden messages in the Bible, illustrate an extreme form of data-dredging. Critics have pointed out that similar procedures will also detect hidden messages in *War and Peace* or *Moby Dick* (McKay et al., 1999).

† Or you should apply a post hoc procedure [see ?TukeyHSD and the multcomp package in R] that corrects for the fact that you are testing a pattern that was not suggested in advance—however, even these procedures apply corrections only for a specific set of possible comparisons, not for all possible patterns that you could have found in your data.

But these injunctions may be too strict for ecologists. Unexpected patterns in the data can inspire you to ask new questions, and it is foolish not to explore your hard-earned data. *Exploratory data analysis* (EDA; Tukey, 1977; Cleveland, 1993; Hoaglin et al., 2000, 2006) is a set of graphical techniques for finding interesting patterns in data. EDA was developed in the late 1970s when computer graphics first became widely available. It emphasizes *robust* and *nonparametric* methods, which make fewer assumptions about the shapes of curves and the distributions of the data and hence are less sensitive to nonlinearity and outliers. Most of the rest of this book will focus on models that, in contrast to EDA, are parametric (i.e., they specify particular distributions and curve shapes) and mechanistic. These methods are more powerful and give more ecologically meaningful answers, but they are also susceptible to being misled by unusual patterns in the data.

The big advantages of EDA are that it gets you looking at and thinking about your data (whereas stepwise approaches are often substitutes for thought), and that it may reveal patterns that standard statistical tests would overlook because of their emphasis on specific models. However, EDA isn't a magic formula for interpreting your data without the risk of data dredging. Only common sense and caution can keep you in the zone between ignoring interesting patterns and overinterpreting them. It's useful to write down a list of the ecological patterns you're looking for and how they relate your ecological questions *before* you start to explore your data, so that you can distinguish among (1) patterns you were initially looking for, (2) unanticipated patterns that answer the same questions in different ways, and (3) interesting (but possibly spurious) patterns that suggest new questions.

The rest of this chapter describes how to get your data into R and how to make some basic graphs in order to search for expected and unexpected patterns. The text covers both philosophy and some nitty-gritty details. The supplement at the end of the chapter gives a sample session and more technical details.

2.2 Getting Data into R

2.2.1 Preliminaries

ELECTRONIC FORMAT

Before you can analyze your data you have to get them into R. Data come in a variety of formats—in ecology, most are either plaintext files (space- or comma-delimited) or Excel files.* R prefers plaintext files with "white space" (arbitrary numbers of tabs or spaces) or commas between columns. Text files are less structured and may take up more disk space than more specialized formats, but they are the lowest common denominator of file formats and so can be read by almost anything (and, if necessary, examined and adjusted in any text editor). Since a wide variety of text editors can read plaintext formats, they are unlikely to be made obsolete by changes

---

* Your computer may be set up to open comma-delimited (.csv) files in Excel, but underneath they are just text files.

in technology (you could say they're already obsolete), and less likely to be made unusable by corruption of a few bits of the file; only hard copy is better.*

R is platform-agnostic. While text files do have very slightly different formats on Unix, Microsoft Windows, and Macintosh operating systems, R handles these differences. If you later save data sets or functions in R's own format (using save to save and load to load them), you will be able to exchange them freely across platforms.

Many ecologists keep their data in Excel spreadsheets. The read.xls function in the gdata package allows R to read Excel files directly, but the best thing to do with an Excel file (if you have access to a copy of Excel, or if you can open it in an alternative spreadsheet program) is to save the worksheet you want as a .csv (comma-separated values) file. Saving as a .csv file will also force you to go into the worksheet and clean up any random cells that are outside of the main data table—R won't like these. If your data are in some more exotic form (e.g., within a GIS or database system), you'll have to figure out how to extract them from that particular system into a text file. There are ways of connecting R directly with databases or GIS systems, but they're beyond the scope of this book. If you have trouble exporting data or you expect to have large quantities of data (e.g., more than tens of thousands of observations) in one of these exotic forms, read the R *Data Import/Export Manual*, which is accessible through Help in the R menus.

METADATA

Metadata is the information that describes the properties of a data set: the names of the variables, the units they were measured in, when and where the data were collected, etc. R does not have a structured system for maintaining metadata, but it does allow you to include a good deal of this metadata within your data file, and it is good practice to keep as much of this information as possible associated with the data file. Some tips on metadata in R:

- Column names are the first row of the data set. Choose names that compromise between convenience (you will be typing these names a lot) and clarity; larval_density or larvdens is better than either x or larval_density_per_m3_in_ponds. Use underscores or dots to separate words in variable names, not spaces. Begin names with a letter, not a number.
- R will ignore any information on a line following a #.[†] I usually use this comment character to include general metadata at the beginning of my data file, such as the data source, units, and so forth—anything that can't easily be encoded in the variable names. I also use comments before, or at the ends of, particular lines in the data set that might need annotation, such as the circumstances surrounding questionable data points. You can't use # to make a comment in the middle of a line: use a comment like # pH calibration failed at the end of the line to indicate that a particular field in that line is suspect.

---

* Unless your data are truly voluminous, you should also save a hard-copy, archival version of your data (Gotelli and Ellison, 2004).

[†] In a text file: when using read.csv you need to add comment.char="#" if you add metadata in this way.

- If you have other metadata that can't easily be represented in plaintext format (such as a map), you'll have to keep it separately. You can reference the file in your comments, keep a separate file that lists the location of data and metadata, or use a system like Morpho (from `ecoinformatics.org`) to organize it.

Whatever you do, make sure that you have some workable system for maintaining your metadata. Eventually, your R scripts—which document how you read in your data, transformed it, and drew conclusions from it—will also become a part of your metadata. As mentioned in Chapter 1, this is one of the advantages of R over (say) Excel: after you've done your analysis, *if you were careful to document your work sufficiently as you went along*, you will be left with a set of scripts that will allow you to verify what you did; make minor modifications and rerun the analysis; and apply the same or similar analyses to future data sets.

SHAPE

Just as important as electronic or paper format is the organization or *shape* of your data. Most of the time, R prefers that your data have a single *record* (typically a line of data values) for each individual observation. This basically means that your data should usually be in "long" (or "indexed") format. For example, the first few lines of the seed removal data set look like this, with a line giving the number of seeds present for each station/date combination:

	station	date	dist	species	seeds
1	1	1999-03-23	25	psd	5
2	1	1999-03-27	25	psd	5
3	1	1999-04-03	25	psd	5
4	2	1999-03-23	25	uva	5
5	2	1999-03-27	25	uva	5
6	2	1999-04-03	25	uva	5

Because each station has seeds of only one species and can be at only a single distance from the forest, these values are repeated for every date. During the first two weeks of the experiment no seeds of psd or uva were taken by predators, so the number of seeds remained at the initial value of 5.

Alternatively, you will often come across data sets in "wide" format, like this:

	station	species	dist	seeds.1999-03-23	seeds.1999-03-27
1	1	psd	25	5	5
2	2	uva	25	5	5
3	3	pol	25	5	4
4	4	dio	25	5	5
5	5	cor	25	5	4
6	6	abz	25	5	5

(I kept only the first two date columns in order to make this example narrow enough to fit on the page.)

Long format takes up more room, especially if you have data (such as dist above, the distance of the station from the edge of the forest) that apply to each station independent of sample date or species (which therefore have to be repeated many times in the data set). However, you'll find that this format is typically what statistical packages request for analysis.

You can read data into R in wide format and then convert it to long format. R has several different functions—reshape and stack/unstack in the base package, and melt/cast/recast in the reshape package*—that will let you switch data back and forth between wide and long formats. Because there are so many different ways to structure data, and so many different ways you might want to aggregate or rearrange them, software tools designed to reshape arbitrary data are necessarily complicated (Excel's pivot tables, which are also designed to restructure data, are as complicated as reshape).

- stack and unstack are simple but basic functions—stack converts from wide to long format and unstack from long to wide; they aren't very flexible.
- reshape is very flexible and preserves more information than stack/unstack, but its syntax is tricky: if long and wide are variables holding the data in the examples above, then

```
> reshape(wide, direction = "long", timevar = "date",
+     varying = 4:5)
> reshape(long, direction = "wide", timevar = "date",
+     idvar = c("station", "dist", "species"))
```

  convert back and forth between them. In the first case (wide to long) we specify that the time variable in the new long-format data set should be date and that columns 4–5 are the variables to collapse. In the second case (long to wide) we specify that date is the variable to expand and that station, dist, and species should be kept fixed as the identifiers for an observation.
- The reshape package contains the melt, cast, and recast functions, which are similar to reshape but sometimes easier to use, e.g.,

```
> library(reshape)
> recast(wide, formula = ... ~ ., id.var = c("station",
+     "dist", "species"))
> recast(long, formula = station + dist + species ~
+     ..., id.var = c("station", "dist", "species",
+     "date"))
```

  in the formulas above, . . . denotes "all other variables" and . denotes "nothing," so the formula . . .~. means "separate out by all variables" (long format) and station+dist+species~. . . means "separate out by station, distance, and species, put the values for each date on one line."

In general you will have to look carefully at the examples in the documentation and play around with subsets of your data until you get it reshaped exactly the way you want. Alternatively, you can manipulate your data in Excel, either with pivot tables or by brute force (cutting and pasting). In the long run, learning to reshape data will pay off, but for a single project it may be quicker to use brute force.

---

* If you don't know what a package is, go back and read about them in the R supplement for Chapter 1.

2.2.2 Reading in Data

BASIC R COMMANDS

The basic R commands for reading in a data set, once you have it in a long-format text file, are `read.table` for space-separated data and `read.csv` for comma-separated data. If there are no complications in your data, you should be simply be able to say (e.g.)

```
> data = read.table("mydata.dat", header = TRUE)
```

(if your file is actually called `mydata.dat` and includes a first row with the column names) to read your data in (as a *data frame*; see p. 35) and assign it to the variable `data`.

Reading in files presents several potential complications, which are more fully covered in the R supplement: (1) telling R where to look for data files on your computer system; (2) checking that every line in the file has the same number of variables, or *fields*—R won't read it otherwise; and (3) making sure that R reads all your variables as the right data types (discussed in the next section).

2.3 Data Types

When you read data into a computer, the computer stores those data as some particular data *type*. This is partly for efficiency—it's more efficient to store numbers as strings of bits rather than as human-readable character strings—but its main purpose is to maintain a sort of metadata about variables, so the computer knows what to do with them. Some operations make sense only with particular types—what should you get when you try to compute 2+"A"? "2A"? If you try to do something like this in Excel, you get an error code—#VALUE!; if you do it in R, you get the message `Error ... non-numeric argument to binary operator.`*

Computer packages vary in how they deal with data. Some lower-level languages like C are *strongly typed*; they insist that you specify exactly what type every variable should be and require you to convert variables between types (say integer and real, or floating-point) explicitly. Languages or packages like R or Excel are looser; they try to guess what you have in mind and convert variables between types (*coerce*) automatically as appropriate. For example, if you enter 3/25 into Excel, it automatically converts the value to a date—March 25 of the current year.

R makes similar guesses as it reads in your data. By default, if every entry in a column is a valid number (e.g., 234, -127.45, 1.238e3 [computerese for 1.238×10^3]), then R guesses the variable is numeric. Otherwise, it makes it a *factor*—an indexed list of values used to represent categorical variables, which I will describe in more detail shortly. Thus, any error in a numeric variable (extra decimal point, included letter, etc.) will lead R to classify that variable as a factor rather than a number. R also has a detailed set of rules for dealing with missing values (internally

---

* The + symbol is called a "binary operator" because it is used to combine two values.

represented as NA, for Not Available). If you use missing-value codes (such as * or -9999) in your data set, you have to tell R about it or it will read them naively as strings or numbers.

While R's standard rules for guessing about input data are pretty simple and allow you only two options (numeric or factor), there are a variety of ways for specifying more detail either as R reads in your data or after it has read them in; these are covered in more detail in the accompanying material.

2.3.1 Basic Data Types

R's basic (or *atomic*) data types are integer, numeric (real numbers), logical (TRUE or FALSE), and character (alphanumeric strings). (There are a few more, such as complex numbers, that you probably won't need.) At the most basic level, R organizes data into *vectors* of one of these types, which are just ordered sets of data. Here are a couple of simple (numeric and character) vectors:

```
> 1:5
[1] 1 2 3 4 5
> c("yes", "no", "maybe")
[1] "yes"   "no"    "maybe"
```

More complicated data types include dates (Date) and factors (factor). Factors are R's way of dealing with categorical variables. A factor's underlying structure is a set of (integer) levels along with a set of the labels associated with each level.

One advantage of using these more complex types, rather than converting your categorical variables to numeric codes, is that it's much easier to remember the meaning of the levels as you analyze your data, for example, north and south rather than 0 and 1. Also, R can often do the right things with your data automatically if it knows what types they are (this is an example of crude-versus-sophisticated where a little more sophistication may be useful). Much of R's built-in statistical modeling software depends on these types to do the right analyses. For example, the command lm(y~x) (meaning "fit a linear model of y as a function of x," analogous to SAS's PROC GLM) will do an ANOVA if x is categorical (i.e., stored as a factor) or a linear regression if x is numeric. If you want to analyze variation in population density among sites designated with integer codes (e.g., 101, 227, 359) and haven't specified that R should interpret the codes as categorical rather than numeric values, R will try to fit a linear regression rather than doing an ANOVA. Many of R's plotting functions will also do different things depending on what type of data you give them. For example, R can automatically plot date axes with appropriate labels. To repeat, data types are a form of metadata; the more information about the meaning of your data that you can retain in your analysis, the better.

2.3.2 Data Frames and Matrices

R can organize data at a higher level than simple vectors. A *data frame* is a table of data that combines vectors (columns) of different types (e.g., character, factor, and numeric data). Data frames are a hybrid of two simpler data structures: *lists*, which can mix arbitrary types of data but have no other structure, and *matrices*, which are structured by rows and columns but usually contain only one data type (typically numeric). When treating the data frame as a list, there are a variety of different ways of extracting columns of data from the data frame to work with:

```
> SeedPred[[3]]
> SeedPred[["species"]]
> SeedPred$species
```

all extract the third column (a factor containing species abbreviations) from the data frame SeedPred. You can also treat the data frame as a matrix and use square brackets [] to extract (e.g.) the third column:

```
> SeedPred[, 3]
> SeedPred[, "species"]
```

or rows 1 through 10

```
> SeedPred[1:10, ]
```

(SeedPred[i,j] extracts the matrix element in row(s) i and column(s) j; leaving the columns or rows specification blank, as in SeedPred[i,] or SeedPred[,j], takes row i (all columns) or column j (all rows) respectively.) A few operations, such as transposing or calculating a variance-covariance matrix, work only with matrices (not with data frames); R will usually convert (*coerce*) data frames to matrices automatically when it makes sense to, but you may sometimes have to use as.matrix to manually convert a data frame to a matrix.*

2.3.3 Checking Data

Now suppose you've decided on appropriate types for all your data and told R about it. Are the data you've read in actually correct, or are there still typographical or other errors?

summary

First check the results of summary. For a numeric variable summary will list the minimum, first quartile, median, mean, third quartile, and maximum. For a factor it will list the numbers of observations with each of the first six factor levels, then the number of remaining observations. (Use table on a factor to see the numbers of observations at all levels.) It will list the number of NAs for all types.

---

*Matrices and data frames can appear identical but behave differently. If x is a data frame, either colnames(x) or names(x) will tell you the column names. If x has a column called a, either x$a or x[["a"]] or x[,"a"] will retrieve it. If x is a matrix, you must use colnames(x) to get the column names and x[,"a"] to retrieve a column (the other commands will give errors). Use is.data.frame or class to tell matrices and data frames apart.

For example:

```
> summary(SeedPred[, 1:4])
```

```
     station              dist          species              date
1      :     74      10:5883      abz     :1480      Min.    :1999-03-23
2      :     74      25:5920      cd      :1480      1st Qu. :1999-05-23
3      :     74                   cor     :1480      Median  :1999-07-24
4      :     74                   dio     :1480      Mean    :1999-07-25
5      :     74                   pol     :1480      3rd Qu. :1999-09-28
6      :     74                   psd     :1480      Max.    :1999-11-28
(Other):11359                     (Uther) :2923
```

(To keep the output short, I'm looking at the first four columns of the data frame only: summary(SeedPred) would summarize the whole thing.)

Check the following points:

- Is the total number of observations right? For factors, is the number of observations in each level right?
- Do the summaries of the numeric variables—mean, median, etc.—look reasonable? Are the minimum and maximum values about what you expected?
- Are the numbers of NAs in each column reasonable? If not (especially if you have extra mostly NA columns), you may want to go back a few steps and use count.fields to identify rows with extra fields.

str

The str command tells you about the **structure** of an R variable: it is slightly less useful than summary for dealing with data, but it may come in handy later for figuring out more complicated R variables. Applied to a data frame, it tells you the total number of observations (rows) and variables (columns) and prints out the names and classes of each variable along with the first few observations in each variable.

```
> str(SeedPred)
```

```
'data.frame':   11803 obs. of  9 variables:
 $ station  : Factor w/ 160 levels "1","2","3","4",..: 1 1 1 1 1 1 1
                1 1 1 ...
 $ dist     : Factor w/ 2 levels "10","25": 1 1 1 1 1 1 1 1 1 1 ...
 $ species  : Factor w/ 8 levels "abz","cd","cor",..: 7 7 7 7 7 7 7
                7 7 7 ...
 $ date     : Class 'Date'  num [1:11803] 10675 10678 10685 10692
                10699 ...
 $ seeds    : int  5 5 5 5 0 0 0 0 0 0 ...
 $ tcum     : num  0 3 10 17 24 31 39 46 53 60 ...
 $ tint     : num  NA 3 7 7 7 8 7 7 7 ...
 $ taken    : int  NA 0 0 0 5 0 0 0 0 0 ...
 $ available: int  NA 5 5 5 5 0 0 0 0 0 ...
```

class

The class command prints out the class (numeric, factor, Date, logical, etc.) of a variable. class(SeedPred) gives "data.frame"; sapply(SeedPred, class) applies class to each column of the data individually.

```
> class(SeedPred)
```

```
[1] "data.frame"
```

```
> sapply(SeedPred, class)
```

```
   station       dist    species      date      seeds        tcum
  "factor"   "factor"   "factor"    "Date"  "integer"   "numeric"
      tint      taken  available
 "numeric"  "integer"  "integer"
```

head

The head command just prints out the beginning of a data frame; by default it prints the first six rows, but head(data,10) (e.g.) will print out the first 10 rows.

```
> head(SeedPred)
```

```
  station dist species       date seeds tcum tint taken available
1       1   10     psd 1999-03-25     5    0   NA    NA        NA
2       1   10     psd 1999-03-28     5    3    3     0         5
3       1   10     psd 1999-04-04     5   10    7     0         5
4       1   10     psd 1999-04-11     5   17    7     0         5
5       1   10     psd 1999-04-18     0   24    7     5         5
6       1   10     psd 1999-04-25     0   31    7     0         0
```

The tail command prints out the end of a data frame.

table

table is R's command for cross-tabulation; you can use it to check that you have appropriate numbers of observations in different factor combinations.

```
> table(SeedPred$station, SeedPred$species)
```

	abz	cd	cor	dio	mmu	pol	psd	uva
1	0	0	0	0	0	0	74	0
2	0	0	0	0	0	0	0	74
3	0	0	0	0	0	74	0	0
4	0	0	0	74	0	0	0	0
5	0	0	74	0	0	0	0	0
6	74	0	0	0	0	0	0	0

(just the first six lines are shown): apparently, each station has seeds of only a single species. The $ extracts variables from the data frame SeedPred, and table says we want to count the number of instances of each combination of station and species; we could also do this with a single factor or with more than two.

DEALING WITH NAs

Missing values are a nuisance, but a fact of life. Throwing out or ignoring missing values is tempting, but it can be dangerous. Ignoring missing values can bias your analyses, especially if the pattern of missing values is not completely random. R is conservative by default and assumes that, for example, 2+NA equals NA—if you don't know what the missing value is, then the sum of it and any other number is also unknown. Almost any calculation you make in R will be contaminated by NAs, which is logical but annoying. Perhaps most difficult is that you can't just do what comes naturally and say (e.g.) x=x[x!=NA] to remove values that are NA from a variable, because even comparisons to NA result in NA!*

- You can use the special function is.na to count the number of NA values (sum(is.na(x))) or to throw out the NA values in a vector (x=x[!is.na(x)]).
- Functions such as mean, var, sd, sum (and some others) have an optional na.rm argument: na.rm=TRUE drops NA values before doing the calculation. Otherwise if x contains any NAs, mean(x) will result in NA and sd(x) will give an error about missing observations.
- To convert NA values to a particular value, use x[is.na(x)]=value; e.g., to set NAs to zero x[is.na(x)]=0, or to set NAs to the mean value x[is.na(x)]=mean(x,na.rm=TRUE). *Don't do this unless you have a very good, and defensible, reason.*
- na.omit will drop NAs from a vector (na.omit(x)), but it is also smart enough to do the right thing if x is a data frame instead, and throw out all the cases (rows) where *any* variable is NA; however, this may be too stringent if you are analyzing a subset of the variables. For example, you might have a really unreliable soil moisture meter that produces lots of NAs, but you don't need to throw away all of these data points while you're analyzing the relationship between light and growth. (complete.cases returns a logical vector that says which rows have no NAs; if x is a data frame, na.omit(x) is equivalent to x[complete.cases(x),].)
- Calculations of covariance and correlation (cov and cor) have more complicated options: use="all.obs", use="complete.obs", or use="pairwise.complete.obs". all.obs uses all of the data (but the answer will contain NAs every time either variable contains one); complete.obs uses only the observations for which *none* of the variables are NA (but may thus leave out a lot of data); and pairwise.complete.obs computes the pairwise covariance/correlations using the observations where both of each particular pair of variables are non-NA (but may lead in some cases to incorrect estimates).

As you discover errors in your data, you may have to go back to your original data set to correct errors and then reenter them into R (using the commands you have saved, of course). Or you can change a few values in R, e.g.,

```
> SeedPred[24, "species"] = "mmu"
```

---

*!= means "not equal to"; in general (but not for NAs), x[x!=y] will select values of x that are not equal to y.

to change the species in the 24th observation from psd to mmu. Whatever you do, document this process as you go along, and always maintain your original data set in its original, archival, form, even including data you think are errors (this is easier to remember if your original data set is in the form of field notebooks). Keep a log of what you modify so conflicting versions of your data don't confuse you.

2.4 Exploratory Data Analysis and Graphics

The next step in checking your data is to graph them, which leads on naturally to exploring patterns. Graphing is the best way to understand not only data, but also the models that you fit to data; as you develop models you should graph the results frequently to make sure you understand how the model is working.

R gives you complete control of all aspects of graphics (Figure 1.7) and lets you save graphics in a wide range of formats. The only major nuisance of doing graphics in R is that R constructs graphics as though it were drawing on a static page, not by adding objects to a dynamic scene. You generally specify the positions of all graphics on the command line, not with the mouse (although the locator and identify functions can be useful). Once you tell R to draw a point, line, or piece of text there is no way to erase or move it. The advantage of this procedure, like logging your data manipulations, is that you have a complete record of what you did and can easily recreate the picture with new data.

R actually has two different coexisting graphics systems. The base graphics system is cruder and simpler, while the lattice graphics system (in the lattice package) is more sophisticated and complex. Both can create scatterplots, box-and-whisker plots, histograms, and other standard graphical displays. Lattice graphics do more automatic processing of your data and produce prettier graphs, but the commands are harder to understand and customize. In the realm of 3D graphics, there are several more options, at different stages of development. Base graphics and lattice graphics both have some 3D capabilities (persp in base, wireframe and cloud in lattice); the scatterplot3d package builds on base to draw 3D point clouds; the rgl package (still under development) allows you to rotate and zoom the 3D coordinate system with the mouse; and the ggobi package is an interface to a system for visualizing multidimensional point data.

2.4.1 Seed Removal Data: Discrete Numeric Predictors, Discrete Numeric Responses

As described in Chapter 1, the seed removal data set from Duncan and Duncan (2000) gives information on the rate at which seeds were removed from experimental stations set up in a Ugandan grassland. Seeds of eight species were set out at stations along two transects different distances from the forest and monitored every few days for more than eight months. We have already seen a subset of these data in a brief example, but we haven't really examined the details of the data set. There are a total of 11,803 observations, each containing information on the station number (station), distance in meters from the forest edge (dist), the species

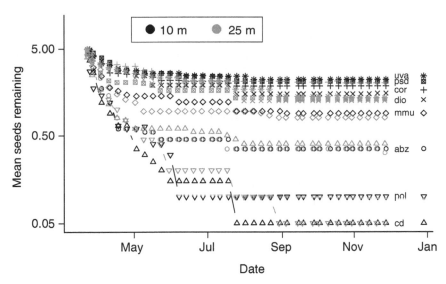

Figure 2.1 Seed removal data: mean seeds remaining by species over time. Functions: (main plot) `matplot`, `matlines`; (annotation) `axis`, `axis.Date`, `legend`, `text`, `points`.

code (`species`),* the date sampled (`date`), and the number of seeds present (`seeds`). The remaining columns in the data set are derived from the first five: the cumulative elapsed time (in days) since the seeds were put out (`tcum`); the time interval (in days) since the previous observation (`tint`); the number of seeds removed since the previous observation (`taken`); and the number of seeds present at the previous observation (`available`).

2.4.1.1 DECREASE IN NUMBERS OVER TIME

The first thing to look at is the mean number of seeds remaining over time (Figure 2.1). I plotted the mean on a logarithmic scale; if seeds were removed at a constant per capita rate (a reasonable null hypothesis), the means should decrease exponentially over time and the lines should be straight on a log scale. (It's much easier to see differences from linearity than to tell whether a curve is decreasing faster or slower than exponentially.) They are not: the seeds that remain after July appear to be taken at a much slower rate. (See the R supplement, p. 63, for the code to create the figure.)

Figure 2.1 also reveals differences among species larger than the differences between the two distances from the forest. However, it also seems that some species may have a larger difference between distances from the forest; *C. durandii* (cd, △) disappears 10 times faster near than far from the forest. Like all good graphics, the figure raises many questions (only some of which can be answered from the data at hand): Is the change in disappearance rate indicated by the flattening out of the

---

* `abz`=*Albizia grandibracteata*, `cd`=*Celtis durandii*, `cor`=*Cordia abyssinica*, `dio`=*Diospyros abyssinica*, `mmu`=*Mimusops bagshawei*, `pol`=*Polyscias fulva*, `psd`=*Pseudospondias microcarpa*, `uva`=*Uvariopsis congensis*.

curves driven by the elapsed time since the seeds were set out, the season, or the declining density of seeds? Or is there variation within species, such that predators take all the tasty seeds at a station and leave the nontasty ones? Is the change in rate a gradual decrease or an abrupt change? Does it differ among species? Are the overall differences in removal rate among species, between distances from the forest, and their interaction (i.e., the fact that cd appears to be more sensitive to differences in distance) real or just random fluctuations? Are they related to seed mass or some other known characteristic of the species?

2.4.1.2 NUMBER TAKEN OUT OF NUMBER AVAILABLE

Plotting the mean number remaining over time shows several facets of the data (elapsed time, species, distance from edge) and asks and answers important ecological questions, but it ignores another facet—the variability or *distribution* of the number of seeds taken. To explore this facet, I'll now look at the patterns of the number of seeds taken as a function of the number available.

The simplest starting point is to plot the number taken between each pair of samples (on the y axis) as a function of the number available (on the x axis). If x and y are numeric variables, plot(x,y) draws a scatterplot. Here we use plot(SeedPred$available,SeedPred$taken). The lattice package equivalent would be xyplot(taken~available,data=SeedPred). The scatterplot turns out not to be very informative in this case (try it and see!); all the repeated points in the data overlap, so that all we see in the plot is that any number of seeds up to the number available can be taken.

One quick-and-dirty way to get around this problem is to use the jitter command, which adds a bit of random variation so that the data points don't all land in exactly the same place: Figure 2.2a shows the results, which are ugly but do give some idea of the patterns.

sizeplot, from the plotrix package, deals with repeated data points by making the area of plotting symbols proportional to the number of observations falling at a particular point (Figure 2.2b; in this case I've used the text command to add text to the circles with the actual numbers from cross-tabulating the data by number available and number taken (t1=table(SeedPred$available,SeedPred$taken)). More generally, *bubble plots* superimpose a third variable on an *x-y* scatterplot by changing symbol sizes: in R, you can either use the symbols command or just set cex to a vector in a plot command (e.g., plot(x,y,cex=z) plots y vs. x with symbol sizes proportional to z). sizeplot is a special-case bubble plot; it counts the number of points with identical x and y values and makes the area of the circle proportional to that number. If (as in this case) these x and y values come from a cross-tabulation, two other ways to plot the data are a *mosaic plot* (e.g., mosaicplot(t1), mosaicplot(available~taken,data=SeedPred) or mosaicplot(available~taken,data=SeedPred,subset=taken>0)) or a *balloon plot* (balloonplot in the gplots package: balloonplot(t1)). You could also try dotchart(t1); *dot charts* are an invention of W. Cleveland that perform approximately the same function as bar charts. (Try these and see for yourself.)

R is *object-oriented*, which in this context means that it will try to "do the right thing" when you ask it to do something with a variable. For example, if you

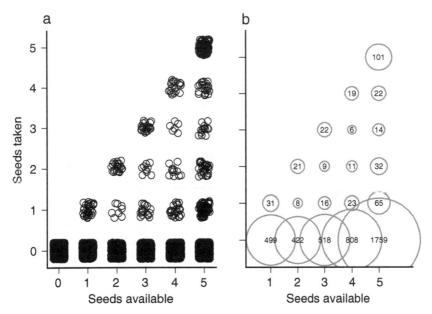

Figure 2.2 (a) Jittered scatterplot of number of seeds taken as a function of number of seeds available: all species and dates combined. (b) Bubble plot of combined seed removal data (`sizeplot`: (0,0) category dropped for clarity).

simply say `plot(t1)`, R knows that `t1` is a two-way table, and it will plot something reasonably sensible—in this case the mosaic plot mentioned above.

Bar plots are another way to visualize the distribution of number of seeds taken (Figure 2.3). The `barplot` command can plot either a vector (as single bars) or a matrix (as stacked bars, or as grouped sets of bars). Here we want to plot groups of stacked bars, one group for each number of available seeds. The only remaining trick here is that `barplot` plots each *column* of the matrix as a group, whereas we want our bar plot grouped by number available, which are the *rows* of our table. We could go back and recalculate `table(taken,available)`, which would switch the order of rows and columns. However, it's easier to use the transpose command `t` to exchange rows and columns of the table.

I also decided to put the plot on a logarithmic scale, since the data span a wide range of numbers of counts. Since the data contain zeros, taking logarithms of the raw data may cause problems; since they are count data, it is reasonable to add 1 as an offset. I decided to use logarithms base 10 (`log10`) rather than natural logarithms (`log`) since I find them easier to interpret. (Many of R's plot commands, including `barplot`, have an argument `log` that can be used to specify that the x, y, or both axes are logarithmic (`log="x"`, `log="y"`, `log="xy"`)—this has the advantage of plotting an axis with the original, more interpretable values labeled but unevenly spaced. In this particular case the figure is slightly prettier the way I've done it.)

The main conclusions from Figures 2.2 and 2.3 and the table, which have really shown essentially the same thing in four different ways, are that (1) the number of seeds taken increases as the number of seeds available increases (this is not surprising);

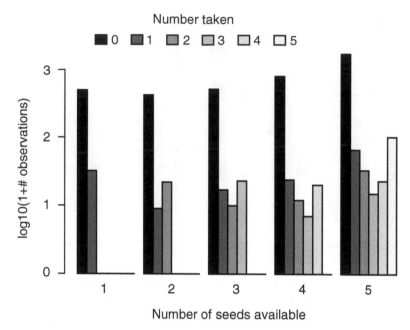

Figure 2.3 Bar plot of observations of number of seeds taken, subdivided by number available: `barplot(t(log10(t1+1)), beside=TRUE)`.

(2) the *distribution* of number of seeds taken is bimodal (has two peaks) with modes at zero and at the total number of seeds available—all or nothing; (3) the distribution of the number of seeds taken looks roughly constant as the number of seeds available increases. Observation 2 in particular starts to suggest some ecological questions: it makes sense for there to be a mode at zero (when seed predators don't find the seeds at all) and one away from zero (when they do), but why would seed predators take either few or many but not an intermediate number? Perhaps this pattern, which appears at the level of the whole data set, emerges from variability among low- and high-vulnerability sites or species, or perhaps it has something to do with the behavior of the seed predators.

Yet another graphical approach would be to try to visualize these data in three dimensions, as a 3D bar plot or "lollipop plot" (adding stems to a 3D scatterplot to make it easier to locate the points in space; Figure 2.4). 3D graphics do represent a wide new range of opportunities for graphing data, but they are often misused and sometimes actually convey less information than a carefully designed 2D plot; it's hard to design a really good 3D plot. To present 3D graphics in print you also have to pick a single viewpoint, although this is not an issue for exploratory graphics. Finally, R's 3D capabilities are less well developed than those of MATLAB or Mathematica (although the rgl package, which is used in Figure 2.4 and has been partially integrated with the Rcmdr and vegan packages, is under rapid development). A package called ggobi allows you to explore scatterplots of high-dimensional/multivariate data sets.

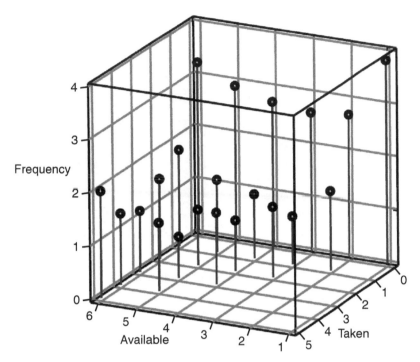

Figure 2.4 3D graphics: lollipop plot produced in rgl (plot3d(...,type="s") to plot spheres, followed by plot3d(...,type="h") to plot stems).

2.4.1.3 FRACTION OF SEEDS TAKEN

It may make more sense to work with the *fraction of seeds taken*, and to see how this varies with number available: Is it constant? Or does the fraction of seeds taken increase with the density of seeds (predator attraction) or decrease (predator saturation) or vary among species?

```
> frac.taken = SeedPred$taken/SeedPred$available
```

Plotting the fraction taken directly (e.g., as a function of number available: plot(SeedPred$available,frac.taken)) turns out to be uninformative, since all of the possible values (e.g. 0/3, 1/3, 2/3, 1) appear in the data set and so there is lots of overlap; we could use sizeplot or jitter again, or we could compute the mean fraction taken as a function of species, date, and number of seeds available.

Suppose we want to calculate the mean fraction taken for each number of seeds available. The command

```
> mean.frac.by.avail = tapply(frac.taken, available,
+     mean, na.rm = TRUE)
```

computes the mean fraction taken (frac.taken) for each different number of seeds available (available: R temporarily converts available into a factor for this purpose). (The tapply command is discussed in more detail in the R supplement.)

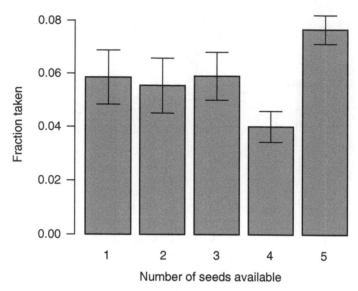

Figure 2.5 Bar plot with error bars: mean fraction taken as a function of number available: barplot2 (mean.frac.by.avail, plot.CI=TRUE,...).

We can also use `tapply` to calculate the standard errors, $\sigma/\sqrt{n}$:

```
> n.by.avail = table(available)
> sd.by.avail = tapply(frac.taken, available, sd,
+      na.rm = TRUE)
> se.by.avail = sd.by.avail/sqrt(n.by.avail)
```

I'll use a variant of `barplot`, `barplot2` (from the `gplots` package), to plot these values with standard errors. R does not supply error-bar plotting as a built-in function, but you can use the `barplot2` (gplots package) or `plotCI` (gplots or plotrix package) function to add error bars to a plot (see the R supplement).

While a slightly larger fraction of available seeds is removed when 5 seeds are available, there is not much variation overall (Figure 2.5). We can use `tapply` to cross-tabulate by species as well; the following commands would show a bar plot of the fraction taken for each combination of number available and species:

```
> mean.frac.by.avail.sp = tapply(frac.taken, list(available,
+      species), mean, na.rm = TRUE)
> mean.frac.by.avail.sp = na.omit(mean.frac.by.avail.sp)
> barplot(mean.frac.by.avail.sp, beside = TRUE)
```

It's often better to use a *box plot* (or *box-and-whisker plot*) to compare continuous data in different groups. Box plots show more information than bar plots, and they show it in a robust form (see p. 49 for an example). However, in this case the box plot is dominated by zeros and so is not very informative.

One more general plotting strategy is to use *small multiples* (Tufte, 2001), breaking the plot into an array of similar plots comparing patterns at different levels (by species, in this case). To make small multiples in base graphics, I would use `par(mfrow=c(row,col))` to divide the plot region into a grid with `row` rows and `col` columns and then draw a plot for each level separately. The `lattice` package handles small multiples automatically, and elegantly. In this case, I used the command

```
> nz = subset(SeedPred, taken > 0)
```

to separate out the cases where at least 1 seed was removed, and then

```
> barchart(table(nz$available, nz$species, nz$taken),
+        stack = FALSE)
```

to plot bar charts showing the distribution of the number of seeds taken for each number available, subdivided by species. (`barchart(...,stack=FALSE)` is the lattice equivalent of `barplot(...,beside=TRUE)`.) In other contexts, the lattice package uses a vertical bar | to denote a small-multiple plot. For example, `bwplot(frac.taken~available|species)` would draw an array of box plots, one for each species, of the fraction of seeds taken as a function of the number available (see p. 49 for another example).

Figure 2.6 shows that the all-or-nothing distribution seen in Figure 2.3 is not just an artifact of lumping all the species together, but holds up at the individual species level. The patterns are slightly different, since in Figure 2.3 we chose to handle the large number of zero cases by log-transforming the number of counts (to compress the range of number of counts), while here we have just dropped the zero cases. Nevertheless, it is still more likely that a small or large fraction of the available seeds will disappear, rather than an intermediate fraction.

We could ask many more questions about these data.

- Is the length of time available for removal important? Although most stations were checked every 7 days, the interval ranged from 3 to 10 days (`table(tint)`). Would separating the data by `tint`, or standardizing to a removal rate (`tint/taken`), show any new patterns?
- Do the data contain more information about the effects of distance from the forest? Would any of Figures 2.2–2.6 show different patterns if we separated the data by distance?
- Do the seed removal patterns vary *along* the transects (remember that the stations are spaced every 5 m along two transects)? Are neighboring stations more likely to be visited by predators? Are there gradients in removal rate from one end of the transect to the other?

However, you may be getting tired of seeds by now. The remaining examples in this chapter show more kinds of graphs and more techniques for rearranging data.

2.4.2 Tadpole Predation Data

The next example data set describes the survival of tadpoles of an African treefrog, *Hyperolius spinigularis*, in field predation trials conducted in large tanks. Vonesh and Bolker (2005) present the full details of the experiment; the goal was to understand the trade-offs that *H. spinigularis* face between avoiding predation in the egg stage (eggs are attached to tree leaves above ponds, and are exposed to predation by other frog species and by parasitoid flies) and in the larval stage (tadpoles drop into the water and are exposed to predation by many aquatic organisms including larval

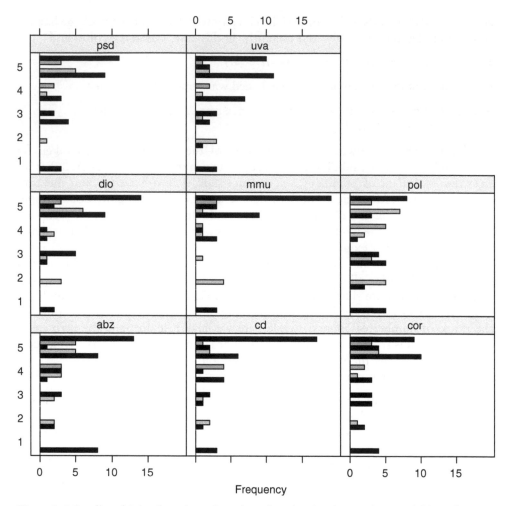

Figure 2.6 Small multiples: bar plots of number of seeds taken by number available and species (barchart(frac.taken|species)).

dragonflies). In particular, juveniles may face a trade-off between hatching earlier (and hence smaller) to avoid egg predators and surviving as tadpoles, since smaller tadpoles are at higher risk from aquatic predators.* Here, we're just going to look at the data as an example of dealing with continuous predictor variables (i.e., exploring how predation risk varies with tadpole size and density).

Since reading in these data is straightforward, we'll take a shortcut and use the data command to pull the data into R from the emdbook package. Three data sets correspond to three different experiments:

- ReedfrogPred: Results of a factorial experiment that quantified the number of tadpoles surviving for 16 weeks (surv: survprop gives the proportion

---

* In fact, the study found that smaller, earlier-hatched tadpoles manage to compensate for this risk by growing faster through the size range in which they are vulnerable to aquatic predators.

surviving) with and without predators (pred), with three different tadpole densities (density), at two different initial tadpole sizes (size).

- ReedfrogSizepred: Data from a more detailed experiment on the effects of size (TBL, tadpole body length) on survival over 3 days (Kill, number killed out of 10).
- ReedfrogFuncresp: Data from a more detailed experiment on the effects of initial tadpole density (Initial) on the number killed over 14 days (Killed).

2.4.2.1 FACTORIAL PREDATION EXPERIMENT (ReedfrogPred)

What are the overall effects of predation, size, density, and their interactions on survival? Figure 2.7 uses boxplot(propsurv~size*density*pred) to display the experimental results (bwplot is the lattice equivalent of boxplot). Box plots show more information than bar plots. In general, you should prefer box plots to bar plots unless you are particularly interested in comparing values to zero (bar plots are anchored at zero, which emphasizes this comparison).

Specifically, the line in the middle of each box represents the median; the ends of the boxes ("hinges") are the first and third quartiles (approximately; see ?boxplot.stats for gory details); the "whiskers" extend to the most extreme data point in either direction that is within a factor of 1.5 of the hinge; any points beyond the whiskers (there happen to be none in Figure 2.7) are considered outliers and are plotted individually. It's clear from the picture that predators significantly lower survival (not surprising). Density and tadpole size also have effects, and may interact (the effect of tadpole size in the predation treatment appears larger at high densities).* The order of the factors in the box plot formula doesn't really change the answers, but it does change the order in which the bars are presented, which emphasizes different comparisons. In general, you should organize bar plots and other graphics to focus attention on the most important or most interesting question: in this case, the effect of predation is so big and obvious that it's good to separate predation from no-predation first so we can see the effects of size and density. I chose size*density*pred to emphasize the effects of size by putting the big- and small-tadpole bars within a density treatment next to each other; density*size*pred would emphasize the effects of density instead.

Box plots are also implemented in the lattice package:

```
> bwplot(propsurv ~ density | pred * size,
+       data = ReedfrogPred, horizontal = FALSE)
```

gives a box plot. Substituting dotplot for bwplot would produce a dotplot instead, which shows the precise value for each experimental unit—good for relatively small data sets like this one, although in this particular example several points fall on top of each other in the treatments where there was high survival.

---

* An analysis of variance on the arcsine-square root transformed proportion surviving (Table 1 in Vonesh and Bolker (2005)) identifies significant effects of density, predator, density × predator and size × predator interactions (i.e., density and size matter only when predators are present), but not a significant density × size × predator interaction. Either the apparent increase in size effect at high densities in the presence of a predator is by chance alone, or the statistical test was not powerful enough to distinguish it from chance.

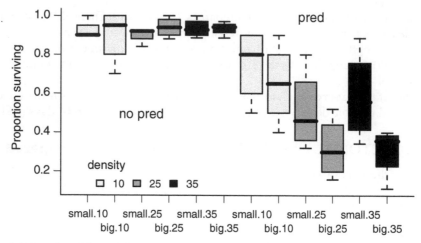

Figure 2.7 Results of factorial experiment on *H. spinigularis* predation: boxplot(propsurv~ size*density*pred,data=ReedfrogPred).

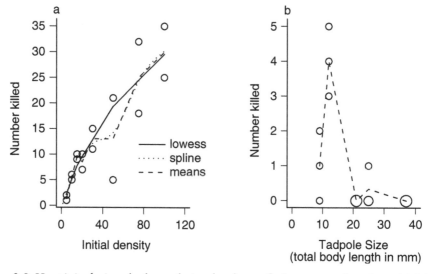

Figure 2.8 *H. spinigularis* tadpole predation by dragonfly larvae as a function of (a) initial density of tadpoles (b) initial size of tadpoles.

2.4.2.1 EFFECTS OF DENSITY AND TADPOLE SIZE

Once the factorial experiment had established the qualitative effects of density and tadpole size on predation, Vonesh ran more detailed experiments to explore the ecological mechanisms at work: how, precisely, do density and size affect predation rate, and what can we infer from these effects about tadpole life history choices?

Figure 2.8 shows the relationship between (a) initial density and (b) tadpole size and the number of tadpoles killed by aquatic predators. The first relationship shows the predator *functional response*—how the total number of prey eaten increases, but saturates, as prey density increases. The second relationship demonstrates a *size refuge*—small tadpoles are protected because they are hidden or ignored by predators, while large tadpoles are too big to be eaten or big enough to escape predators.

Questions about the functional relationship between two continuous variables, asking how one ecological variable affects another, are very common in ecology. Chapter 3 will present a wide variety of plausible mathematical functions to describe such relationships. When we do exploratory data analysis, on the other hand, we want ways of "connecting the dots" that are plausible but that don't make too many assumptions. Typically we're interested in smooth, continuous functions. For example, we think that a small change in initial density should not lead to an abrupt change in the number of tadpoles eaten.

The pioneers of exploratory data analysis invented several recipes to describe such smooth relationships.

- R incorporates two slightly different implementations of *robust locally weighted regression* (`lowess` and `loess`). This algorithm runs linear or quadratic regressions on successive chunks of the data to produce a smooth curve. `lowess` has an adjustable smoothness parameter (in this case the proportion of points included in the "neighborhood" of each point when smoothing) that lets you choose curves ranging from smooth lines that ignore a lot of the variation in the data to wiggly lines that pass through every point; in Figure 2.8a, I used the default value (`lines(lowess(Initial,Killed))`).
- Figure 2.8a also shows a *spline* fit to the data which uses a series of cubic curves to fit the data. Splines also have a smoothing parameter, the *degrees of freedom* or number of different piecewise curves fitted to the data; in this case I set the degrees of freedom to 5 (the default here would be 2) to get a slightly more wiggly curve (`smooth.spline(Initial, Killed,df=5)`).
- Simpler possibilities include just drawing a straight line between the mean values for each initial density (using `tapply(Killed,Initial,mean)` to calculate the means and `unique(Initial)` to get the nonrepeated values of the initial density), or plotting the results of a linear or quadratic regression of the data (not shown; see the R supplement). I plotted straight lines between the means in Figure 2.8b because local robust regression and splines worked poorly.

To me, these data present fewer intriguing possibilities than the seed removal data—primarily because they represent the results of a carefully targeted experiment, designed to answer a very specific question, rather than a more general set of field observations. The trade-off is that there are fewer loose ends; in the end we were actually able to use the detailed information about the shapes of the curves to explain why small tadpoles experienced higher survival, despite starting out at an apparent disadvantage.

2.4.3 Damselfish Data

The next example comes from Schmitt et al.'s (1999) work on a small reef fish, the three-spotted damselfish (*Dascyllus trimaculatus*), in French Polynesia. Like many reef fish, *Dascyllus*'s local population dynamics are *open*. Pelagic larval fish immigrate from outside the area, settling when they arrive on sea anemones. Schmitt et al. were interested in understanding how the combination of larval supply (settler density), density-independent mortality, and density-dependent mortality determines local population densities.

The data are observations of the numbers of settlers found on previously cleared anemones after settlement pulses and observations of the number of subadults recruiting (surviving after 6 months) in an experiment where densities were artificially manipulated.

The settlement data set, `DamselSettlement`, includes 600 observations at 10 sites, across 6 different settlement pulses in 2 years. Each observation records the site at which settlement was observed (`site`), the month (`pulse`), and the number (`obs`) and density per 0.1 m^2 (`density`) of settling larvae. The first recruitment data set, `DamselRecruitment`, gives the anemone area in 0.1 m^2 (`area`), the initial number of settlers introduced (`init`), and the number of recruits (subadults surviving after 6 months: `surv`). The second recruitment data set, `DamselRecruitment_sum`, gives information on the recruitment according to target densities (the densities the experimenters were trying to achieve), rather than the actual experimental densities, and summarizes the data by category. It includes the target settler density (`settler.den`), the mean recruit density in that category after 6 months (`surv.den`), and the standard error of recruit density (`SE`).

2.4.3.1 DENSITY-RECRUITMENT EXPERIMENT

The relationship between settler density and recruit density (Figure 2.9) is ecologically interesting but does not teach us many new graphical or data analysis tricks. I did plot the x axis on a log scale, which shows the low-density data more clearly but makes it harder to see whether the curve fits any of the standard ecological models (e.g., purely density-independent survival would produce a straight line on a regular (linear) scale). Nevertheless, we can see that the number recruiting at high densities levels off (evidence of density-dependent survival) and there is even a suggestion of overcompensation—a decreasing density of recruits at extreme densities.

Settlement Data

The reef fish data also provide us with information about the variability in settlement density across space and time. Schmitt et al. lumped all of these data together, to find out how the distribution of settlement density affects the relative importance of density-independent and density-dependent factors (Figure 2.10).

Figure 2.10 shows a histogram of the settlement densities. Histograms (`hist` in basic graphics or `histogram` in lattice graphics) resemble bar plots but are designed for continuous rather than discrete distributions. They break the data up into evenly spaced categories and plot the number or proportion of data points that fall into each bin. You can use histograms for discrete data, if you're careful to set the breaks between integer values (e.g., at `seq(0,100,by=0.5)`), but `plot(table(x))` and `barplot(table(x))` are generally better. Although histograms are familiar to most ecologists, *kernel density estimators* (`density`: Venables and Ripley, 2002), which produce a smooth estimate of the probability density rather than breaking the counts into discrete categories, are often better than histograms—especially for

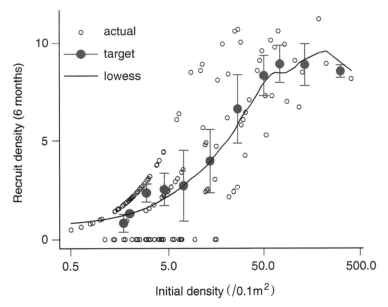

Figure 2.9 Recruit (subadult) *D. trimaculatus* density after 6 months, as a function of experimentally manipulated settler density. Black points show actual densities and survivorship; gray points with error bars show the recruit density, ± 1 SE, by the target density category; line is a lowess fit.

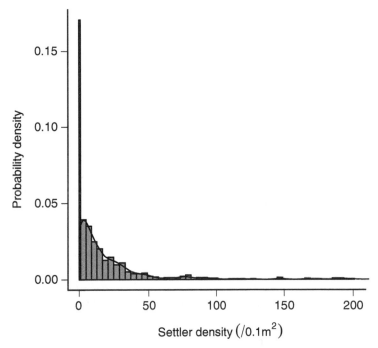

Figure 2.10 Overall distribution of settlement density of *D.trimaculatus* across space and time (only values < 200/(0.1 m^2); 8 values excluded): hist and lines(density(...)).

large data sets. While any form of binning (including kernel density estimation) requires some choice about how finely versus coarsely to subdivide or smooth the data, density estimators have a better theoretical basis for understanding this trade-off. It is also simpler to superimpose densities graphically to compare distributions. The only case where I prefer histograms to densities is when I am interested in the distribution near a boundary such as zero, when density estimation can produce artifacts. Estimating the density and adding it to Figure 2.10 was as simple as `lines(density(setdens))`.

The zero-settlement events are shown as a separate category by using `breaks=c(0,seq(1,200,by=4))`. Rather than plot the number of counts in each category, the *probability density* is shown using `prob=TRUE`, so that the area in each bar is proportional to the number of counts. Perhaps the most striking feature of the histogram is the large number of zeros, but this aspect is downplayed by the original histogram in Schmitt et al. (1999), which plots the zero counts separately but fails to increase the height of the bar to compensate for its narrower width. The zero counts seem to fall into a separate category; ecologically, one might wonder why there are so many zeros, and whether there are any covariates that would predict where and when there were no settlers. Depending on your ecological interests, you also might want to replot the histogram without the zeros to focus attention on the shape of the rest of the distribution.

The histogram also shows that the distribution is very wide (one might try plotting a histogram of $\log(1 + x)$ to compress the distribution). In fact, I excluded the eight largest values from the histogram. (R's histogram function does not have a convenient way to lump "all larger values" into the last bar, as in Schmitt et al.'s original figure.) The first part of the distribution falls off smoothly (once we ignore the zeros), but there are enough extremely large values to make us ask both what is driving these extreme events and what effects they may be having.

Schmitt et al. did not explore the distribution of settlement across time and space. We could use

```
> bwplot(log10(1 + density) ~ pulse | site,
+        data = DamselSettlement, horizontal = FALSE)
```

to plot box-and-whisker plots of settlement divided by pulse, with small multiples for each site, for the damselfish settlement data. We can also use a *pairs plot* (`pairs`) or *scatterplot matrix* (`splom` in the `lattice` package) to explore the structure of multivariate data (many predictor variables, many response variables, or both; Figure 2.11). The pairs plot shows a table of *x-y* plots, one for each pair of variables in the data set. In this case, I've used it to show the correlations between settlement to a few of the different sites in Schmitt et al.'s data set (each site contains multiple reefs where settlement is counted). Because the `DamselSettlement` data set is in long form, we first have to reshape it so that we have a separate variable for each site:

```
> library(reshape)
> x2 = melt(DamselSettlement, measure.var = "density")
> x3 = cast(x2, pulse + obs ~ ...)
```

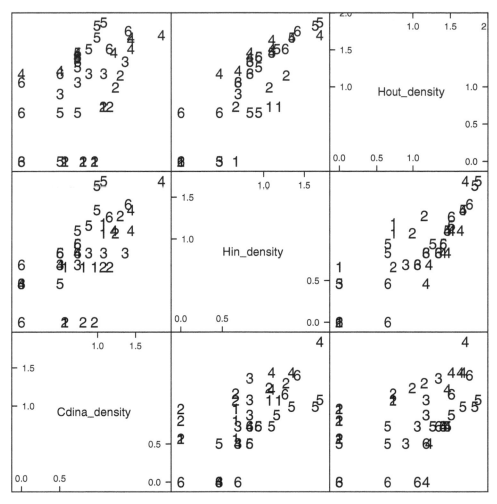

Figure 2.11 Scatterplot matrix of settlement to three selected reefs (logarithm$(1+x)$ scale), with points numbered according to pulse: splom(log10(1+x3[,3:5]),groups=x3$pulse, pch=as.character(1:6)).

The first few rows and columns of the reshaped data set look like this:

	pulse	obs	Cdina_density	Hin_density	Hout_density	...
1	1	1	2.7	0.0	0	...
2	1	2	2.7	0.0	0	...
3	1	3	2.7	0.0	0	...
4	1	4	2.7	3.6	0	...

and we can now use pairs(log10(1+x3[,3:5])) (or splom(log10(1+x3[,3:5]))
to use lattice graphics) to produce the scatterplot matrix (Figure 2.11).

2.4.4 Goby Data

We can explore the effect of density on survival in more detail with another data set on reef fish survivorship, this one on the marine gobies *Elacatinus prochilos* and *E. evelynae* in St. Croix (Wilson, 2004). Like damselfish, larval marine gobies also immigrate to a local site, although these species settle on coral heads rather than on anemones. Wilson experimentally manipulated density in a series of experiments across several years; she replaced fish as they died in order to maintain the local density experienced by focal individuals.*

Previous experiments and observations suggested that patch reefs with higher natural settlement rate have lower mortality rates, once one accounts for the effects of density. Thus reefs with high natural settlement rates were deemed to be of putatively high "quality," and Wilson took the natural settlement rate as an index of quality in subsequent experiments in which she manipulated density.

Reading from a comma-separated file, specifying that the first four columns are factors and the last four are numeric:

```
> gobydat = read.csv("GobySurvival.csv",
+     colClasses = c(rep ("factor", 4), rep("numeric", 4)))
```

Left to its own devices, R would have guessed that the first two columns (experiment number and year) were numeric rather than factors. I could then have converted them back to factors via gobydat$exper=factor(gobydat$exper) and gobydat$year=factor(gobydat$year).

R has an attach command that gives direct access to the columns in a data frame: if we say

```
> attach(gobydat)
```

we can then refer to year, exper, d1 rather than gobydat$year, gobydat$exper, gobydat$d1, and so forth. attach can make your code easier to read, but it can also be confusing; see p. 63 for some warnings.

For each individual monitored, the data give the experiment number (exper: five separate experiments were run between 2000 and 2002) and information about the year and location of the experiment (year, site); information about the location (coral head: head) of each individual and the corresponding density (density) and quality (qual) of the coral head; and the fate of the individual—the last day it was observed (d1) and the first day it was *not* seen (d2, set to 70 if the fish was present on the last sampling day of the experiment). (In survival analysis literature, individuals that are still alive when the study ends are called *right-censored*). Since juvenile gobies of these species rarely disperse, we will assume that a fish that disappears has died.

Survival data are challenging to explore graphically, because each individual provides only a single discrete piece of information (its time of death or disappearance).

---

* Unlike the rest of the data sets in the book, I did not include this one in the emdbook package, since all the analyses have not yet been published. I will include them as soon as they become available; please feel free to contact me (BMB) in the meanwhile if you would like access to them.

In this case we will approximate time of death as halfway between the last time an individual was observed and the first time it was not observed:

```
> meansurv = (d1 + d2)/2
```

For visualization purposes, it will be useful to define low- and high-density and low- and high-quality categories. We will use the ifelse(val,a,b) command to assign value a if val is TRUE or b if val is FALSE, and the factor command to make sure that level low is listed before high even though it is alphabetically after it.

```
> dens.cat = ifelse(density > median(density), "high",
+        "low")
> dens.cat = factor(dens.cat, levels - c("low", "high"))
> qual.cat = ifelse(qual > median(qual), "high", "low")
> qual.cat = factor(qual.cat, levels = c("low", "high"))
```

Figure 2.12 shows an xyplot of the mean survival value, jittered and divided into low- and high-quality categories, with linear-regression lines added to each subplot. There is some mild evidence that mean survival declines with density at low-quality sites, but much of the pattern is driven by the fish with meansurv of > 40 (which are all fish that survived to the end of the experiment) and by the large cluster of short-lived fish at low quality and high densities (> 10).

Let's try calculating and plotting the mortality rate over time, and the proportion surviving over time (the *survival curve*), instead.

Starting by taking all the data together, we would calculate these values by first tabulating the number of individuals disappearing in each time interval:

```
> survtab = table(meansurv)
> survtab
```

```
  meansurv
   1.5    2.5    3.5    5    6    7    9    9.5    10    11    40.5    41
   137    113    17     8    14   3    5    13     4     3     26      26
```

To calculate the number of individuals that disappeared on or after a given time, reverse the table (rev) and take its cumulative sum (cumsum):

```
> csurvtab = cumsum(rev(survtab))
> csurvtab
```

```
   41    40.5    11    10    9.5    9    7    6    5    3.5    2.5    1.5
   26    52      55    59    72     77   80   94   102   119    232    369
```

Reversing the vector again sorts it into order of increasing time:

```
> csurvtab = rev(csurvtab)
```

To calculate the proportional mortality at each time step, divide the number disappearing by the total number still present (I have rounded to two digits):

```
> survtab/csurvtab
```

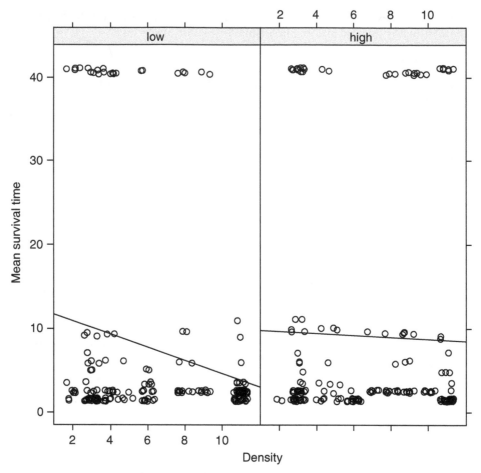

Figure 2.12 Mean survival time as a function of density, divided by quality (background settlement) category: xyplot.

```
meansurv
1.5   2.5   3.5   5     6     7     9     9.5   10    11    40.5  41
0.37  0.49  0.14  0.08  0.15  0.04  0.06  0.18  0.07  0.05  0.50  1.00
```

Figure 2.13 plots the proportion dying and survival curves by quality/density category. The plot of proportion dying is very noisy but does suggest that the disappearance rate starts relatively high ($\approx$ 50% per observation period) and then decreases (the end of the experiment gets *very* noisy, and was left off the plot). The survival curve is clearer. Since it is plotted on a logarithmic scale, the leveling off of the curves is an additional indication that the mortality rate decreases with time (constant mortality would lead to exponential decline, which would appear as a straight line on a logarithmic graph). As expected, the low-quality, high-density treatment has the lowest proportion surviving, with the other three treatments fairly closely clustered and not in the expected order (we would expect the high-quality, low-density treatment to have the highest survivorship). Don't forget to clean up with detach (gobydat).

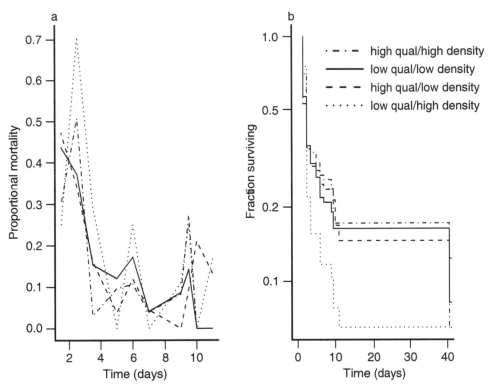

Figure 2.13 Goby survival data: proportional mortality and fraction surviving over time, for different quality/density categories.

2.5 Conclusion

This chapter has given an overview and examples of how to get data into R and start plotting it in various ways to ask ecological questions. I overlooked a variety of special kinds of data (e.g., circular data such as directional data or daily event times; highly multivariate data; spatial data and maps; compositional data, where the sum of proportions in different categories adds to 1.0); Table 2.1 gives some ideas for handling these data types, but you may also have to search elsewhere, for example, using RSiteSearch("circular") to look for information on circular statistics.

2.6 R Supplement

All of the R code in the supplements is available from http://press.princeton .edu/titles/8709.html in an electronic format that will be easier to cut and paste from, in case you want to try it out for yourself (you should).

TABLE 2.1
Summary of Graphical Procedures

Predictors	Response	Plot Choices
Single categorical	Single categorical	`table`, `barplot` , `dotchart`, `barchart` [L], `dotplot` [L]
Multiple categorical	Single categorical	as above, plus `mosaicplot`, small multiples (`par(mfrow)`/`par(mfcol)` or `lattice` plots), `sizeplot` [plotrix] or 3D histogram [scatterplot3d, rgl]
Circular	Categorical	`rose.diag` [CircStats]
Circular	Continuous	`polar.plot` [plotrix]
None	Compositional	`barplot(...,beside=FALSE)`, `barchart(...,stack=TRUE)` [L], `ternaryplot` [vcd], `triax.plot` [plotrix]
Single categorical	Multiple continuous	`stars`
None or single categorical	Single continuous	`boxplot`, `bwplot` [L], violin plots (`bwplot(...,panel=panel.violin)` [L]), `vioplot` [vioplot], `stripplot` [L], `barplot2` [gplot] for error bars
Continuous + categorical	Single continuous	scatterplot (`plot` , `xyplot` [L]) with categories indicated by plotting symbols (`pch`), color (`col`), size (`cex`) or (in `lattice`) groups argument
Single continuous	Single continuous	`plot` , `xyplot` [L]; `lowess`, `supsmu`, `smooth.spline` for curves; `plotCI` [gplots or plotrix] for error bars
Multiple continuous	Multiple continuous	conditioning plots (`coplot` or `lattice` plots), 3D scatter- or lollipop plots (`cloud` [L], `scatterplot3d` [scatterplot3d] or `plot3d` [rgl])
Continuous (time or 1D space)	Continuous	`plot`/`xyplot` with `type="l"` or `type="b"`
Continuous (2D space)	Continuous	`image`, `contour`, `persp`, `kde2d` [MASS], `wireframe` [L], `surface3d` [rgl], `maps` package, `maptools` package, `sp` package

Square brackets denote functions in packages; [L] denotes functions in the `lattice` package.

2.6.1 Clean, Reshape, and Read In Data

To let R know where your data files are located, you have a few choices:

- Spell out the *path*, or file location, explicitly. (Use a single forward slash to separate folders (e.g., `"c:/Users/bolker/My Documents/R/script.R"`); this works on all platforms.)

- Use `filename=file.choose()`, which will bring up a dialog box to let you choose the file and set `filename`. This works on all platforms but is useful only on Windows and MacOS.
- Use menu commands to change your working directory to wherever the files are located: `File/Change dir` (Windows) or `Misc/Change Working Directory` (Mac).
- Change your working directory to wherever the files are located using the `setwd` (set working directory) function, e.g., `setwd("c:/temp")`.

Changing your working directory is more efficient in the long run. You should save all the script and data files for a particular project in the same directory and switch to that directory when you start work.

The seed removal data were originally stored in two separate Excel files, one for the 10-m transect and one for the 25-m transect. After a couple of preliminary errors I decided to include `na.strings="?"` (to turn question marks into NAs) and `comment=""` (to deal with a # character in the column names—although I could also have edited the Excel file to remove it):

```
> dat_10 = read.csv("duncan_10m.csv", na.strings = "?",
+      comment = "")
> dat_25 = read.csv("duncan_25m.csv", na.strings = "?",
+      comment = "")
```

`str` and `summary` originally showed that I had some extra columns and rows: row 160 of `dat_10` and columns 40–44 of `dat_25` were junk. I could have gotten rid of them this way:

```
> dat_10 = dat_10[1:159, ]
> dat_25 = dat_25[, 1:39]
```

(I could also have used *negative* indices to drop specific rows or columns: `dat_10 [-160,]` and `dat_25[-(40:44),]` would have the same effect).

Now we reshape the data, specifying `id.var=1:2` to preserve the first two columns, station and species, as identifier variables:

```
> library(reshape)
> dat_10_melt = melt(dat_10, id.var = 1:2)
```

Convert the third column to a date, using `paste` to append 1999 to each date (`sep="."` separates the two pasted strings with a period):

```
> date_10 = paste(dat_10_melt[, 3], "1999", sep = ".")
```

Then use `as.Date` to convert the string to a date (%d means day, %b means three-letter month abbreviation, and %Y means four-digit year; check `?strptime` for more date format details). Because the column names in the Excel file began with numbers (e.g., 25-Mar), R automatically added an X to the beginning as well as converting dashes to periods (X25.Mar)—include this in the format string:

```
> dat_10_melt[, 3] = as.Date(date_10, format = "X%d.%b.%Y")
```

Finally, rename the columns.:

```
> names(dat_10_melt) = c("station", "species", "date",
+      "seeds")
```

Do the same for the 25-m transect data:

```
> dat_25_melt = melt(dat_25, id.var = 1:2)
> date_25 = paste(dat_25_melt[, 3], "1999", sep = ".")
> dat_25_melt[, 3] = as.Date(date_25, format = "X%d.%b.%Y")
> names(dat_25_melt) = c("station", "species", "date",
+      "seeds")
```

We've finished cleaning up and reformatting the data. Now we would like to calculate some derived quantities: specifically, tcum (elapsed time from the first sample), tint (time since previous sample), taken (number removed since previous sample), and available (number available at previous sample). We'll split the data frame up into a separate chunk for each station:

```
> split_10 = split(dat_10_melt, dat_10_melt$station)
```

for loops are a general way of executing similar commands many times. A for loop runs for every value in a vector.

```
for (var in vec) {
  commands
}
```

runs the R commands inside the curly brackets once for each element of vec, each time setting the variable var to the corresponding element of vec. The most frequent use of for loops is to run a set of commands n times by making vec equal 1:n.

For each data chunk (corresponding to the data from one station), we want to calculate (1) the cumulative time elapsed by subtracting the first date from all the dates; (2) the time interval since the previous observation by taking the difference of successive dates (with diff) and putting an NA at the beginning; (3) the number of seeds lost since the previous observation by taking the *negative* of the difference of successive numbers of seeds and prepending an NA; and (4) the number of seeds available at the previous observation by prepending NA and dropping the last element. We then put the new derived variables together with the original data and reassign them.

The for loop below does all these calculations for each chunk by executing each statement inside the curly brackets {}, setting i to each value between 1 and the number of stations:

```
> for (i in 1:length(split_10)) {
+      x = split_10[[i]]
+      tcum = as.numeric(x$date - x$date[1])
+         #(1) cumulative time
+      tint = as.numeric(c(NA, diff(x$date)))
+         #(2) time interval
+      taken = c(NA, -diff(x$seeds)) #(3) seeds taken
+      available = c(NA, x$seeds[-nrow(x)]) #(4) seeds
```

```
+           available
+       split_10[[i]] = data.frame(x, tcum, tint, taken,
+           available)
+ }
```

Now we want to stick all of the chunks of the data frame back together. rbind (for row **bind**) combines columns, but normally we would say rbind(x,y,z) to combine three matrices or data frames with the same number of columns. If, as in this case, we have a *list* of matrices that we want to combine, we have to use do.call("rbind",list) to apply rbind to the list:

```
> dat_10 = do.call("rbind", split_10)
```

The approach shown here is also useful when you have individuals or stations that have data recorded only for the first observation of the individual. In some cases you can also do these manipulations by working with the data in wide format.

Do the same for the 25-m data (not shown).

Create new data frames with an extra column that gives the distance from the forest (rep is the R command to **rep**eat values); then stick them together.

```
> dat_10 = data.frame(dat_10, dist = rep(10, nrow(dat_10)))
> dat_25 = data.frame(dat_25, dist = rep(25, nrow(dat_25)))
> SeedPred = rbind(dat_10, dat_25)
```

Convert station and distance from numeric to factors:

```
> SeedPred$station = factor(SeedPred$station)
> SeedPred$dist = factor(SeedPred$dist)
```

Reorder columns:

```
> SeedPred = SeedPred[, c("station", "dist", "species",
+     "date", "seeds", "tcum", "tint", "taken", "available")]
```

2.6.2 Plots: Seed Data

2.6.2.1 MEAN NUMBER REMAINING WITH TIME

Attach the seed removal (predation) data:

```
> attach(SeedPred)
```

Using attach can make your code easier to read, since you don't have to put SeedPred$ in front of the column names, but it's important to realize that attaching a data frame makes a local copy of the variables. Changes that you make to these variables are *not* saved in the original data frame, which can be very confusing. Therefore, it's best to use attach only after you've finished modifying your data. attach can also be confusing if you have columns with the same name in two different attached data frames: use search to see where R is looking for variables. It's best to attach just one data frame at a time—and make sure to detach it when you finish.

Separate out the 10-m and 25-m transect data from the full seed removal data set:

```
> SeedPred_10 = subset(SeedPred, dist == 10)
> SeedPred_25 = subset(SeedPred, dist == 25)
```

The `tapply` (for table **apply**, pronounced "t apply") function splits a vector into groups according to the list of factors provided, then *applies* a function (e.g., `mean` or `sd`) to each group. To split the data on numbers of seeds present by `date` and `species` and take the mean (`na.rm=TRUE` says to drop NA values):

```
> s10_means = tapply(SeedPred_10$seeds,
+      list(SeedPred_10$date, SeedPred_10$species),
+      mean, na.rm = TRUE)
> s25_means = tapply(SeedPred_25$seeds,
+      list(SeedPred_25$date, SeedPred_25$species),
+      mean, na.rm = TRUE)
```

`matplot` ("**matrix plot**") plots all the columns of a matrix against a single *x* variable. Use it to plot the 10-m data on a log scale (`log="y"`) with both lines and points (`type="b"`), in black (`col=1`), with plotting characters (`pch`) 1 through 8, with solid lines (`lty=1`). Use `matlines` ("**matrix lines**") to add the 25-m data in gray. (`lines` and `points` are the base graphics commands to add lines and points to an existing graph.)

```
> matplot(s10_means, log = "y", type = "b", col = 1,
+      pch = 1:8, lty = 1)
> matlines(s25_means, type = "b", col = "gray", pch = 1:8,
+      lty = 1)
```

2.6.2.2 SEED DATA: DISTRIBUTION OF NUMBER TAKEN VERSUS AVAILABLE

Jittered plot:

```
> plot(jitter(SeedPred$available), jitter(SeedPred$taken))
```

Bubble plot:

```
> library(plotrix)
> sizeplot(SeedPred$available, SeedPred$taken, scale = 0.5,
+      pow = 0.5, xlim = c(-2, 6), ylim = c(-2, 5))
```

This plot differs from Figure 2.2 because I don't exclude cases where there are no seeds available. (I use `xlim` and `ylim` to extend the axes slightly.) `scale` and `pow` can be tweaked to change the size and scaling of the symbols.

To plot the numbers in each category, I use `text`, `row` to get row numbers, and `col` to get column numbers; I subtract 1 from the row and column numbers to plot values starting at zero.

```
> t1 = table(SeedPred$available, SeedPred$taken)
> text(row(t1) - 1, col(t1) - 1, t1)
```

Or you can use `balloonplot` from the gplots package:

```
> library(gplots)
> balloonplot(t1)
```

Finally, you can use the default mosaic plot, either using the default `plot` command on the existing tabulation

```
> plot(t1)
```

or using `mosaicplot` with a formula based on the columns of `SeedPred`:

```
> mosaicplot(~available + taken, data = SeedPred)
```

Bar plot:

```
> barplot(t(log10(t1 + 1)), beside = TRUE,
+     xlab = "Available", ylab = "log10(1+# observations)")
```

or

```
> barplot(t(t1 + 1), log = "y", beside = TRUE,
+     xlab = "Available", ylab = "1+# observations")
```

Bar plot of mean fraction taken:

```
> mean.frac.by.avail = tapply(frac.taken, available,
+     mean, na.rm = TRUE)
> n.by.avail = table(available)
> se.by.avail = tapply(frac.taken, available, sd,
+     na.rm = TRUE)/sqrt(n.by.avail)
> barplot2(mean.frac.by.avail, plot.ci = TRUE,
+     ci.l = mean.frac.by.avail - se.by.avail,
+     ci.u = mean.frac.by.avail + se.by.avail,
+     xlab = "Number available", ylab = "Fraction taken")
```

Bar plot of mean fraction taken *by species*—in this case we use `barplot`, saving the *x* locations of the bars in a variable b, and then add the confidence intervals with `plotCI`:

```
> library(plotrix)
> frac.taken = SeedPred$taken/SeedPred$available
> mean.frac.avail.by.species = tapply(frac.taken,
+     list(available, species), mean, na.rm = TRUE)
> n.avail.by.species = table(available, species)
> se.avail.by.species = tapply(frac.taken, list(available,
+     species), sd, na.rm = TRUE)/sqrt(n.avail.by.species)
> b = barplot(mean.frac.avail.by.species, beside = TRUE)
> plotCI(b, mean.frac.avail.by.species,
+     se.avail.by.species, add = TRUE, pch = ".", gap = FALSE)
```

3D Plots

Using t1 from above, define the *x*, *y*, and *z* variables for the plot:

```
> avail = row(t1)[t1 > 0]
> taken = col(t1)[t1 > 0] - 1
> freq = log10(t1[t1 > 0])
```

The scatterplot3d library is simpler to use, but less interactive—once the plot is drawn you can't change the viewpoint. Plot -avail and -taken to reverse the order of the axes and use type="h" (originally named for a "high-density" plot in R's 2D graphics) to draw lollipops:

```
> library(scatterplot3d)
> scatterplot3d(-avail, -taken, freq, type = "h", angle = 50,
+      pch = 16)
```

With the rgl library, first plot spheres (type="s") hanging in space:

```
> library(rgl)
> plot3d(avail, taken, freq, lit = TRUE, col.pt = "gray",
+      type = "s", size = 0.5, zlim = c(0, 4))
```

Then add stems and grids to the plot:

```
> plot3d(avail, taken, freq, add = TRUE, type = "h",
+      size = 4, col = gray(0.2))
> grid3d(c("x+", "y-", "z"))
```

Use the mouse to move the viewpoint until you like the result.

2.6.2.3 HISTOGRAM/SMALL MULTIPLES

Using lattice graphics, as in the text:

```
> histogram(~frac.taken | species, xlab = "Fraction taken")
```

or with base graphics:

```
> op = par(mfrow = c(3, 3))
> for (i in 1:length(levels(species))) {
+      hist(frac.taken[species == levels(species)[i]],
+           xlab = "Fraction taken", main = "", col = "gray")
+ }
> par(op)
```

op stands for "old parameters" (you can name this variable anything you want). Saving the old parameters in this way and using par(op) at the end of the plot restores the original graphical parameters.

Clean up:

```
> detach(SeedPred)
```

2.6.3 Tadpole Data

As mentioned in the text, reading in the data was fairly easy in this case: read.table(...,header=TRUE) and read.csv worked without any tricks. I take a shortcut, therefore, to load these data sets from the emdbook library:

```
> data(ReedfrogPred)
> data(ReedfrogFuncresp)
> data(ReedfrogSizepred)
```

2.6.3.1 BOX PLOT OF FACTORIAL EXPERIMENT

The box plot is fairly easy:

```
> graycols = rep(rep(gray(c(0.4, 0.7, 0.9)), each = 2), 2)
> boxplot(propsurv ~ size * density * pred,
+       data = ReedfrogPred, col = graycols)
```

Play around with the order of the factors to see what the different plots tell you.

graycols specifies the colors of the bars to mark the different density treatments. gray(c(0.4,0.7,0.9)) produces a vector of three colors; rep(gray(c(0.4,0.7,0.9)),each=2) repeats each color twice (for the big and small treatments within each density treatment; and rep(rep(gray(c(0.4,0.7,0.9)), each=2),2) repeats the whole sequence twice (for the no-predator and predator treatments).

2.6.3.2 FUNCTIONAL RESPONSE VALUES

First attach the functional response data:

```
> attach(ReedfrogFuncresp)
```

A simple x-y plot, with an extended x axis and some axis labels:

```
> plot(Initial, Killed, xlim = c(0, 100), ylab = "Number
+       killed", xlab = "Initial density")
```

Adding the lowess fit (lines is the general command for adding lines to a plot: points is handy too):

```
> lines(lowess(Initial, Killed))
```

Calculate mean values and corresponding initial densities, add to the plot with a different line type:

```
> meanvals = tapply(Killed, Initial, mean)
> densvals = unique(Initial)
> lines(densvals, meanvals, lty = 3)
```

Fit a spline to the data using the smooth.spline command:

```
> lms = smooth.spline(Initial, Killed, df = 5)
```

To add the spline curve to the plot, I have to use `predict` to calculate the predicted values for a range of initial densities, then add the results to the plot:

```
> ps = predict(lms, x = 0:100)
> lines(ps, lty = 2)
```

Equivalently, I could use the `lm` function with `ns` (natural spline), which is a bit more complicated in this case but has more general uses:

```
> library(splines)
> lm1 = lm(Killed ~ ns(Initial, df = 5),
+       data = ReedfrogSizepred)
> p1 = predict(lm1, newdata = data.frame(Initial = 1:100))
> lines(p1, lty = 2)
```

Finally, I could do linear or quadratic regression (I need to use `I(Initial^2)` to tell R I really want to fit the square of the initial density); adding the lines to the plot would follow the procedure above.

```
> lm2 = lm(Killed ~ Initial, data = ReedfrogSizepred)
> lmq = lm(Killed ~ Initial + I(Initial^2),
+       data = ReedfrogSizepred)
```

Clean up:

```
> detach(ReedfrogFuncresp)
```

The (tadpole size) vs. (number killed) plot follows similar lines, although I did use `sizeplot` because there were overlapping points.

2.6.4 Damselfish data

2.6.4.1 SURVIVORS AS A FUNCTION OF DENSITY

Load and attach data:

```
> data(DamselRecruitment)
> data(DamselRecruitment_sum)
> attach(DamselRecruitment)
> attach(DamselRecruitment_sum)
```

Plot surviving vs. initial density (scaled to per $0.1\,m^2$); use `plotCI` to add the summary data by target density; and add a `lowess`-smoothed curve to the plot:

```
> plot(init/area * 1000, surv/area * 1000, log = "x")
> plotCI(settler.den, surv.den, SE, add = TRUE, pch = 16,
+     col = "darkgray", gap = 0)
> lines(lowess(init.dens, surv.dens))
```

Clean up:

```
> detach(DamselRecruitment)
> detach(DamselRecruitment_sum)
```

2.6.4.2 DISTRIBUTION OF SETTLEMENT DENSITY

Plot the histogram (normally one would specify freq=FALSE to plot probabilities rather than counts, but the uneven breaks argument makes this happen automatically).

```
> attach(DamselSettlement)
> hist(density[density < 200], breaks = c(0, seq(1,
+      201, by = 4)), col = "gray", xlab = "",
+      ylab = "Prob. density")
> lines(density(density[density < 200], from = 0))
```

The last command is potentially confusing because density is both a data vector (settlement density) and a built-in R command (kernel density estimator), but R can tell the difference.
Some alternatives to try:

```
> hist(log(1 + density))
> hist(density[density > 0], breaks = 50)
```

(you can use breaks either to specify particular breakpoints or to give the total number of bins to use).
If you really want to lump all the large values together:

```
> h1 = hist(density, breaks = c(0, seq(1, 201, by = 4),
+      500), plot = FALSE)
> b = barplot(h1$counts, space = 0)
> axis(side = 1, at = b, labels = h1$mids)
```

These commands (1) use hist to calculate the number of counts in each bin without plotting anything; (2) use barplot to plot the values (ignoring the uneven width of the bins!), with space=0 to squeeze them together; and (3) add a custom x axis.
Box-and-whisker plots:

```
> bwplot(log10(1 + density) ~ pulse | site,
+      data = DamselSettlement, horizontal = FALSE)
```

Other variations to try:

```
> densityplot(~density, groups = site,
+      data = DamselSettlement, xlim = c(0, 100))
> bwplot(density ~ site, horizontal = FALSE,
+      data = DamselSettlement)
> bwplot(density ~ site | pulse, horizontal = FALSE,
+      data = DamselSettlement)
> bwplot(log10(1 + density) ~ site | pulse,
+      data = DamselSettlement,
+      panel = panel.violin, horizontal = FALSE)
> boxplot(density ~ site * pulse)
```

Scatterplot matrices: first reshape the data.

```
> library(reshape)
> x2 = melt(DamselSettlement, measure.var = "density")
> x3 = cast(x2, pulse + obs ~ ...)
```

Scatterplot matrix of columns 3 to 5 (sites Cdina, Hin, and Hout)—using base graphics:

```
> pairs(log10(1 + x3[, 3:5]))
```

Using lattice graphics:

```
> splom(log10(1 + x3[, 3:5]), groups = x3$pulse,
+       pch = as.character(1:6), col = 1)

> detach(DamselSettlement)
```

2.6.5 Goby Data

Plotting mean survival by density subdivided by quality category:

```
> attach(gobydat)
> xyplot(jitter(meansurv, factor = 2) ~ jitter(density, 2)
+       | qual.cat, xlab = "Density", ylab = "Mean
+       survival time")
```

The default amount of jittering is too small, so factor=2 doubles it; see ?jitter for details.

2.6.5.1 LATTICE PLOTS WITH SUPERIMPOSED LINES AND CURVES

To add "extras" like extra points, linear regression lines, or loess fits to lattice graphics, you have to write a new panel function, combining a a default lattice panel function (usually called panel.xxx, e.g., panel.xyplot, panel.densityplot) with components from ?panel.functions. For example, here is a panel function that plots an *x-y* plot and adds a linear regression line:

```
> panel1 = function(x, y) {
+     panel.xyplot(x, y)
+     panel.lmline(x, y)
+ }
```

Then call the original lattice function with the new panel function:

```
> xyplot(jitter(meansurv, factor = 2) ~ jitter(density, 2)
+       | qual.cat, xlab = "Density", ylab = "Mean
+       survival time", panel = panel1)
> detach(gobydat)
```

2.6.5.2 PLOTTING SURVIVAL CURVES

First set up categories for different combinations of quality and density by using interaction, and count the number of observations in each combination.

```
> intcat = interaction(qual.cat, dens.cat)
> cattab = table(intcat)
```

Tabulate the number disappearing at each time in each category:

```
> survtab = table(meansurv, intcat)
```

Reverse order and calculate the cumulative sum by column (margin 2):

```
> survtab = survtab[nrow(survtab):1, ]
> csurvtab = apply(survtab, 2, cumsum)
```

Divide each column (survival curve per category) by the total number for that category:

```
> cnsurvtab = sweep(csurvtab, 2, cattab, "/")
```

Calculate the fraction disappearing at each time:

```
> fracmort = survtab/csurvtab
```

Extract the time coordinate:

```
> days = as.numeric(rownames(csurvtab))
```

Plot survival curves by category:

```
> matplot(days, cnsurvtab, type = "s", xlab = "Time (days)",
+     ylab = "Proportion of cohort surviving", log = "y")
```

3 Deterministic Functions for Ecological Modeling

This chapter first covers the mathematical tools and R functions that you need in order to figure out the shape and properties of a mathematical function from its formula. It then presents a broad range of frequently used functions and explains their general properties and ecological uses.

3.1 Introduction

You've now learned how to start exploring the patterns in your data. The methods introduced in Chapter 2 provide only *qualitative* descriptions of patterns: when you first explore your data, you don't want to commit yourself to any particular description of those patterns. To tie the patterns to ecological theory, however, we often want to use particular mathematical functions to describe the deterministic patterns in the data. Sometimes phenomenological descriptions, intended to describe the pattern as simply and accurately as possible, are sufficient. Whenever possible, however, it's better to use mechanistic descriptions with meaningful parameters, derived from a theoretical model that you or someone else has invented to describe the underlying processes driving the pattern. (Remember from Chapter 1 that the same function can be either phenomenological or mechanistic depending on context.) In any case, you need to know something about a wide range of possible functions, and need even more to learn (or remember) how to discover the properties of a new mathematical function. This chapter first presents a variety of analytical and computational methods for finding out about functions, and then goes through a "bestiary" of useful functions for ecological modeling. The chapter uses differential calculus heavily. If you're rusty, it would be a good idea to look at the appendix for some reminders.

For example, look again at the data introduced in Chapter 2 on predation rate of tadpoles as a function of tadpole size (Figure 3.1). We need to know what kinds of functions might be suitable for describing these data. The data are humped in the middle and slightly skewed to the right, probably reflecting the balance between small tadpoles' ability to hide from (or be ignored by) predators and large tadpoles' ability to escape them or be too big to swallow. What functions could fit this pattern? What

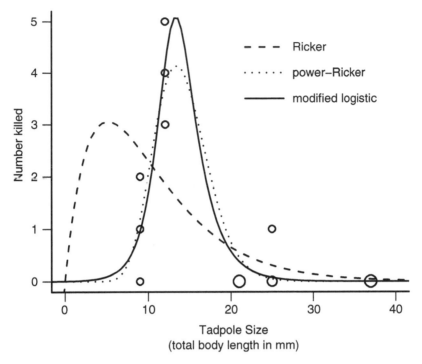

Figure 3.1 Tadpole predation as a function of size, with some possible functions fitted to the data.

do their parameters mean in terms of the shapes of the curves? In terms of ecology? How do we "eyeball" the data to obtain approximate parameter values, which we will need as a starting point for more precise estimation and as a check on our results?

The Ricker function, $y = axe^{-bx}$, is a standard choice for hump-shaped ecological patterns that are skewed to the right, but Figure 3.1 shows that it doesn't fit well. Two other choices, the power-Ricker (Persson et al., 1998) and a modified logistic equation (Vonesh and Bolker, 2005), fit pretty well; later in the chapter we will explore some strategies for modifying standard functions to make them more flexible.

3.2 Finding Out about Functions Numerically

3.2.1 Calculating and Plotting Curves

You can use R to experiment numerically with different functions. It's better to experiment numerically *after* you've got some idea of the mathematical and ecological meanings of the parameters; otherwise you may end up using the computer as an expensive guessing tool. Having some idea what the parameters of a function mean will allow you to eyeball your data to get a rough idea of the appropriate values, and to tweak the parameters intelligently when necessary. Nevertheless, I'll show you first some of the ways that you can use R to compute and draw pictures of functions so that you can sharpen your intuition as we go along.

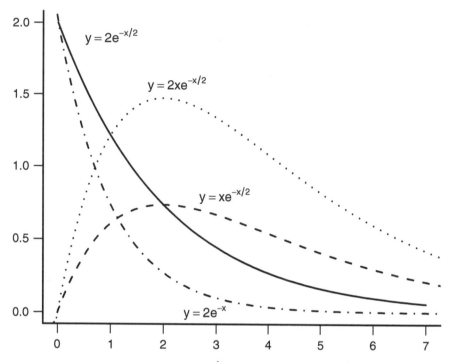

Figure 3.2 Negative exponential ($y = ae^{-bx}$) and Ricker ($y = axe^{-bx}$) functions: curve.

As examples, I'll use the (negative) exponential function, ae^{-bx} (R uses exp(x) for the exponential function e^x) and the Ricker function, axe^{-bx}. Both are very common in ecological modeling.

As a first step, you can simply use R as a calculator to plug values into functions, e.g., 2.3*exp(1.7*2.4). Since most functions in R operate on vectors (or "are vectorized," ugly as the expression is), you can calculate values for a range of inputs or parameters with a single command.

Next simplest, you can use the curve function to have R compute and plot values for a range of inputs: use add=TRUE to add curves to an existing plot (Figure 3.2). (Remember the differences between mathematical and R notation: the exponential is ae^{-bx} or $a\exp(-bx)$ in math notation, but it's a*exp(-b*x) in R. Using math notation in a computer context will give you an error. Using computer notation in a math context is just ugly.)

If you want to keep the values of the function and do other things with them, you may want to define your own vector of x values (with seq: call it something like xvec) and then use R to compute the values (e.g., xvec = seq(0,7,length=100); y = a * exp(-b * xvec)).

If the function you want to compute *does not* work on a whole vector at once, then you can't use either of the above recipes. The easiest shortcut in this case, and a worthwhile thing to do for other reasons, is to write your own small R function that computes the value of the function for a given input value, then use sapply to run the function on all of the values in your x vector. When you write such an R function, you would typically make the input value (x) be the first argument, followed by all of the other parameters. Often typing and assigning *default values* to the other parameters saves time; in the following example, both *a* and *b* have default values of 1.

```
> ricker = function(x, a = 1, b = 1) {
+     a * x * exp(-b * x)
+ }
> yvals = sapply(xvec, ricker)
```

(In this case, since `ricker` uses vectorized operations only, `ricker(xvec)` would work just as well.)*

3.2.2 Plotting Surfaces

Functions of two (or more) variables are a bit more complicated to handle: R's range of 3D graphics is more limited, it is harder to vectorize operations over two different parameters, and you may want to compute the value of the function so many times that you have to worry about computational efficiency (this is our first hint of the so-called *curse of dimensionality*, which will come back to haunt us later).

Base R doesn't have exact multidimensional analogues of `curve` and `sapply`, but the emdbook package supplies some: `curve3d` and `apply2d`. The `apply2d` function takes an *x* vector and a *y* vector and computes the value of a function for all of the combinations, while `curve3d` does the same thing for surfaces that `curve` does for curves: it computes the function value for a range of values and plots it.[†] The basic function for plotting surfaces in R is `persp`. You can also use `image` or `contour` to plot 2D graphics, or `wireframe` [lattice package], or `persp3d` [rgl package] as alternatives to `persp`. With `persp` and `wireframe`, you may want to play with the viewing point for the 3D perspective (modify `theta` and `phi` for `persp` and `screen` for `wireframe`); the `rgl` package lets you use the mouse to move the viewpoint.

For example, Vonesh and Bolker (2005) combined size- and density-dependent tadpole mortality risk by using a modified logistic function of size as in Figure 3.1 to compute an attack rate $\alpha(s)$, then assuming that per capita mortality risk declines with density *N* as $\alpha(s)/(1 + \alpha(s)HN)$, where *H* is the handling time (i.e., a Holling type II functional response). Suppose we already have a function `attackrate` that computes the attack rate as a function of size. Then our mortality risk function would be

```
> mortrisk = function(N, size, H = 0.84) {
+     a <- attackrate(size)
+     a/(1 + a * N * H)
+ }
```

The `H=0.84` in the function definition sets the default value of the handling time parameter.

---

* The definition of "input values" and "parameters" is flexible. You can also compute the values of the function for a fixed value of *x* and a range of one of the parameters, e.g., `ricker(1,a=c(1.1,2.5,3.7))`.

[†] For simple functions you can use the built-in `outer` function, but `outer` requires vectorized functions: `apply2d` works around this limitation.

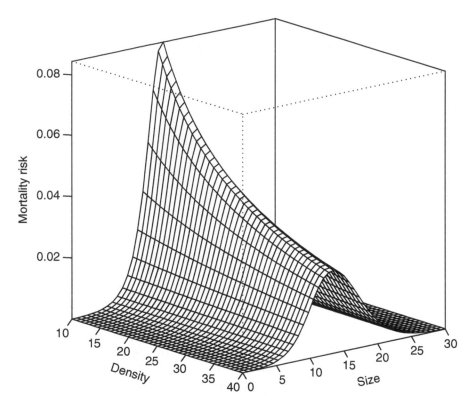

Figure 3.3 Perspective plot for the mortality risk function used in Vonesh and Bolker (2005): `curve3d(mortrisk(N=x, size=y),to=c(40,30),theta=50)`.

3.3 Finding Out about Functions Analytically

Exploring functions numerically is quick and easy, but limited. To fully understand a function's properties, you must explore it analytically—i.e., you have to analyze its equation mathematically. To do this and then translate your mathematical intuition into ecological intuition, you must remember some algebra and calculus. In particular, this section will explain how to take limits at the ends of the range of the function; understand the behavior in the middle of the range; find critical points; understand what the parameters mean and how they affect the shape of the curve; and approximate the function near an arbitrary point (Taylor expansion). These tools will probably tell you everything you need to know about a function.

3.3.1 Taking Limits: What Happens at Either End?

FUNCTION VALUES

You can take the limit of a function as x gets large ($x \to \infty$) or small ($x \to 0$, or $x \to -\infty$ for a function that makes sense for negative x values). The basic principle is to throw out lower-order terms. As x grows, it will eventually grow much larger than the largest constant term in the equation. Terms with larger powers of x will dwarf smaller powers, and exponentials will dwarf any power. If x is very small, then you apply the same logic in reverse; constants are bigger than (positive) powers of x, and negative powers ($x^{-1} = 1/x$, $x^{-2} = 1/x^2$, etc.) are bigger than any constants. (Negative exponentials go to 1 as x approaches zero and 0 as x approaches ∞.) Exponentials are stronger than powers: $x^{-n}e^x$ eventually gets big and $x^n e^{-x}$ eventually gets small as x increases, no matter how big n is.

Our examples of the exponential and the Ricker function are almost too simple: we already know that the negative exponential function approaches 1 (or a, if we are thinking about the form ae^{-bx}) as x approaches 0 and 0 as x becomes large. The Ricker is slightly more interesting: for $x = 0$ we can calculate the value of the function directly (to get $a \cdot 0 \cdot e^{-b \cdot 0} = 0 \cdot 1 = 0$) or argue qualitatively that the e^{-bx} part approaches 1 and the ax part approaches zero (and hence the whole function approaches zero). For large x we have a concrete example of the $x^n e^{-x}$ example given above (with $n = 1$) and use our knowledge that exponentials always win to say that the e^{-bx} part should dominate the ax part to bring the function down to zero in the limit. (When you are doing this kind of qualitative reasoning you can almost always ignore the constants in the equation.)

As another example, consider the Michaelis-Menten function ($f(x) = ax/(b+x)$). We see that as x gets large we can say that $x \gg b$, no matter what b is ($\gg$ means "is much greater than"), so $b + x \approx x$, so

$$\frac{ax}{b+x} \approx \frac{ax}{x} = a : \tag{3.3.1}$$

the curve reaches a constant value of a. As x gets small, $b \gg x$, so

$$\frac{ax}{b+x} \approx \frac{ax}{b} : \tag{3.3.2}$$

the curve approaches a straight line through the origin, with slope a/b. As x goes to zero you can see that the value of the function is exactly zero $(a \times 0)/(b+0) = 0/b = 0$).

For more difficult functions that contain a fraction whose numerator and denominator both approach zero or infinity in some limit (and thus make it hard to find the limiting value), you can try *L'Hôpital's Rule*, which says that the limit of the function equals the limit of the ratio of the derivatives of the numerator and the denominator:

$$\lim \frac{a(x)}{b(x)} = \lim \frac{a'(x)}{b'(x)}. \tag{3.3.3}$$

($a'(x)$ and $\frac{da}{dx}$ are alternative notations for the derivative of a with respect to x.)

DERIVATIVES

As well as knowing the limits of the function, we also want to know how the function increases or decreases toward them: the limiting slope. Does the function shoot up or down (a derivative that "blows up" to positive or negative infinity), change linearly (a derivative that reaches a positive or negative constant limiting value), or flatten out (a derivative with limit 0)? To figure this out, we need to take the derivative with respect to x and then find its limit at the edges of the range.

The derivative of the exponential function $f(x) = ae^{-bx}$ is easy (if it isn't, review the appendix): $f'(x) = -abe^{-bx}$. When $x = 0$ this becomes $-ab$, and when x gets large the e^{-bx} part goes to zero, so the answer is zero. Thus (as you may already have known), the slope of the (negative) exponential is negative at the origin ($x = 0$) and the curve flattens out as x gets large.

The derivative of the Ricker is only a little harder (use the product rule):

$$\frac{d(axe^{-bx})}{dx} = (a \cdot e^{-bx} + ax \cdot -be^{-bx}) = (a - abx) \cdot e^{-bx} = a(1 - bx)e^{-bx}. \quad (3.3.4)$$

At zero, this is easy to compute: $a(1 - b \cdot 0)e^{-b \cdot 0} = a \cdot 1 \cdot 1 = a$. As x goes to infinity, the $(1 - bx)$ term becomes negative (and large in magnitude) and the e^{-bx} term goes toward zero, and we again use the fact that exponentials dominate linear and polynomial functions to see that the curve flattens out, rather than becoming more and more negative and crashing toward negative infinity. (In fact, we already knew that the curve approaches zero, so we could also have deduced that the curve must flatten out and the derivative must approach zero.)

In the case of the Michaelis-Menten function it's easy to figure out the slope at zero (because the curve becomes approximately $(a/b)x$ for small x), but in some cases you might have to take the derivative first and then set x to 0. The derivative of $ax/(b+x)$ is (using the quotient rule)

$$\frac{(b+x) \cdot a - ax \cdot 1}{(b+x)^2} = \frac{ab + ax - ax}{(b+x)^2} = \frac{ab}{(b+x)^2} \quad (3.3.5)$$

which (as promised) is approximately a/b when $x \approx 0$ (following the rule that $(b + x) \approx b$ for $x \approx 0$). Using the quotient rule often gives you a complicated denominator, but when you are only looking for points where the derivative is zero, you can calculate when the numerator is zero and ignore the denominator (assuming it is not zero at the same points where the numerator is).

3.3.2 *What Happens in the Middle? Scale Parameters and Half-Maxima*

It's also useful to know what happens in the middle of a function's range.

For *unbounded* functions (functions that increase to ∞ or decrease to $-\infty$ at the ends of their range), such as the exponential, we may not be able to find special points in the middle of the range, although it's worth trying out special cases such as $x = 1$ (or $x = 0$ for functions that range over negative and positive values) just to see if they have simple and interpretable answers.

In the exponential function ae^{-bx}, b is a *scale parameter*. In general, if a parameter appears in a function in the form of bx or x/c, its effect is to scale the curve along the x axis—stretching it or shrinking it, but keeping the qualitative shape the same. If the scale parameter is in the form bx, then b has inverse-x units (if x is a time measured in hours, then b is a rate per hour with units hour^{-1}): such a parameter might

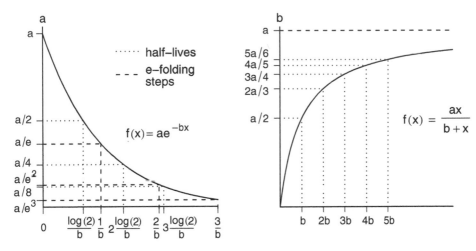

Figure 3.4 (a) Half-lives and *e*-folding times for a negative exponential function. (b) Half-maximum and characteristic scales for a Michaelis-Menten function.

sometimes be called a *rate parameter* instead of a scale parameter. If the expression is in the form x/c, then c has the same units as x, and we can call c a *characteristic scale*. Mathematicians often choose the form bx because it looks cleaner, while ecologists may prefer x/c because it's easier to interpret the parameter when it has the same units as x. Mathematically, the two forms are equivalent, with $b = 1/c$; this is an example of changing the *parameterization* of a function (see p. 81).

For the negative exponential function ae^{-bx}, the characteristic scale $1/b$ is also sometimes called the *e-folding time* (or *e*-folding distance if x measures distance rather than time). The value of the function drops from a at $x = 0$ to $ae^{-1} = a/e$ when $x = 1/b$, and drops a further factor of $e = 2.718\ldots \approx 3$ every time x increases by $1/b$ (Figure 3.4). Exponential-based functions can also be described in terms of the *half-life* (for decreasing functions) or *doubling time* (for increasing functions), which is $T_{1/2} = \ln 2/b$. When $x = T_{1/2}$, $y = a/2$, and every time x increases by $T_{1/2}$ the function drops by another factor of 2.

For the Ricker function, we already know that the function is zero at the origin and approaches zero as x gets large. We also know that the derivative is positive at zero and negative (but the curve is flattening out, so the derivative is increasing toward zero) as x gets large. We can deduce* that the derivative must be zero and the function must reach a peak somewhere in the middle; we will calculate the location and height of this peak in the next section.

For functions that reach an asymptote, like the Michaelis-Menten, it's useful to know when the function gets "halfway up"—the *half-maximum* is a point on the x axis, not the y axis. We figure this out by finding the asymptote ($= a$ for this parameterization of the Michaelis-Menten function) and solving $f(x_{1/2}) = \text{asymptote}/2$ for $x_{1/2}$. In this case

$$\frac{ax_{1/2}}{b + x_{1/2}} = \frac{a}{2}$$

---

* Because the Ricker function has continuous derivatives.

$$ax_{1/2} = \frac{a}{2} \cdot (b + x_{1/2})$$

$$\left(a - \frac{a}{2}\right) x_{1/2} = \frac{ab}{2}$$

$$x_{1/2} = \frac{2}{a} \cdot \frac{ab}{2} = b.$$

The half-maximum b is the characteristic scale for the Michaelis-Menten; we can see this by dividing the numerator and denominator by b to get $f(x) = a \cdot (x/b)/(1 + x/b)$. As x increases by half-maximum units (from $x_{1/2}$ to $2x_{1/2}$ to $3x_{1/2}$), the function first reaches 1/2 its asymptote, then 2/3 its asymptote, then 3/4 ... (Figure 3.4).

We can calculate the half-maximum for any function that starts from zero and reaches an asymptote, although it may not be a simple expression.

3.3.3 Critical Points and Inflection Points

We may also be interested in the *critical points*—maxima and minima—of a function. To find the critical points of f, remember from calculus that they occur where $f'(x) = 0$; calculate the derivative, solve it for x, and plug that value for x into $f(x)$ to determine the value (peak height or trough depth) at that point.* The exponential function is *monotonic*: it is always either increasing or decreasing depending on the sign of b (its slope is always either positive or negative for all values of x)—so it never has any critical points.

The Michaelis-Menten curve is also monotonic: we figured out above that its derivative is $ab/(b + x)^2$. Since the denominator is squared, the derivative is always positive. (Strictly speaking, this is true only if $a > 0$. Ecologists are usually sloppier than mathematicians, who are careful to point out all the assumptions behind a formula (like $a > 0$, $b > 0$, $x \geq 0$). I'm acting like an ecologist rather than a mathematician, assuming parameters and x values are positive unless otherwise stated.) While remaining positive, the derivative decreases to zero as $x \to \infty$ (because $ab/(b + x)^2 \approx ab/x^2 \to 0$); such a function is called *saturating*.

We already noted that the Ricker function, axe^{-bx}, has a peak in the middle somewhere: where is it? Using the product rule:

$$\frac{d(axe^{-bx})}{dx} = 0$$

$$ae^{-bx} + ax(-be^{-bx}) = 0$$

$$(1 - bx)ae^{-bx} = 0.$$

The left-hand side can be zero only if $1 - bx = 0$, $a = 0$ (a case we're ignoring as ecologists) or $e^{-bx} = 0$. The exponential part e^{-bx} is never equal to 0, so we simply solve $(1 - bx) = 0$ to get $x = 1/b$. Plugging this value of x back into the equation

---

* The derivative is also zero at *saddle points*, where the function temporarily flattens on its way up or down.

tells us that the height of the peak is $(a/b)e^{-1}$. (You may have noticed that the peak location, $1/b$, is equal to the characteristic scale for the Ricker equation.)

3.3.4 Understanding and Changing Parameters

Once you know something about a function (its value at zero or other special points, value at ∞, half-maximum, slope at certain points, and the relationship of these values to the parameters), you can get a rough idea of the meanings of the parameters. You will find, alas, that scientists rarely stick to one parameterization. Reparameterization seems like an awful nuisance—why can't everyone just pick one set of parameters and stick to it?—but, even setting aside historical accidents that make different fields adopt different parameterizations, different parameterizations are useful in different contexts. Different parameterizations have different mechanistic interpretations. For example, we'll see shortly that the Michaelis-Menten function can be interpreted (among other possibilities) in terms of enzyme reaction rates and half-saturation constants or in terms of predator attack rates and handling times. Some parameterizations make it easier to estimate parameters by eye. For example, half-lives are easier to see than e-folding times, and peak heights are easier to see than slopes. Finally, some sets of parameters are strongly correlated, making them harder to estimate from data. For example, if you write the equation of a line in the form $y = ax + b$, the estimates of the slope a and the intercept b are negatively correlated, but if you instead say $y = a(x - \bar{x}) + \bar{y}$, estimating the mean value of y rather than the intercept, the estimates are uncorrelated. You just have to brush up your algebra and learn to switch among parameterizations.

We know the following things about the Michaelis-Menten function $f(x) = ax/(b + x)$: the value at zero $f(0) = 0$; the asymptote $f(\infty) = a$; the initial slope $f'(0) = a/b$; and the half-maximum (the characteristic scale) is b.

You can use these characteristics to crudely estimate the parameters from the data. Find the asymptote and the x value at which y reaches half of its maximum value, and you have a and b. (You can approximate these values by eye, or use a more objective procedure such as taking the mean of the last 10% of the data to find the asymptote.) Or you can estimate the asymptote and the initial slope ($\Delta y/\Delta x$), perhaps by linear regression on the first 20% of the data, and then use the algebra $b = a/(a/b) = $ asymptote/(initial slope) to find b.

Equally important, you can use this knowledge of the curve to translate among algebraic, geometric, and mechanistic meanings. When we use the Michaelis-Menten in community ecology as the Holling type II functional response, its formula is $P(N) = \alpha N/(1 + \alpha H N)$, where P is the predation rate, N is the density of prey, α is the attack rate, and H is the handling time. In this context, the initial slope is α and the asymptote is $1/H$. Ecologically, this makes sense because at low densities the predators will consume prey at a rate proportional to the attack rate ($P(N) \approx \alpha N$) while at high densities the predation rate is entirely limited by handling time ($P(N) \approx 1/H$). It makes sense that the high-density predation *rate* is the inverse of the handling *time*: if a predator needs half an hour to handle (capture, swallow, digest, etc.) a prey, and needs essentially no time to locate a new one (since the prey density is very high), then the predation rate is $1/(0.5$ hour$) = 2$/hour. The half-maximum in this parameterization is $1/(\alpha H)$.

On the other hand, biochemists usually parameterize the function more as we did above, with a maximum rate v_{max} and a half-maximum K_m: as a function of concentration C, $f(C) = v_{max}C/(K_m + C)$.

As another example, recall the following facts about the Ricker function $f(x) = axe^{-bx}$: the value at zero $f(0) = 0$; the initial slope $f'(0) = a$; the horizontal location of the peak is at $x = 1/b$; and the peak height is $a/(be)$. The form we wrote above is algebraically simplest, but parameterizing the curve in terms of its peak location (let's say $p = 1/b$) might be more convenient: $y = axe^{-x/p}$. Fisheries biologists often use another parameterization, $R = Se^{-a_3 - bS}$, where $a_3 = -\log a$ (Quinn and Deriso, 1999).*

3.3.5 Transformations

Beyond changing the parameterization, you can also change the scales of the x and y axes, or in other words *transform* the data. For example, in the Ricker example just given ($R = Se^{-a_3 - bS}$), if we plot $-\log(R/S)$ against S, we get the line $-\log(R/S) = a_3 + bS$, which makes it easy to see that a_3 is the intercept and b is the slope.

Log transformations of x or y or both are common because they make exponential relationships into straight lines. If $y = ae^{-bx}$ and we log-transform y, we get $\log y = \log a - bx$ (a *semi-log* plot). If $y = ax^b$ and we log-transform both x and y, we get $\log y = \log a + b \log x$ (a *log-log* plot).

Another example: if we have a Michaelis-Menten curve and plot x/y against x, the relationship is

$$x/y = \frac{x}{ax/(b+x)} = \frac{b+x}{a} = \frac{1}{a} \cdot x + \frac{b}{a},$$

which represents a straight line with slope $1/a$ and intercept b/a.

All of these transformations are called *linearizing transformations*. Researchers often used them in the past to fit straight lines to data when computers were slower. Linearizing is not recommended when another alternative such as non-linear regression is available, but transformations are still useful. Linearized data are easier to eyeball, so you can get rough estimates of slopes and intercepts by eye, and deviations are easier to see from linearity than from (e.g.) an exponential curve. Log-transforming data on geometric growth of a population lets you look at *proportional* changes in the population size (a doubling of the population is always represented by the same distance on the y axis). Square-root-transforming data on variances lets you look at standard deviations, which are measured in the same units as the original data and may thus be easier to understand.

The *logit* or *log-odds* function, $\text{logit}(x) = \log(x/(1-x))$ (`qlogis` in R) is another common linearizing transformation. If x is a probability, then $x/(1-x)$ is the ratio of the probability of occurrence (x) to the probability of nonoccurrence ($1-x$), which is called the *odds* (e.g., a probability of 0.1 or 10% corresponds to odds of

---

* Throughout this book I use $\log(x)$ to mean the natural logarithm of x (also called $\ln(x)$ or $\log_e(x)$). If you need a refresher on logarithms, see the appendix.

$0.1/0.9 = 1/9$). The logit transformation makes a logistic curve, $y = e^{a+bx}/(1 + e^{a+bx})$, into a straight line:

$$y = e^{a+bx}/(1 + e^{a+bx})$$

$$(1 + e^{a+bx})y = e^{a+bx}$$

$$y = e^{a+bx}(1 - y)$$

$$\frac{y}{1 - y} = e^{a+bx}$$

$$\log\left(\frac{y}{1 - y}\right) = a + bx$$

(3.3.6)

3.3.6 Shifting and Scaling

Another way to change or extend functions is to shift or scale them. For example, let's see how we can manipulate the simplest form of the Michaelis-Menten function (Figure 3.5). The function $y = x/(1 + x)$ starts at 0, increases to 1 as x gets large, and has a half-maximum at $x = 1$.

- We can stretch, or scale, the x axis by dividing x by a constant—this means you have to go farther on the x axis to get the same increase in y. If we substitute x/b for x everywhere in the function, we get $y = (x/b)/(1 + x/b)$. Multiplying the numerator and denominator by b shows us that $y = x/(b + x)$, so b is just the half-maximum, which we identified before as the characteristic scale. In general a parameter that we multiply or divide x by is called a *scale parameter* because it changes the horizontal scale of the function.
- We can stretch or scale the y axis by multiplying the whole right-hand side by a constant. If we use a, we have $y = ax/(b + x)$, which as we saw above moves the asymptote from 1 to a.
- We can shift the whole curve to the right or the left by subtracting or adding a constant *location parameter* from x throughout; subtracting a positive constant from x shifts the curve to the right. Thus, $y = a(x - c)/(b + (x - c))$ hits $y = 0$ at c rather than zero. (You may want in this case to specify that $y = 0$ if $y < c$—otherwise the function may behave badly [try curve(x/(x-1),from=0,to=3) to see what might happen].)
- We can shift the whole curve up or down by adding or subtracting a constant on the right-hand side: $y = a(x - c)/(b + (x - c)) + d$ would start from $y = d$, rather than zero, when $x = c$ (the asymptote also moves up to $a + d$).

These recipes can be used with any function. For example, Emlen (1996) wanted to describe a relationship between the prothorax and the horn length of horned beetles where the smallest beetles in his sample had a constant, but nonzero, horn length. He added a constant to a generalized logistic function to shift the curve up from its usual zero baseline.

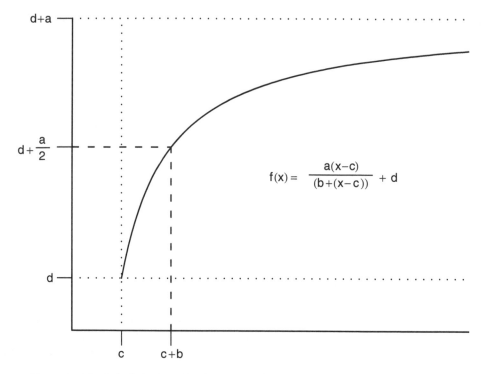

Figure 3.5 Scaled, shifted Michaelis-Menten function $y = a(x-c)/((x-c)+b)+d$.

3.3.7 Taylor Series Approximation

Taylor approximation or the *Taylor series* is the single most useful, and used, application of calculus for an ecologist. Two particularly useful applications of Taylor approximation are understanding the shapes of goodness-of-fit surfaces (Chapter 6) and the delta method for estimating errors in estimation (Chapter 7).

The Taylor series allows us to approximate a complicated function near a point we care about, using a simple function—a polynomial with a few terms, say a quadratic curve. All we have to do is figure out the slope (first derivative) and curvature (second derivative) at that point. Then we can construct a parabola that matches the complicated curve in the neighborhood of the point we are interested in. (In reality the Taylor series goes on forever—we can approximate the curve more precisely with a cubic, then a fourth-order polynomial, and so forth—but in practice ecologists never go beyond a quadratic expansion.)

Mathematically, the Taylor series says that, near a given point x_0,

$$f(x) \approx f(x_0) + \frac{df}{dx}\bigg|_{x_0} \cdot (x - x_0) + \frac{d^2 f}{dx^2}\bigg|_{x_0} \cdot \frac{(x - x_0)^2}{2} + \cdots$$

$$+ \frac{d^n f}{dx^n}\bigg|_{x_0} \cdot \frac{(x - x_0)^n}{n!} + \cdots \qquad (3.3.7)$$

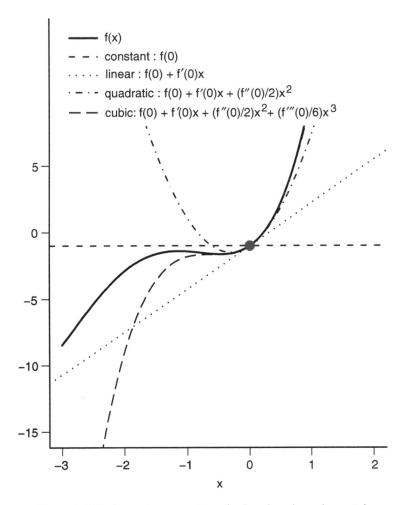

Figure 3.6 Taylor series expansion of a fourth-order polynomial.

(the notation $\frac{df}{dx}\big|_{x_0}$ means "the derivative evaluated at the point $x = x_0$," and $n!$ denotes a factorial—see note p. 106). Taylor *approximation* just means dropping terms past the second or third.

Figure 3.6 shows a function and the constant, linear, quadratic, and cubic approximations (Taylor expansion using one to four terms). The linear approximation is bad but the quadratic fit is good very near the center point, and the cubic accounts for some of the asymmetry in the function. In this case one more term would match the function exactly, since it is actually a fourth-degree polynomial.

THE EXPONENTIAL FUNCTION

The Taylor expansion of the exponential, e^{rx}, around $x = 0$ is $1 + rx + (rx)^2/2 + (rx)^3/(2 \cdot 3)$.... Remembering this fact rather than working it out every time may save you time in the long run—for example, to understand how the Ricker function works for small x we can substitute $(1 - bx)$ for e^{-bx} (dropping all but the first two terms!) to get $y \approx ax - abx^2$. This tells us immediately that the function starts linear but begins to curve downward right away.

THE LOGISTIC CURVE

Calculating Taylor approximations is often tedious (all those derivatives), but we usually try to do it at some special point where a lot of the complexity goes away (such as $x = 0$ for a logistic curve).

The general form of the logistic (p. 95) is $e^{a+bx}/(1 + e^{a+bx})$, but doing the algebra will be simpler if we take the special case $a = 0$ and divide numerator and denominator by e^{bx} to get $f(x) = 1/(1 + e^{-bx})$. Taking the Taylor expansion around $x = 0$:

- $f(0) = 1/2$.
- $f'(x) = \dfrac{be^{-bx}}{(1+e^{-bx})^2}$ (writing the formula as $(1 + e^{-bx})^{-1}$ and using the power rule and the chain rule twice) so $f'(0) = (b \cdot 1)/((1 + 1)^2) = b/4$.*
- Using the quotient rule and the chain rule:

$$f''(0) = \left. \frac{(1+e^{-bx})^2(-b^2 e^{-bx}) - (be^{-bx})(2(1+e^{-bx})(-be^{-bx}))}{(1+e^{-bx})^4} \right|_{x=0}$$

$$= \frac{(1+1)^2(-b^2) - (b)(2(1+1)(-b))}{(1+1)^4}$$

$$= \frac{(-4b^2) + (4b^2)}{16}$$

$$= 0. \tag{3.3.8}$$

R will actually compute simple derivatives for you (using D; see p. 101), but it won't simplify them at all. If you just need to compute the numerical value of the derivative for a particular b and x, it may be useful, but you'll often miss general answers by doing it this way (e.g., in the above case that $f''(0)$ is zero for any value of b).

Stopping to interpret the answer we got from all that tedious algebra: we find out that the slope of a logistic function around its midpoint is $b/4$, and its curvature (second derivative) is zero. This means that the midpoint is an *inflection point* (where there is no curvature, or where the curve switches from being concave to convex), as you might have known already. It also means that near the inflection point, the logistic can be closely approximated by a straight line. (For y near zero, exponential growth is a good approximation; for y near the asymptote, exponential approach to the asymptote is a good approximation.)

---

*We calculate $f'(x)$ and evaluate it at $x = 0$. We *don't* calculate the derivative of $f(0)$, because $f(0)$ is a constant value (1/2 in this case) and its derivative is zero.

3.4 Bestiary of Functions

The remainder of the chapter describes different families of functions that are useful in ecological modeling; Table 3.1 gives an overview of their qualitative properties. This section includes little R code, although the formulas should be easy to translate into R. You should skim through this section on the first reading to get an idea of what functions are available. If you begin to feel bogged down, you can skip ahead and use the section for reference as needed.

3.4.1 Functions Based on Polynomials

3.4.1.1 POLYNOMIAL FUNCTIONS

A polynomial is a function of the form $y = \sum_{i=0}^{n} a_i x^i$.

Examples

- Linear: $f(x) = a + bx$, where a is the intercept (value when $x = 0$) and b is the slope. (You know this, right?)
- Quadratic: $f(x) = a + bx + cx^2$. The simplest nonlinear model.
- Cubics and higher-order polynomials: $f(x) = \sum_{i=0}^{n} a_i x^i$. The *order* or *degree* of a polynomial is the highest power that appears in it (so, e.g., $f(x) = x^5 + 4x^2 + 1$ is fifth-order).

Advantages

Polynomials are easy to understand. They are easy to reduce to simpler functions (*nested* functions) by setting some of the parameters to zero. High-order polynomials can fit arbitrarily complex data.

Disadvantages

On the other hand, polynomials are often hard to justify mechanistically (can you think of a reason an ecological relationship should be a cubic polynomial?). They don't level off as x goes to $\pm\infty$—they always go to $-\infty$ or ∞ as x gets large. Extrapolating polynomials often leads to nonsensically large or negative values. High-order polynomials can be unstable: following Forsythe et al. (1977) you can show that extrapolating a high-order polynomial from a fit to U.S. census data from 1900 to 2000 predicts a population crash to zero around 2015!

It is sometimes convenient to parameterize polynomials differently. For example, we could reparameterize the quadratic function $y = a_0 + a_1 x + a_2 x^2$ as $y = a + c(x - b)^2$ (where $a_0 = a + cb^2$, $a_1 = -2cb$, $a_2 = c$). It's now clear that the curve has its minimum (if $c > 0$) at $x = b$ (because $c(x - b)^2$ is zero there and positive everywhere else), that $y = a$ at the minimum, and that c governs how fast the curve increases away from its minimum. Polynomials can be particularly simple if some of their coefficients are zero: for example, $y = bx$ (a line through the origin, or

TABLE 3.1
Qualitative Properties of Bestiary Functions

Function	Range	Left End	Right End	Middle
Polynomials				
Line	$\{-\infty, \infty\}$	$y \to \pm\infty$, constant slope	$y \to \pm\infty$, constant slope	monotonic
Quadratic	$\{-\infty, \infty\}$	$y \to \pm\infty$, accelerating	$y \to \pm\infty$, accelerating	single max/min
Cubic	$\{-\infty, \infty\}$	$y \to \pm\infty$, accelerating	$y \to \pm\infty$, accelerating	up to 2 max/min
Piecewise polynomials				
Threshold	$\{-\infty, \infty\}$	flat	flat	breakpoint
Hockey stick	$\{-\infty, \infty\}$	flat or linear	flat or linear	breakpoint
Piecewise linear	$\{-\infty, \infty\}$	linear	linear	breakpoint
Rational				
Hyperbolic	$\{0, \infty\}$	$y \to \infty$ or finite	$y \to 0$	decreasing
Michaelis-Menten	$\{0, \infty\}$	$y = 0$, linear	asymptote	saturating
Holling type III	$\{0, \infty\}$	$y = 0$, accelerating	asymptote	sigmoid
Holling type IV ($c < 0$)	$\{0, \infty\}$	$y = 0$, accelerating	asymptote	hump-shaped
Exponential-based				
Negative exponential	$\{0, \infty\}$	y finite	$y \to 0$	decreasing
Monomolecular	$\{0, \infty\}$	$y = 0$, linear	asymptote	saturating
Ricker	$\{0, \infty\}$	$y = 0$, linear	$y \to 0$	hump-shaped
logistic	$\{0, \infty\}$	y small, accelerating	asymptote	sigmoid
Power-based				
Power law	$\{0, \infty\}$	y small or $\to \infty$	$y \to 0$ or $\to \infty$	monotonic
von Bertalanffy	like logistic			
Gompertz	like logistic			
Shepherd	like Ricker			
Hassell	like Ricker			
Nonrectangular hyperbola	like Michaelis-Menten			

direct proportionality) or $y = cx^2$. Where a polynomial actually represents proportionality or area, rather than being an arbitrary fit to data, you can often simplify in this way.

The advantages and disadvantages just listed all concern the mathematical and phenomenological properties of polynomials. Sometimes linear and quadratic polynomials do actually make sense in ecological settings. For example, a population or resource that accumulates at a constant rate from outside the system will grow linearly with time. The rates of ecological or physiological processes (e.g., metabolic cost or resource availability) that depend on an organism's skin surface or mouth area will be a quadratic function of linear measurements of its size (e.g., snout-to-vent length or height).

3.4.1.2 PIECEWISE POLYNOMIAL FUNCTIONS

You can make polynomials (and other functions) more flexible by using them as components of *piecewise* functions. In this case, different functions apply over different ranges of the predictor (x) variable. (See p. 101 for information on using R's `ifelse` function to build piecewise functions.)

Examples

- The simplest piecewise function is a simple *threshold model*—$y = a_1$ if x is less than some threshold T, and $y = a_2$ if x is greater. Hilborn and Mangel (1997) use a threshold function in an example of the number of eggs a parasitoid lays in a host as a function of how many she has left (her "egg complement"), although the original researchers used a logistic function instead (Rosenheim and Rosen, 1991).
- The *hockey-stick* function (Bacon and Watts, 1971, 1974) is a combination of a constant and a linear piece, typically either flat and then increasing linearly or linearly increasing and then suddenly hitting a plateau. Hockey-stick functions have a long history in ecology, at least as far back as the definition of the Holling type I functional response, which is supposed to represent foragers like filter feeders that can continually increase their uptake rate until they suddenly hit a maximum (Jeschke et al., 2004). Hockey-stick models have recently become popular in fisheries modeling, for modeling stock-recruitment curves (Barrowman and Myers, 2000), and in ecology, for detecting edges in landscapes (Toms and Lesperance, 2003).* Under the name of *self-excitable threshold autoregressive* (SETAR) models, such functions have been used to model density dependence in population dynamic models of lemmings (Framstad et al., 1997), feral sheep (Grenfell et al., 1998), and moose (Post et al., 2002); in another population dynamic context, Brannström and Sumpter (2005) call them *ramp* functions.
- Threshold functions are flat (i.e., the slope is zero) on both sides of the breakpoint, and hockey sticks are flat on one side. More general piecewise linear

---

* It is surely only a coincidence that so much significant work on hockey-stick functions has been done by Canadians.

functions have nonzero slopes on both sides of the breakpoint s_1:

$$y = a_1 + b_1 x$$

for $x < s_1$ and

$$y = (a_1 + b_1 s_1) + b_2(x - s_1)$$

for $x > s_1$. (The extra complications in the formula for $x > s_1$ ensure that the function is continuous.)

- *Cubic splines* are a general-purpose tool for fitting curves to data. They are *piecewise cubic* functions that join together smoothly at transition points called *knots*. They are typically used as purely phenomenological curve-fitting tools, when you want to fit a smooth curve to data but don't particularly care about interpreting its ecological meaning (Wood, 2001, 2006). Splines have many of the useful properties of polynomials (adjustable complexity or smoothness; simple basic components) without their instability.

Advantages

Piecewise functions make sense if you believe there could be a biological switch point. For example, in optimal behavior problems theory often predicts sharp transitions among different behavioral strategies (Hilborn and Mangel, 1997, ch. 4). Organisms might decide to switch from growth to reproduction, or to migrate between locations, when they reach a certain size or when resource supply drops below a threshold. Phenomenologically, using piecewise functions is a simple way to stop functions from dropping below zero or increasing indefinitely when such behavior would be unrealistic.

Disadvantages

Piecewise functions present some special technical challenges for parameter fitting, which probably explains why they have gained attention only recently. Using a piecewise function means that the rate of change (the derivative) changes suddenly at some point. Such a discontinuous change may make sense, for example, if the last prey refuge in a reef is filled, but transitions in ecological systems usually happen more smoothly. When thresholds are imposed phenomenologically to prevent unrealistic behavior, it may be better to go back to the original biological system to try to understand what properties of the system would actually stop (e.g.) population densities from becoming negative: would they hit zero suddenly, or would a gradual approach to zero (perhaps represented by an exponential function) be more realistic?

3.4.1.3 RATIONAL FUNCTIONS: POLYNOMIALS IN FRACTIONS

Rational functions are ratios of polynomials, $(\sum a_i x^i)/(\sum b_j x^j)$.

Examples

- The simplest rational function is the *hyperbolic* function, a/x; it is often used in models of plant competition, to fit seed production as a function of plant

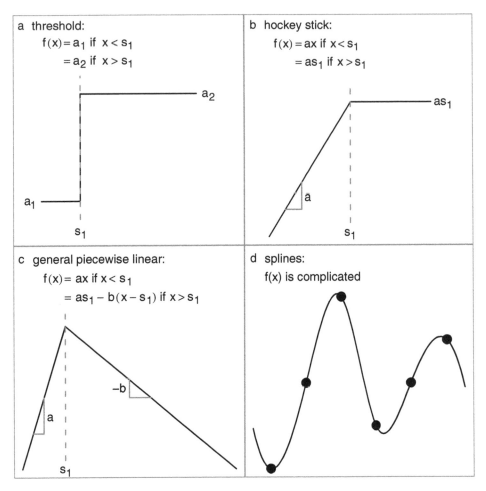

a threshold:
$$f(x) = a_1 \text{ if } x < s_1$$
$$= a_2 \text{ if } x > s_1$$

b hockey stick:
$$f(x) = ax \text{ if } x < s_1$$
$$= as_1 \text{ if } x > s_1$$

c general piecewise linear:
$$f(x) = ax \text{ if } x < s_1$$
$$= as_1 - b(x - s_1) \text{ if } x > s_1$$

d splines:
f(x) is complicated

Figure 3.7 Piecewise polynomial functions: the first three (threshold, hockey stick, general piecewise linear) are all piecewise linear. Splines are piecewise cubic; the equations are complicated and are usually handled by software (see ?spline and ?smooth.spline).

density. A mechanistic explanation might be that if resources per unit area are constant, the area available to a plant for resource exploitation might be proportional to 1/density, which would translate (assuming uptake, allocation, etc., all stay the same) into a hyperbolically decreasing amount of resource available for seed production. A better-behaved variant of the hyperbolic function is $a/(b+x)$, which doesn't go to infinity when $x = 0$ (Pacala and Silander, 1987, 1990).

- The next most complicated, and probably the most famous, rational function is the *Michaelis-Menten* function: $f(x) = ax/(b+x)$. Michaelis and Menten introduced it in the context of enzyme kinetics; it is also known, by other names, in resource competition theory (as the Monod function), predator-prey dynamics (Holling type II functional response), and fisheries biology (Beverton-Holt model). It starts at 0 when $x = 0$ and approaches an asymptote at a as x gets large. The only major caveat with this function is that it takes

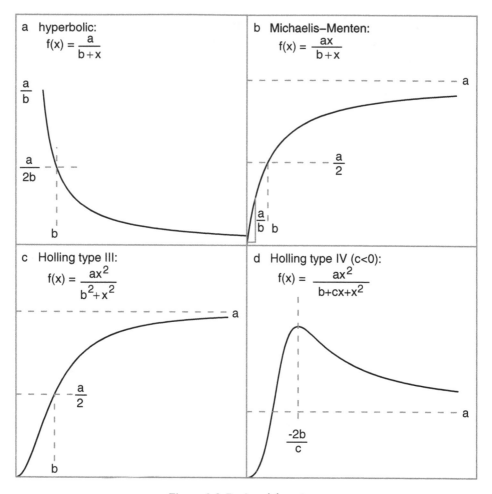

Figure 3.8 Rational functions.

surprisingly long to approach its asymptote: $x/(1+x)$, which is halfway to its asymptote when $x = 1$, still reaches only 90% of its asymptote when $x = 9$. The Michaelis-Menten function can be parameterized in terms of any two of the asymptote, half-maximum, initial slope, or their inverses.

The mechanism behind the Michaelis-Menten function in biochemistry and ecology (Holling type II) is similar; as substrate (or prey) become more common, enzymes (or predators) have to take a larger and larger fraction of their time handling rather than searching for new items. In fisheries, the Beverton-Holt stock-recruitment function comes from assuming that over the course of the season the mortality rate of young-of-the-year is a linear function of their density (Quinn and Deriso, 1999).

- We can go one more step, from a linear to a quadratic function in the denominator, and define a function sometimes known as the *Holling type III* functional response: $f(x) = ax^2/(b^2 + x^2)$. This function is *sigmoid*, or S-shaped. The asymptote is at a; its shape is quadratic near the origin, starting from zero with slope zero and curvature a/b^2; and its half-maximum is at $x = b$. It can occur mechanistically in predator-prey systems because of predator switching from rare to common prey, predator aggregation, and spatial and other forms of heterogeneity (Morris, 1997).

- Some ecologists have extended this family still further to the *Holling type IV functional response*: $f(x) = ax^2/(b + cx + x^2)$. Turchin (2003) derives this function (which he calls a "mechanistic sigmoidal functional response") by assuming that the predator attack rate in the Holling type II functional response is itself an increasing Michaelis-Menten function of prey density— that is, predators prefer to pursue more abundant prey. In this case, $c > 0$. If $c < 0$, then the Holling type IV function is *unimodal*, or "hump-shaped," with a maximum at intermediate prey density. Ecologists have used this version of the Holling type IV phenomenologically to describe situations where predator interference or induced prey defenses lead to decreased predator success at high predator density (Holt, 1983; Collings, 1997; Wilmshust et al., 1999; Chen, 2004). Whether c is negative or positive, the Holling type IV reaches an asymptote at a as $x \to \infty$. If $c < 0$, then it has a maximum at $x = -2b/c$.
- More complicated rational functions are potentially useful but rarely used in ecology. The (unnamed) function $y = (a + bx)/(1 + cx)$ has been used to describe species-area curves (Flather, 1996; Tjørve, 2003).

Advantages

Like polynomials, rational functions are very flexible (you can always add more terms in the numerator or denominator) and simple to compute; unlike polynomials, they can reach finite asymptotes at the ends of their range. In many cases, rational functions make mechanistic sense, arising naturally from simple models of biological processes such as competition or predation.

Disadvantages

Rational functions can be complicated to analyze because the quotient rule makes their derivatives complicated. Like the Michaelis-Menten function they approach their asymptotes very slowly, which makes estimating the asymptote difficult— although this problem really says more about the difficulty of getting enough data rather than about the appropriateness of rational functions as models for ecological systems. Section 3.4.3 shows how to make rational functions even more flexible by raising some of their terms to a power, at the cost of making them even harder to analyze.

3.4.2 Functions Based on Exponential Functions

3.4.2.1 SIMPLE EXPONENTIALS

The simplest examples of functions based on exponentials are the exponential growth (ae^{bx}) or decay (ae^{-bx}) and saturating exponential growth functions (*monomolecular*, $a(1 - e^{-bx})$). The monomolecular function (so named because it represents the buildup over time of the product of a single-molecule chemical reaction) is also

- The *catalytic curve* in infectious disease epidemiology, where it represents the change over time in the fraction of a cohort that has been exposed to disease (Anderson and May, 1991).
- The simplest form of the *von Bertalanffy* growth curve in organismal biology and fisheries, where it arises from the competing effects of changes in catabolic and metabolic rates with changes in size (Essington et al., 2001).
- The *Skellam model* in population ecology, giving the number of offspring in the next year as a function of the adult population size in the current year when competition has a particularly simple form (Skellam, 1951; Brännström and Sumpter, 2005).

These functions have two parameters, the multiplier a, which expresses the starting or final size depending on the function, and the exponential rate b or "*e*-folding time" $1/b$ (the time needed to reach e times the initial value, or the initial value divided by e, depending whether b is positive or negative). The *e*-folding time can be expressed as a half-life or doubling time ($\log(2)/b$) as well. Such exponential functions arise naturally from any compounding process where the population loses or gains a constant proportion per unit time; one example is *Beers' Law* for the decrease in light availability with depth in a vegetation canopy (Teh, 2006).

The differences in shape between an exponential-based function and its rational-function analogue (e.g., between the monomolecular and the Michaelis-Menten function) are usually subtle. Unless you have a lot of data you're unlikely to be able to distinguish from the data which fits better, and you will instead have to choose on the basis of which one makes more sense mechanistically, or possibly which is more convenient to compute or analyze (Figure 3.9).

3.4.2.2 COMBINATIONS OF EXPONENTIALS WITH OTHER FUNCTIONS

Ricker Function

The Ricker function, $ax \exp(-bx)$, is a common model for density-dependent population growth; if per capita fecundity decreases exponentially with density, then overall population growth will follow the Ricker function. It starts off growing linearly with slope a and has its maximum at $x = 1/b$; it's similar in shape to the generalized Michaelis-Menten function ($RN/(1 + (aN)^b)$). It is very widely used as a phenomenological model for ecological variables that start at zero, increase to a peak, and decrease gradually back to zero.

Several authors (Hassell, 1975; Royama, 1992; Brännström and Sumpter, 2005) have derived Ricker equations for the dependence of offspring number on density, assuming that adults compete with each other to reduce fecundity; Quinn and Deriso (1999, p. 89) derive the Ricker equation in a fisheries context, assuming that young-of-year compete with each other and increase mortality (e.g., via cannibalism).

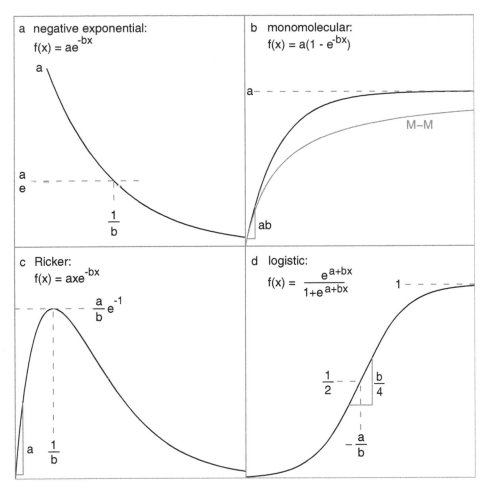

Figure 3.9 Exponential-based functions. "M-M" in the monomolecular figure is the Michaelis-Menten function with the same asymptote and initial slope.

Logistic Function

Two parameterizations of the logistic function are widely used. The first,

$$y = \frac{e^{a+bx}}{1 + e^{a+bx}} \tag{3.4.1}$$

(or equivalently $y = 1/(1 + e^{-(a+bx)})$) comes from a statistical or phenomenological context. The function goes from 0 at $-\infty$ to 1 at $+\infty$. The location parameter a shifts the curve left or right: the half-maximum, which is also the inflection point, occurs at $x = -a/b$ when the term in the exponent is 0. The scale parameter b controls the steepness of the curve.*

---

* If we reparameterized the function as $\exp(b(x-c))/(1 + \exp(b(x-c)))$, the half-maximum would be at c. Since b is still the steepness parameter, we could then shift and steepen the curve independently.

The second parameterization comes from population ecology:

$$n(t) = \frac{K}{1 + \left(\frac{K}{n_0} - 1\right) e^{-rt}} \tag{3.4.2}$$

where K is the carrying capacity, n_0 the value at $t = 0$, and r the initial per capita growth rate. (The statistical parameterization is less flexible, with only two parameters: it has $K = 1$, $n_0 = e^a/(1 + e^a)$, and $r = b$.)

The logistic is popular both because it's a simple sigmoid function (although its rational analogue the Holling type III functional response is also simple) and because it's the solution to one of the simplest population-dynamic models, the *logistic equation*:

$$\frac{dn}{dt} = rn\left(1 - \frac{n}{K}\right), \tag{3.4.3}$$

which says that per capita growth rate $((dn/dt)/n)$ decreases linearly from a maximum of r when n is much less than K to zero when $n = K$. Getting from the logistic equation (3.4.3) to the logistic function (3.4.2) involves solving the differential equation by integrating by parts, which is tedious but straightforward (see any calculus book, e.g., Adler (2004)).

In R you can write out the logistic function yourself, using the `exp` function, as `exp(x)/(1+exp(x))`, or you can use the `plogis` function. The *hyperbolic tangent* (`tanh`) function is another form of the logistic. Its range extends from -1 as $x \to -\infty$ to 1 as $x \to \infty$ instead of from 0 to 1.

Gompertz Function

The *Gompertz* function, $f(x) = e^{-ae^{-bx}}$, is an alternative to the logistic function. Similar to the logistic, it is accelerating at $x = 0$ and exponentially approaches 1 as x gets large, but it is asymmetric—the inflection point or change in curvature occurs $1/e \approx 1/3$ of the way up to the asymptote, rather than halfway up. In this parameterization the inflection point occurs at $x = 0$; you may want to shift the curve c units to the right by using $f(x) = e^{-ae^{b(x-c)}}$. If we derive the curves from models of organismal or population growth, the logistic assumes that growth decreases linearly with size or density while the Gompertz assumes that growth decreases exponentially.

3.4.3 Functions Involving Power Laws

So far the polynomials involved in our rational functions have been simple linear or quadratic functions. Ecological modelers sometimes introduce an arbitrary (fractional) power as a parameter (x^b) instead of using only integer values (e.g., x, x^2, x^3); using power laws in this way is often a phenomenological way to vary the shape of a curve, although these functions may also have mechanistic derivations.

Here are some categories of power-law functions.

- Simple power laws $f(x) = ax^b$ (for noninteger b; otherwise the function is just a polynomial; Figure 3.10a) often describe allometric growth (e.g., reproductive

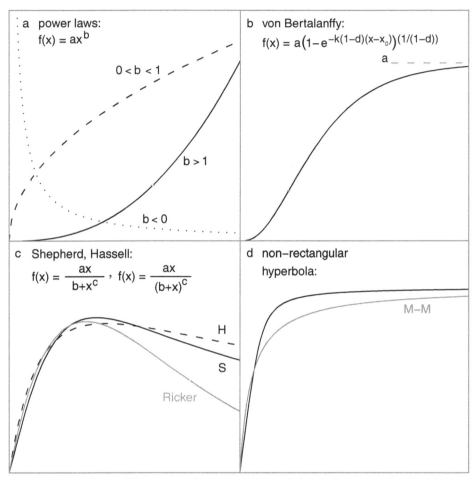

Figure 3.10 Power-based functions. The lower left panel shows the Ricker function for comparison with the Shepherd and Hassell functions. The lower right shows the Michaelis-Menten function (M–M) for comparison with the nonrectangular hyperbola.

biomass as a function of diameter at breast height (Niklas, 1993) or mass as a function of tarsus length in birds); or quantities related to metabolic rates (Etienne et al., 2006a); or properties of landscapes with fractal geometry (Halley et al., 2004); or species-area curves (Tjørve, 2003).

- The generalized form of the von Bertalanffy growth curve, $f(x) = a(1 - \exp(-k(1-d)(x-x_0)))^{1/(1-d)}$ (Figure 3.10b), allows for energy assimilation to change as a function of mass (i.e., assimilation = mass^d). The parameter d is often taken to be 2/3, assuming that energy assimilation is proportional to area (length^2) and mass is proportional to volume (length^3) (Quinn and Deriso, 1999).

- A generalized form of the Michaelis-Menten function, $f(x) = ax/(b + x^c)$ (Figure 3.10c), describes ecological competition (Maynard-Smith and Slatkin, 1973; Brännström and Sumpter, 2005). This model reduces to the standard Michaelis-Menten curve when $c = 1$; $0 < c < 1$ corresponds to "contest" (undercompensating) competition, while $c > 1$ corresponds to "scramble" (overcompensating) competition (the function has maximum for finite densities if $c > 1$). In fisheries, this model is called the *Shepherd function*. Quinn and Deriso (1999) show how the Shepherd function emerges as a generalization of the Beverton-Holt function when the density-dependent mortality coefficient is related to the initial size of the cohort.

- A related function, $f(x) = ax/(b+x)^c$, is known in ecology as the *Hassell* competition function (Hassell, 1975; Brännström and Sumpter, 2005); it is similar to the Shepherd/Maynard-Smith/Slatkin model in allowing Michaelis-Menten ($c = 1$), undercompensating ($c < 1$), or overcompensating ($c > 1$) dynamics.

- Persson et al. (1998) used a generalized Ricker equation, $y = A(\frac{x}{x_0} \exp{(1 - \frac{x}{x_0})})^\alpha$, to describe the dependence of attack rate y on predator body mass x (Figure 3.1 shows the same curve, but as a function of *prey* body mass). In fisheries, Ludwig and Walters proposed this function as a stock-recruitment curve (Quinn and Deriso, 1999). Bellows (1981) suggested a slightly different form of the generalized Ricker, $y = x \exp{(r(1 - (a/x)^\alpha))}$ (the power is inside the exponent instead of outside), to model density-dependent population growth.

- Emlen (1996) used a generalized form of the logistic, $y = a + b/(1 + c \exp{(-dx^\alpha)})$ extended to allow both a nonzero intercept (via the a parameter, discussed above under "Scaling and Shifting") and more flexibility in the shape of the curve (via the power exponent α).

- The *nonrectangular hyperbola* (Figure 3.10), based on first principles of plant physiology, describes the photosynthetic rate P as a function of light availability I:

$$P(I) = \frac{1}{2\theta} \left(\alpha I + p_{max} - \sqrt{(\alpha I + p_{max})^2 - 4\theta \alpha I p_{max}} \right),$$

where α is photosynthetic efficiency (and initial slope); p_{max} is the maximum photosynthetic rate (and asymptote); and θ is a sharpness parameter. In the limit as $\theta \to 0$, the function becomes a Michaelis-Menten function; in the limit as $\theta \to 1$, it becomes piecewise linear (Thornley, 2002).

Advantages

Functions incorporating power laws are flexible, especially since the power parameter is usually added to an existing model that already allows for changes in location, scale, and curvature. In many mechanistically derived power-law functions the value of the exponent comes from intrinsic geometric or allometric properties of the system and hence does not have to be estimated from data.

Disadvantages

Many different mechanisms can lead to power-law behavior (Mitzenmacher, 2003). It can be tempting but is often misguided to reason backward from an observed pattern to infer something about the meaning of a particular estimated parameter.

Despite the apparent simplicity of the formulas, estimating exponents from data can be numerically challenging—leading to poorly constrained or unstable estimates. The exponent of the nonrectangular hyperbola, for example, is notoriously difficult to estimate from reasonable-size data sets (Thornley, 2002). (We will see another example when we try to fit the Shepherd model to data in Chapter 5.)

3.4.4 Other Possibilities

Of course, there is no way I can enumerate all the functions used even within traditional population ecology, let alone fisheries, forestry, ecosystem, and physiological ecology. Haefner (1996, pp. 90–96) gives an alternative list of function types, focusing on functions used in physiological and ecosystem ecology, while Turchin (2003, Table 4.1, p. 81) presents a variety of predator functional response models. Some other occasionally useful categories are:

- *Curves based on other simple mathematical functions*: For example, trigonometric functions like sines and cosines (useful for fitting diurnal or seasonal patterns), and functions based on logarithms.
- *Generalized or "portmanteau" functions*: These are complex, highly flexible functions that reduce to various simpler functions for particular parameter values. For example, the four-parameter Richards growth model

$$y = \frac{k_1}{\left(1 + \left(\frac{k_1}{k_2} - 1\right) e^{-k_3 k_4 x}\right)^{1/k_4}} \qquad (3.4.4)$$

  includes the monomolecular, Gompertz, von Bertalanffy, and logistic equation as special cases (Haefner, 1996; Damgaard et al., 2002). Schnute (1981) defines a still more generalized growth model.
- *Functions not in closed form*: Sometimes it's possible to define the dynamics of a population, but not to find an analytical formula (what mathematicians would call a "closed-form solution") that describes the resulting population density.

  - The *theta-logistic* or *generalized logistic* model (Richards, 1959; Nelder, 1961; Thomas et al., 1980; Sibly et al., 2005) generalizes the logistic equation by adding a power (θ) to the logistic growth equation given above (3.4.3):

$$\frac{dn}{dt} = rn\left(1 - \left(\frac{n}{K}\right)^{\theta}\right). \qquad (3.4.5)$$

    When $\theta = 1$ this equation reduces to the logistic equation, but when $\theta \neq 1$ there is no closed-form solution for $n(t)$—i.e., no solution we can write down in mathematical notation. You can use the `odesolve` library in R to solve the differential equation numerically and get a value for a particular set of parameters.
  - The *Rogers random-predator equation* (Rogers, 1972; Juliano, 1993) describes the numbers of prey eaten by predators, or the numbers of prey

remaining after a certain amount of time in situations where the prey population becomes depleted. Like the theta-logistic, the Rogers equation has no closed-form solution, but it can be written in terms of a mathematical function called the *Lambert W function* (Corless et al. 1996). (See `?lambertW` in the `emdbook` package.)

3.5 Conclusion

The first part of this chapter showed (or reminded you of) some basic tools for understanding the mathematical functions used in ecological modeling—slopes, critical points, derivatives, and limits—and how to use them to figure out the basic properties of functions you come across in your work. The second part of the chapter briefly reviewed some common functions. You will certainly run across others, but the tools from the first part should help you figure out how they work.

3.6 R Supplement

3.6.1 Plotting Functions in Various Ways

Using curve:
   Plot a Michaelis-Menten curve:

```
> curve(2 * x/(1 + x))
```

You do need to specify the parameters: if you haven't defined a and b previously, `curve(a*x/(b+x))` will give you an error. But if you're going to use a function a lot, define a function:

```
> micmen <- function(x, a = 2, b = 1) {
+     a * x/(b + x)
+ }
```

   Now plot several curves (being more specific about the desired x and y ranges; changing colors; and adding a horizontal line (`abline(h=...)`) to show the asymptote).

```
> curve(micmen(x), from = 0, to = 8, ylim = c(0, 10))
> curve(micmen(x, b = 3), add = TRUE, col = 2)
> curve(micmen(x, a = 8), add = TRUE, col = 3)
> abline(h = 8)
```

   Sometimes rather than using *curve* you may want to handle the details yourself. Use `seq` to define a vector of x values:

```
> xvec <- seq(0, 10, by = 0.1)
```

Then use vectorization (yvec=micmen(xvec)) or sapply (yvec=sapply(xvec; micmen)) or a for loop (for i in (1:length(xvec)) {yvec[i] =micmen (xvec[i])}) to calculate the y values. Use plot(xvec,yvec,...), lines(xvec,yvec, ...), etc. (with options you learned in Chapter 2) to produce the graphics.

3.6.2 Piecewise Functions Using ifelse

The ifelse function picks one of two numbers (or values from one of two vectors) depending on a logical condition. For example, a simple threshold function where $y = 1$ if $x < 5$ and $y = 2$ otherwise:

```
> curve(ifelse(x < 5, 1, 2), from = 0, to = 10)
```

or a more complex piecewise linear function:

```
> curve(ifelse(x < 5, 1 + x, 6 - 3 * (x - 5)), from = 0,
+      to = 10)
```

You can also nest ifelse functions to get more than one switching point:

```
> curve(ifelse(x < 5, 1 + x, ifelse(x < 8, 6 - 3 *
+      (x - 5), -3 + 2 * (x - 8))), from = 0, to = 10)
```

3.6.3 Derivatives

You can use D or deriv to calculate derivatives (although R will not simplify the results at all): D gives you a relatively simple answer, while deriv gives you a function that will compute the function and its derivative for specified values of x (you need to use attr(...,"grad") to retrieve the derivative—see below). To use either of these functions, you need to use expression to stop R from trying to interpret the formula.

```
> D(expression(log(x)), "x")
```

1/x

```
> D(expression(x^2), "x")
```

2 * x

Using deriv to plot the logistic and its derivative:

```
> logist = expression(exp(x)/(1 + exp(x)))
> dfun = deriv(logist, "x", function.arg = TRUE)
> xvec = seq(-4, 4, length = 40)
> y = dfun(xvec)
> plot(xvec, y)
> lines(xvec, attr(y, "grad"))
```

Use eval to fill in parameter values in an expression:

```
> d1 = D(expression(a * x/(b + x)), "x")
> d1
a/(b + x) - a * x/(b + x)^2
> eval(d1, list(a = 2, b = 1, x = 3))
[1] 0.125
```

4 Probability and Stochastic Distributions for Ecological Modeling

This chapter continues to review the math you need to fit models to data, moving forward from functions and curves to probability distributions. The first part discusses ecological variability in general terms, then reviews basic probability theory and some important applications, including Bayes' Rule and its application in statistics. The second part reviews how to analyze and understand probability distributions. The third part provides a bestiary of probability distributions, finishing with a short digression on some ways to extend these basic distributions.

4.1 Introduction: Why Does Variability Matter?

For many ecologists and statisticians, noise is just a nuisance—it gets in the way of drawing conclusions from the data. The traditional statistical approach to noise in data was to assume that all variation in the data was normally distributed, or transform the data until it was, and then use classical methods based on the normal distribution to draw conclusions. Some scientists turned to nonparametric statistics, which assume only that the shape of the data distribution is the same in all categories and provide tests of differences in the means or "location parameters" among categories. Unfortunately, such classical nonparametric approaches make it much harder to draw quantitative conclusions from data (rather than simply rejecting or failing to reject null hypotheses about differences between groups).

In the 1980s, as they acquired better computing tools, ecologists began to use more sophisticated models of variability such as generalized linear models (see Chapter 9). Chapter 3 illustrated a wide range of deterministic functions that correspond to deterministic models of the underlying ecological processes. This chapter will illustrate a wide range of models for the stochastic part of the dynamics. In these models, variability isn't just a nuisance but actually tells us something about ecological processes. For example, census counts that follow a negative binomial distribution (p. 124) tell us there is some form of environmental variation or aggregative response among individuals that we haven't taken into account (Shaw and Dobson, 1995).

Remember from Chapter 1 that what we treat as "signal" (deterministic) and what we treat as "noise" (stochastic) depends on the question. The same ecological variability, such as spatial variation in light, might be treated as random variation by a forester interested in the net biomass growth of a forest stand and as a deterministic driving factor by an ecophysiologist interested in the photosynthetic response of individual plants.

Noise affects ecological data in two different ways—as *measurement error* and as *process noise* (this distinction will become important in Chapter 11 when we deal with dynamical models). Measurement error is the variability or "noise" in our measurements, which makes it hard to estimate parameters and make inferences about ecological systems. Measurement error leads to large confidence intervals and low statistical power. Even if we can eliminate measurement error, process noise or process error (often so-called even though it isn't technically an error but a real part of the system) still exists. Variability affects any ecological system. For example, we can observe thousands of individuals to determine the average mortality rate with great accuracy. The fate of a group of a few individuals, however, depends both on the variability in mortality rates of individuals and on the *demographic stochasticity* that determines whether a particular individual lives or dies ("loses the coin toss"). Even though we know the average mortality rate perfectly, our predictions are still uncertain. *Environmental stochasticity*—spatial and temporal variability in (e.g.) mortality rate caused by variation in the environment rather than by the inherent randomness of individual fates—also affects the dynamics. Finally, even if we can minimize measurement error by careful measurement and minimize process noise by studying a large population in a constant environment (i.e., one with low levels of demographic and environmental stochasticity), ecological systems can still amplify variability in surprising ways (Bjørnstad and Grenfell, 2001). For example, a tiny bit of demographic stochasticity at the beginning of an epidemic can trigger huge variation in epidemic dynamics (Rand and Wilson, 1991). Variability also feeds back to change the mean behavior of ecological systems. For example, in the damselfish system described in Chapter 2 the number of recruits in any given cohort is the number of settlers surviving density-dependent mortality, but the average number of recruits is *lower* than expected from an average-sized cohort of settlers because large cohorts suffer disproportionately high mortality and contribute relatively little to the average. This difference is an example of a widespread phenomenon called *Jensen's inequality* (Ruel and Ayres, 1999; Inouye, 2005).

4.2 Basic Probability Theory

To understand stochastic terms in ecological models, you'll have to (re)learn some basic probability theory. To define a probability, we first have to identify the *sample space*, the set of all the possible outcomes that could occur. Then the probability of an event A is the frequency with which that event occurs. A few probability rules are all you need to know:

1. If two events are *mutually exclusive* (e.g., "individual is male" and "individual is female"), then the probability that either occurs (the probability

of *A or B*, or Prob($A \cup B$)) is the sum of their individual probabilities: Prob(male or female) = Prob(male) + Prob(female).

We use this rule, for example, in finding the probability that an outcome is within a certain numeric range by adding up the probabilities of all the different (mutually exclusive) values in the range: for a discrete variable, for example, $P(3 \leq X \leq 5) = P(X = 3) + P(X = 4) + P(X = 5)$.

2. If two events *A* and *B* are not mutually exclusive—the *joint probability* that they occur together, Prob($A \cap B$), is greater than zero—then we have to correct the rule for combining probabilities to account for double-counting:

$$\text{Prob}(A \cup B) = \text{Prob}(A) + \text{Prob}(B) - \text{Prob}(A \cap B).$$

For example, if we are tabulating the color and sex of animals, Prob(blue or male) = Prob(blue) + Prob(male) − Prob(blue male).

3. The probabilities of all possible outcomes of an observation or experiment add to 1.0: Prob(male) + Prob(female) = 1.0.

We will need this rule to understand the form of probability distributions, which often contain a *normalization constant* which ensures that the sum of the probabilities of all possible outcomes is 1.

4. The *conditional probability* of *A* given *B*, Prob($A|B$), is the probability that *A* happens if we know or assume *B* happens. The conditional probability equals

$$\text{Prob}(A|B) = \text{Prob}(A \cap B)/\text{Prob}(B). \tag{4.2.1}$$

For example, continuing the color and sex example:

$$\text{Prob(blue|male)} = \frac{\text{Prob(blue male)}}{\text{Prob(male)}}. \tag{4.2.2}$$

By contrast, we may also refer to the probability of *A* when we make no assumptions about *B* as the *unconditional* probability of *A*: Prob(*A*) = Prob($A|B$)Prob(*B*) + Prob($A|$not *B*)Prob(not *B*). Conditional probability is central to understanding *Bayes' Rule* (Section 4.3).

5. If the conditional probability of *A* given *B*, Prob($A|B$), equals the unconditional probability of *A*, then *A* is *independent* of *B*. Knowing about *B* provides no information about the probability of *A*. Independence implies that

$$\text{Prob}(A \cap B) = \text{Prob}(A)\text{Prob}(B), \tag{4.2.3}$$

which follows from substituting Prob($A|B$) = Prob(*A*) in (4.2.1) and multiplying both sides by Prob(*B*). The probabilities of combinations of independent events are multiplicative.

Multiplying probabilities, or adding log-probabilities ($\log(\text{Prob}(A \cap B)) = \log(\text{Prob}(A)) + \log(\text{Prob}(B))$ if *A* and *B* are independent), is how we find the combined probability of a series of independent observations.

We can immediately use these rules to think about the distribution of seeds taken in the seed removal experiment (Chapter 2). The most obvious pattern in the data is that there are many zeros, probably corresponding to times when no predators visited the station. The sample space for seed disappearance is the number of seeds taken, from 0 to N (the number available). Suppose that when a predator *did* visit the station, which happened with probability v, it had an equal probability of taking any of the possible number of seeds (i.e., a uniform distribution from 0 to N). Since the probabilities must add to 1, this probability (Prob(x taken|predator visit)) is $1/(N+1)$ (0 to N represents $N+1$ different possible events). What is the unconditional probability of x seeds being taken?

If $x > 0$, then only one type of event is possible—the predator visited and took x seeds—with overall probability $v/(N+1)$ (Figure 4.1a).

If $x = 0$, then there are two mutually exclusive possibilities. Either the predator didn't visit (probability $1-v$), or it visited (probability v) and took zero seeds (probability $1/(N+1)$), so the overall probability is

$$\underbrace{(1-v)}_{\text{didn't visit}} + \left(\underbrace{v}_{\text{visited}} \times \underbrace{\frac{1}{N+1}}_{\text{took zero seeds}} \right) = 1-v+\frac{v}{N+1}. \tag{4.2.4}$$

Now make things a little more complicated and suppose that when a predator visits, it decides independently whether or not to take each seed. If the seeds of a given species are all identical, so that each seed is taken with the same probability p, then this process results in a binomial distribution. Using the rules above, the probability of x seeds being taken when each has probability p is p^x. It's also true that $N - x$ seeds are *not* taken, with probability $(1-p)^{N-x}$. Thus the probability is proportional to $p^x \cdot (1-p)^{N-x}$. To get the probabilities of all possible outcomes to add to 1, though, we have to multiply by a normalization constant $N!/(x!(N-x)!)$,[*] or $\binom{N}{x}$. (It's too bad we can't just ignore these ugly normalization factors, which are always the least intuitive parts of probability formulas, but we really need them in order to get the right answers. Unless you are doing advanced calculations, however, you can usually take the formulas for the normalization constants for granted, without trying to puzzle out their meaning.)

Now adding the "predator may or may not visit" layer to this formula, we have a probability

$$\underbrace{(1-v)}_{\text{didn't visit}} + \left(\underbrace{v}_{\text{visited}} \cdot \underbrace{\text{Binom}(0,p,N)}_{\text{took zero seeds}} \right) = (1-v)+v(1-p)^N \tag{4.2.5}$$

if $x = 0$ (since $\binom{N}{0} = 1$, the normalization constant disappears from the second term), or

$$\underbrace{v}_{\text{visited}} \cdot \underbrace{\text{Binom}(x,p,N)}_{\text{took} > 0 \text{ seeds}} = v\binom{N}{x}p^x(1-p)^{N-x} \tag{4.2.6}$$

if $x > 0$ (Figure 4.1b).

This distribution is called the *zero-inflated binomial* (Inouye, 1999; Tyre et al., 2003). With only a few simple probability rules, we have derived a potentially useful

---

[*] $N!$ means $N \cdot (N-1) \cdot \ldots \cdot 2 \cdot 1$, and is referred to as "N factorial."

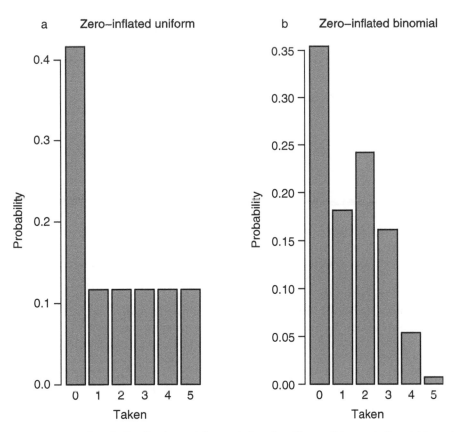

Figure 4.1 Zero-inflated distributions. (a) zero-inflated uniform; (b) zero-inflated binomial. Number of seeds $N = 5$, probability of predator visit $v = 0.7$, binomial probability of individual seed predation $p = 0.4$.

distribution that might describe the pattern of seed predation better than any of the standard distributions we'll see later in this chapter.

4.3 Bayes' Rule

With the simple probability rules defined above we can also derive, and understand, *Bayes' Rule*. Most of the time we will use Bayes' Rule to go from the likelihood Prob($D|H$), the probability of observing a particular set of data D given that a hypothesis H is true (p. 13), to the information we really want, Prob($H|D$)—the probability of our hypothesis H in light of our data D. Bayes' Rule is just a recipe for turning around a conditional probability:

$$P(H|D) = \frac{P(D|H)P(H)}{P(D)}. \tag{4.3.1}$$

Bayes' Rule is general—H and D can be any events, not just hypothesis and data—but it's easier to understand Bayes' Rule when we have something concrete to tie it to.

Deriving Bayes' Rule is almost as easy as remembering it. Rule 4 on p. 105 applied to $P(H|D)$ implies

$$P(H \cap D) = P(H|D)P(D),$$ (4.3.2)

while applying it to $P(D|H)$ tells us

$$P(D \cap H) = P(D|H)P(H).$$ (4.3.3)

But $P(H \cap D) = P(D \cap H)$, so

$$P(H|D)P(D) = P(D|H)P(H)$$ (4.3.4)

and dividing both sides by $P(D)$ gives us (4.3.1).

Equation (4.3.1) says that the probability of the hypothesis given (in light of) the data is equal to the probability of the data given the hypothesis (the *likelihood* associated with H), times the probability of the hypothesis, divided by the probability of the data. There are two problems here: we don't know the probability of the hypothesis, $P(H)$ (isn't that what we were trying to figure out in the first place?), and we don't know the unconditional probability of the data, $P(D)$.

Let's think about the second problem first—our ignorance of $P(D)$. We can calculate an unconditional probability for the data if we have a set of *exhaustive, mutually exclusive* hypotheses: in other words, we assume that one, and only one, of our hypotheses is true. Figure 4.2 shows a geometric interpretation of Bayes' Rule. The gray ellipse represents D, the set of all possibilities that could lead to the observed data.

If one of the hypotheses must be true, then the unconditional probability of observing the data is the sum of the probabilities of observing the data under any of the possible hypotheses. For N different hypotheses H_1 to H_N,

$$P(D) = \sum_{j=1}^{N} P(D \cap H_j)$$

$$= \sum_{j=1}^{N} P(D|H_j)P(H_j).$$ (4.3.5)

In words, the unconditional probability of the data is the sum of the likelihood of each hypothesis ($P(D|H_j)$) times its unconditional probability ($P(H_j)$). In Figure 4.2, taking each wedge (H_j), finding its area of overlap with the gray ellipse ($D \cap H_j$), and summing the area of these "pizza slices" provides the area of the ellipse (D).

Substituting (4.3.5) into (4.3.1) gives the full form of Bayes' Rule for a particular hypothesis H_i when it is one of a mutually exclusive set of hypotheses $\{H_j\}$. The probability of the truth of H_i in light of the data is

$$P(H_i|D) = \frac{P(D|H_i)P(H_i)}{\sum_j P(D|H_j)P(H_j)}.$$ (4.3.6)

In Figure 4.2, having observed the data D means we know that reality lies somewhere in the gray ellipse. The probability that hypothesis 5 is true (i.e., that we are somewhere in the hatched area) is equal to the area of the hatched/shaded

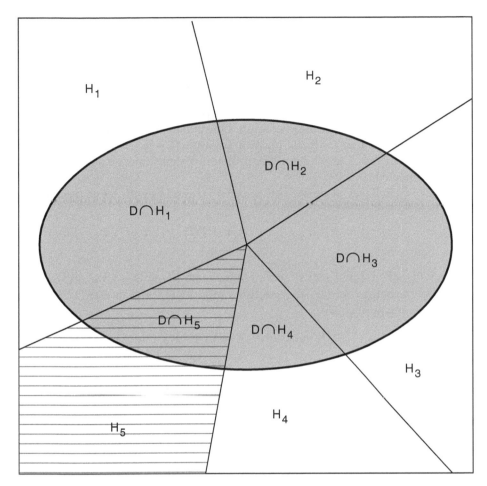

Figure 4.2 Decomposition of the unconditional probability of the observed data (D) into the sum of the probabilities of the intersection of the data with each possible hypothesis ($\sum_{j=1}^{N} D \cap H_j$). The entire gray ellipse in the middle represents D. Each wedge (e.g., the hatched area H_5) represents an alternative hypothesis. The ellipse is divided into "pizza slices" (e.g., $D \cap H_5$, hatched and shaded area). The area of each slice corresponds to $D \cap H_j$, the joint probability of the data D (ellipse) *and* the particular hypothesis H_j (wedge).

"pizza slice" divided by the area of the ellipse. Bayes' Rule breaks this down further by supposing that we know how to calculate the likelihood of the data for each hypothesis—the ratio of the pizza slice divided by the area of the entire wedge (the area of the pizza slice [$D \cap H_5$] divided by the hatched wedge [H_5]). Then we can recover the area of each slice by multiplying the likelihood by the prior (the area of the wedge) and calculate both $P(D)$ and $P(H_5|D)$.

Dealing with the second problem, our ignorance of the unconditional or *prior* probability of the hypothesis $P(H_i)$, is more difficult. In the next section we will simply assume that we have other information about this probability, and we'll revisit the problem shortly in the context of Bayesian statistics. But first, just to practice with Bayes' Rule, we'll explore two simple examples that use Bayes' Rule to manipulate conditional probabilities.

4.3.1 False Positives in Medical Testing

Suppose the unconditional probability of a random person sampled from the population being infected (I) with some deadly but rare disease is one in a million: $P(I) = 10^{-6}$. There is a test for this disease that never gives a false negative result: if you have the disease, you will definitely test positive ($P(+|I) = 1$). However, the test does occasionally give a false positive result. One person in 100 who doesn't have the disease (is uninfected, U) will test positive anyway ($P(+|U) = 10^{-2}$). This sounds like a pretty good test. Let's compute the probability that someone who tests positive is actually infected.

Replace H in Bayes' Rule with "is infected" (I) and D with "tests positive" ($+$). Then

$$P(I|+) = \frac{P(+|I)P(I)}{P(+)}. \qquad (4.3.7)$$

We know $P(+|I) = 1$ and $P(I) = 10^{-6}$, but we don't know $P(+)$, the unconditional probability of testing positive. You are either infected (I) or uninfected (U), so these events are mutually exclusive:

$$P(+) = P(+\cap I) + P(+\cap U). \qquad (4.3.8)$$

Then

$$P(+) = P(+|I)P(I) + P(+|U)P(U) \qquad (4.3.9)$$

because $P(+\cap I) = P(+|I)P(I)$ and similarly for U (4.2.1). We also know that $P(U) = 1 - P(I)$, so

$$
\begin{aligned}
P(+) &= P(+|I)P(I) + P(+|U)(1 - P(I)) \\
&= 1 \times 10^{-6} + 10^{-2} \times (1 - 10^{-6}) \\
&= 10^{-6} + 10^{-2} - 10^{-8} \\
&\approx 10^{-2}.
\end{aligned}
\qquad (4.3.10)
$$

Since 10^{-6} is 10,000 times smaller than 10^{-2}, and 10^{-8} is even tinier, we can neglect them.

Now that we've done the hard work of computing the denominator $P(+)$, we can put it together with the numerator:

$$
\begin{aligned}
P(I|+) &= \frac{P(+|I)P(I)}{P(+)} \\
&\approx \frac{1 \times 10^{-6}}{10^{-2}} \\
&= 10^{-4}.
\end{aligned}
\qquad (4.3.11)
$$

Even though false positives are unlikely, the chance that you are infected if you test positive is still only 1 in 10,000! For a sensitive test (one that produces few false negatives) for a rare disease, the probability that a positive test is detecting a true infection is approximately $P(I)/P$(false positive), which can be surprisingly small.

This false-positive issue also comes up in forensics cases. Assuming that a positive test is significant is called the *base rate fallacy*. It's important to think carefully about the sample population and the true probability of being guilty (or at least having been present at the crime scene) if your DNA matches DNA found at the crime scene.

4.3.2 Bayes' Rule and Liana Infestation

A student of mine used Bayes' Rule as part of a simulation model of liana (vine) dynamics in a tropical forest. He wanted to know the probability that a newly emerging sapling would be in a given "liana class" (L_1 = liana-free, $L_2 - L_3$ = light to moderate infestation, L_4 = heavily infested with lianas). This probability depends on the number of trees nearby that are already infested (N). We have measurements of infestation of saplings from the field, and for each one we know the number of nearby infestations. Thus if we calculate the fraction of individuals in liana class L_i with N nearby infested trees, we get an estimate of $\text{Prob}(N|L_i)$. We also know the overall fractions in each liana class, $\text{Prob}(L_i)$. When we add a new tree to the model, we know the neighborhood infestation N from the model. Thus we can figure out the rules for assigning infestation to a new sapling, $\text{Prob}(L_i|N)$, by using Bayes' Rule to calculate

$$\text{Prob}(L_i|N) = \frac{\text{Prob}(N|L_i)\text{Prob}(L_i)}{\sum_{j=1}^{4} \text{Prob}(N|L_j)\text{Prob}(L_j)}. \tag{4.3.12}$$

For example, suppose we find that a new tree in the model has 3 infested neighbors. Let's say that the probabilities of each liana class (1 to 4) having 3 infested neighbors are $\text{Prob}(3|L_i) = \{0.05, 0.1, 0.3, 0.6\}$ and that the overall fractions of each liana class in the forest (unconditional probabilities) are $L_i = \{0.5, 0.25, 0.2, 0.05\}$. Then the probability that the new tree is heavily infested (i.e., is in class L_4) is

$$\frac{0.6 \times 0.05}{(0.05 \times 0.5) + (0.1 \times 0.25) + (0.3 \times 0.2) + (0.6 \times 0.05)} = 0.21. \tag{4.3.13}$$

We would expect that a new tree with several infested neighbors has a much higher probability of heavy infestation than the overall (unconditional) probability of 0.05. Bayes' Rule allows us to quantify this guess.

4.3.3 Bayes' Rule in Bayesian Statistics

So what does Bayes' Rule have to do with Bayesian statistics?

Bayesians translate likelihood into information about parameter values using Bayes' Rule as given above. The problem is that we have the likelihood $\mathcal{L}$(data|hypothesis), the probability of observing the data given the model (parameters); what we want is Prob(hypothesis|data). After all, we already know what the data are!

4.3.3.1 PRIORS

In the disease testing and the liana examples, we knew the overall, unconditional probability of disease or liana class in the population. When we're doing Bayesian statistics, however, we interpret $P(H_i)$ instead as the *prior probability* of a hypothesis, our belief about the probability of a particular hypothesis *before* we see the data. Bayes' Rule is the formula for updating the prior in order to compute the *posterior probability* of each hypothesis, our belief about the probability of the hypothesis *after* we see the data. Suppose I have two hypotheses A and B and have observed some data D with likelihoods $\mathcal{L}_A = 0.1$ and $\mathcal{L}_B = 0.2$. In other words, the probability of D occurring if hypothesis A is true ($P(D|A)$) is 10%, while the probability of D occurring if hypothesis B is true ($P(D|B)$) is 20%. If I assign the two hypotheses equal prior probabilities (0.5 each), then Bayes' Rule says the posterior probability of A is

$$P(A|D) = \frac{0.1 \times 0.5}{0.1 \times 0.5 + 0.2 \times 0.5} = \frac{0.1}{0.3} = \frac{1}{3} \qquad (4.3.14)$$

and the posterior probability of B is 2/3. However, if I had prior information that said A was twice as probable ($\mathrm{Prob}(A) = 2/3$, $\mathrm{Prob}(B) = 1/3$), then the probability of A given the data would be 0.5 (do the calculation). If you rig the prior, you can get whatever answer you want: e.g., if you assign B a prior probability of 0, then no data will *ever* convince you that B is true (in which case you probably shouldn't have done the experiment in the first place). Frequentists claim that this possibility makes Bayesian statistics open to cheating (Dennis, 1996); however, every Bayesian analysis must clearly state the prior probabilities it uses. If you have good reason to believe that the prior probabilities are not equal, from previous studies of the same or similar systems, then arguably you should *use* that information rather than starting as frequentists do from the ground up every time. (The frequentist-Bayesian debate is one of the oldest and most virulent controversies in statistics (Dennis, 1996; Ellison 1996); I can't possibly do it justice here.)

However, trying so-called *flat* or *weak* or *uninformative* priors—priors that assume you have little information about which hypothesis is true—as a part of your analysis is a good idea, even if you do have prior information (Edwards, 1996). You may have noticed in the first example above that when we set the prior probabilities equal, the posterior probabilities were just equal to the likelihoods divided by the sum of the likelihoods. If all the $P(H_i)$ are equal to the same constant C, then

$$P(H_i|D) = \frac{P(D|H_i)C}{\sum_j P(D|H_j)C} = \frac{\mathcal{L}_i}{\sum_j \mathcal{L}_j} \qquad (4.3.15)$$

where $\mathcal{L}_i$ is the likelihood of hypothesis i.

You may think that setting all the priors equal would be an easy way to eliminate the subjective nature of Bayesian statistics and make everybody happy. Two examples, however, will demonstrate that it's not that easy to say what it means to be completely "objective" or ignorant of which hypothesis is true.

- *Partitioning hypotheses*: Suppose we find a nest missing eggs that might have been taken by a raccoon, a squirrel, or a snake (only). The three hypotheses "raccoon" (R), "squirrel" (Q), and "snake" (S) are our mutually exclusive and exhaustive set of hypotheses for the identity of the predator. If we have

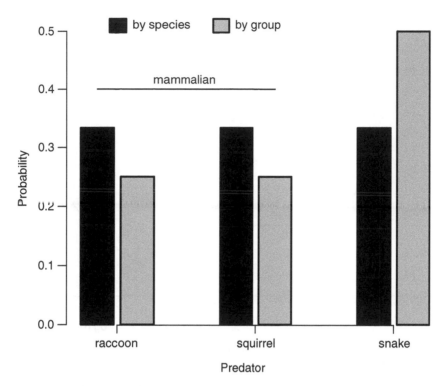

Figure 4.3 The difficulty of defining an uninformative prior for discrete hypotheses. Dark bars are priors that assume predation by each species is equally likely; light bars divide predation by group first, then by species within group.

no other information (e.g., about the local densities or activity levels of different predators), we might choose equal prior probabilities for all three hypotheses. Since there are three mutually exclusive predators, $\text{Prob}(R) = \text{Prob}(Q) = \text{Prob}(S) = 1/3$. Now a friend comes and asks us whether we really believe that mammalian predators are twice as likely to eat the eggs as reptiles ($\text{Prob}(R) + \text{Prob}(Q) = 2\text{Prob}(S)$) (Figure 4.3). What do we do? We might solve this particular problem by setting the probability for snakes (the only reptiles) to 0.5, the probability for mammals ($\text{Prob}(R \cup Q)$) to 0.5, and the probability for raccoons and squirrels equal ($\text{Prob}(R) = \text{Prob}(Q) = 0.25$), but this simple example suggests that such pitfalls are ubiquitous.

• *Changing scales*: A similar problem arises with continuous variables. Suppose we believe that the mass of a particular bird species is between 10 and 100 g, and that no particular value is any more likely than other: the prior distribution is uniform, or flat. That is, the probability that the mass is in some range of width Δm is constant: $\text{Prob}(\text{mass} = m) = 1/90\Delta m$ (so that $\int_{10}^{100} \text{Prob}(m)\,dm = 1$: see p. 116 for more on probability densities).
But is it sensible to assume that the probability that a species' mass is between 10 and 20 is the same as the probability that it is between 20 and 30, or should it be the same as the probability that it is between 20 and 40—that is, would it make more sense to think of the mass distribution on a logarithmic scale?

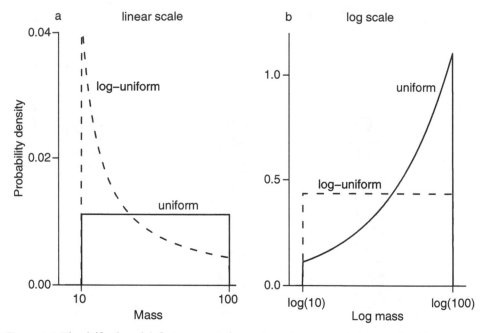

Figure 4.4 The difficulty of defining an uninformative prior on continuous scales. If we assume that the probabilities are uniform on one scale (linear or logarithmic), they must be nonuniform on the other.

If we say that the probability distribution is uniform on a logarithmic scale, then a species is *less* likely to be between 20 and 30 than it is to be between 10 and 20 (Figure 4.4).* Since changing the scale is not really changing anything about the world, just the way we describe it, this change in the prior is another indication that it's harder than we think to say what it means to be ignorant. In any case, many Bayesians think that researchers try too hard to pretend ignorance, and that one really should use what is known about the system. Crome et al. (1996) compare extremely different priors in a conservation context to show that their data really are (or should be) informative to a wide spectrum of stakeholders, regardless of their perspectives.

4.3.3.2 INTEGRATING THE DENOMINATOR

The other challenge with Bayesian statistics, which is purely technical and does not raise any deep conceptual issues, is the problem of adding up the denominator $\sum_j P(H_j)P(D|H_j)$ in Bayes' Rule. If the set of hypotheses (parameters) is continuous, then the denominator is $\int P(h)P(D|h)\,dh$ where h is a particular parameter value.

For example, the binomial distribution says that the likelihood of obtaining 2 heads in 3 (independent, equal-probability) coin flips is $\binom{3}{2}p^2(1-p)$, a function of p.

---

* If the probability is uniform between a and b on the usual, linear scale (Prob(mass $= m$) $= 1/(b-a)\,dm$), then on the log scale it is Prob(log mass $= M$) $= 1/(b-a)e^M\,dM$ [if we change variables to log mass M, then $dM = d(\log m) = 1/m\,dm$, so $dm = m\,dM = e^M\,dM$]. Going the other way, a log-uniform assumption gives Prob(mass $= m$) $= 1/(\log(b/a)m)dm$ on the linear scale.

The likelihood for $p = 0.5$ is therefore 0.375, but to get the posterior probability we have to divide by the probability of getting 2 heads in 3 flips for *any* value of p. Assuming a flat prior, the denominator is $\int_0^1 \binom{3}{2} p^2 (1-p) \, dp = 0.25$, so the posterior probability density of $p = 0.5$ is $0.375/0.25 = 1.5$.*

For the binomial case and other simple probability distributions, it's easy to sum or integrate the denominator either analytically or numerically. If we care only about the *relative* probability of different hypotheses, we don't need to integrate the denominator because it has the same constant value for every hypothesis.

Often, however, we do want to know the absolute probability. Calculating the unconditional probability of the data (the denominator for Bayes' Rule) can be extremely difficult for more complicated problems. Much of current research in Bayesian statistics focuses on ways to calculate the denominator. We will revisit this problem in Chapters 6 and 7, first integrating the denominator by brute-force numerical integration, then looking briefly at a sophisticated technique for Bayesian analysis called Markov chain Monte Carlo.

4.3.4 Conjugate Priors

Using so-called *conjugate priors* makes it easy to do the math for Bayesian analysis. Imagine that we're flipping coins (or measuring tadpole survival or counting numbers of different morphs in a fixed sample) and that we use the binomial distribution to model the data. For a binomial with a per-trial probability of p and N trials, the probability of x successes is proportional (leaving out the normalization constant) to $p^x (1-p)^{N-x}$. Suppose that instead of describing the probability of x successes with a fixed per-trial probability p and number of trials N we wanted to describe the probability of a given per-trial probability p with fixed x and N. We would get Prob(p) proportional to $p^x(1-p)^{N-x}$—*exactly the same formula*, but with a different proportionality constant and a different interpretation. Instead of a discrete probability distribution over a sample space of all possible numbers of successes (0 to N), now we have a continuous probability distribution over all possible probabilities (all values between 0 and 1). The second distribution, for Prob(p), is called the Beta distribution (p. 133) and it is the conjugate prior for the binomial distribution.

Mathematically, conjugate priors have the same structure as the probability distribution of the data. They lead to a posterior distribution with the same mathematical form as the prior, although with different parameter values. Intuitively, you get a conjugate prior by turning the likelihood around to ask about the probability of a parameter instead of the probability of the data.

We'll come back to conjugate priors and how to use them in Chapters 6 and 7.

4.4 Analyzing Probability Distributions

You need the same kinds of skills and intuitions about the characteristics of probability distributions that we developed in Chapter 3 for mathematical functions.

---

* This value is a probability density, not a probability, so it's OK for it to be greater than 1: probability density will be explained on p. 116.

4.4.1 Definitions

DISCRETE

A probability distribution is the set of probabilities on a sample space or set of outcomes. Since this book is about modeling quantitative data, we will always be dealing with sample spaces that are numbers—the number or amount observed in some measurement of an ecological system. The simplest distributions to understand are *discrete* distributions whose outcomes are a set of integers; most of the discrete distributions we deal with describe counting or sampling processes and have ranges that include some or all of the nonnegative integers.

A discrete distribution is most easily described by its distribution function, which is just a formula for the probability that the outcome of an experiment or observation (called a *random variable*) X is equal to a particular value x ($f(x) = \text{Prob}(X = x)$). A distribution can also be described by its cumulative distribution function $F(x)$ (note the uppercase F), which is the probability that the random variable X is less than or equal to a particular value x ($F(x) = \text{Prob}(X \leq x)$). Cumulative distribution functions are most useful for frequentist calculations of tail probabilities, e.g., the probability of getting n or fewer heads in a coin-tossing experiment with a given trial probability.

CONTINUOUS

A probability distribution over a continuous range (such as all real numbers, or the nonnegative real numbers) is called a *continuous* distribution. The cumulative distribution function of a continuous distribution ($F(x) = \text{Prob}(X \leq x)$) is easy to define and understand—it's just the probability that the observed value of a continuous random variable X is smaller than a particular value x in any given observation or experiment. The probability *density function* (the analogue of the distribution function for a discrete distribution), although useful, is more confusing, since the probability of any precise value is zero. You may imagine that a measurement of (say) pH is *exactly* 7.9, but in fact what you have observed is that the pH is between 7.82 and 7.98—if your meter has a precision of ±1%. Thus continuous probability distributions are expressed as probability *densities* rather than probabilities—the probability that random variable X is between x and $x + \Delta x$, divided by Δx ($\text{Prob}(7.82 < X < 7.98)/0.16$, in this case). Dividing by Δx allows the observed probability density to have a well-defined limit as precision increases and Δx shrinks to zero. Unlike probabilities, probability densities can be larger than 1 (Figure 4.5). For example, if the pH probability distribution is uniform on the interval [7,7.1] but zero everywhere else, its probability density is 10 in this range. In practice, we will be concerned mostly with *relative* probabilities or likelihoods, and so the maximum density values and whether they are greater than or less than 1 won't matter much.

4.4.2 Means (Expectations)

The first thing you usually want to know about a distribution is its average value, also called its *mean* or *expectation*.

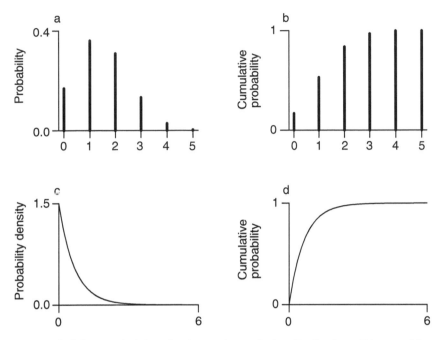

Figure 4.5 Probability, probability density, and cumulative distributions. Discrete (binomial: $N = 5$, $p = 0.3$) (a) probability and (b) cumulative probability distributions. Continuous (exponential: $\lambda = 1.5$) (c) probability density and (d) cumulative probability distributions.

In general the expectation operation, denoted by $E[\cdot]$ (or a bar over a variable, such as $\bar{x}$) gives the "expected value" of a set of data, or a probability distribution, which in the simplest case is the same as its (arithmetic) mean value. For a set of N data values written down separately as $\{x_1, x_2, x_3, \ldots, x_N\}$, the formula for the mean is familiar:

$$E[x] = \frac{\sum_{i=1}^{N} x_i}{N}. \tag{4.4.1}$$

Suppose we have the data tabulated instead, so that for each possible value of x (for a discrete distribution) we have a count of the number of observations (possibly zero, possibly more than 1), which we call $c(x)$. Summing over all of the possible values of x, we have

$$E[x] = \frac{\sum_{i=1}^{N} x_i}{N} = \frac{\sum c(x)x}{N} = \sum \left(\frac{c(x)}{N}\right) x = \sum \text{Prob}(x)x \tag{4.4.2}$$

where $\text{Prob}(x)$ is the discrete probability distribution representing this particular data set. More generally, you can think of $\text{Prob}(x)$ as representing some particular theoretical probability distribution which only approximately matches any actual data set.

We can compute the mean of a continuous distribution as well. First, let's think about grouping (or "binning") the values in a discrete distribution into categories of size Δx. Then if $p(x)$, the density of counts in bin x, is $c(x)/(N\Delta x)$, the formula for the mean becomes $\sum p(x) \cdot x \Delta x$. If we have a continuous distribution with Δx very small, this becomes $\int p(x)x\,dx$. (This is in fact the definition of an integral.) For example, an exponential distribution $p(x) = \lambda \exp(-\lambda x)$ has an expectation or mean value of $\int \lambda \exp(-\lambda x)x\,dx = 1/\lambda$. (You don't need to know how to do this integral analytically: the R supplement will briefly discuss numerical integration in R.)

4.4.3 Variances (Expectation of X^2)

The mean is the expectation of the random variable X itself, but we can also ask about the expectation of functions of X. The first example is the expectation of X^2. We just fill in the value x^2 for x in all of the formulas above: $E[x^2] = \sum \text{Prob}(x)x^2$ for a discrete distribution, or $\int p(x)x^2\,dx$ for a continuous distribution. (We are *not* asking for $\sum \text{Prob}(x^2)x^2$.) The expectation of x^2 is a component of the variance, which is the expected value of $(x - E[x])^2$ or $(x - \bar{x})^2$, or the expected squared deviation around the mean. (We can also show that

$$E[(x - \bar{x})^2] = E[x^2] - (\bar{x})^2 \qquad (4.4.3)$$

by using the rules for expectations that (1) $E[x+y] = E[x]+E[y]$ and (2) if c is a constant, $E[cx] = cE[x]$. The right-hand formula is simpler to compute than $E[(x - \bar{x})^2]$, but more subject to roundoff error.)

Variances are easy to work with because they are additive ($\text{Var}(a+b) = \text{Var}(a) + \text{Var}(b)$ if a and b are uncorrelated), but harder to compare with means since their units are the units of the mean squared. Thus we often use instead the *standard deviation* of a distribution, ($\sqrt{\text{Var}}$), which has the same units as X.

Two other summaries related to the variance are the *variance-to-mean* ratio and the *coefficient of variation* (CV), which is the ratio of the standard deviation to the mean. The variance-to-mean ratio has units equal to the mean; it is used primarily to characterize discrete sampling distributions and compare them to the Poisson distribution, which has a variance-to-mean ratio of 1. The CV is more common and is useful when you want to describe variation that is proportional to the mean. For example, if you have a pH meter that is accurate to $\pm 10\%$, so that a true pH value of x will give measured values that are normally distributed with $2\sigma = 0.1x$,[*] then $\sigma = 0.05\bar{x}$ and the CV is 0.05.

4.4.4 Higher Moments

The expectation of $(x - E[x])^3$ tells you the *skewness* of a distribution or a data set, which indicates whether it is asymmetric around its mean. The expectation $E[(x - E[x])^4]$ measures the *kurtosis*, the "pointiness" or "flatness," of a distribution.[†] These are called the third and fourth *central moments* of the distribution. In

---

[*] Remember that the 95% confidence limits of the normal distribution are approximately $\mu \pm 2\sigma$.
[†] The kurtosis is sometimes scaled by Var^2, or by $3\,\text{Var}^2$.

general, the nth moment is $E[x^n]$, and the nth central moment is $E[(x - \bar{x})^n]$; the mean is the first moment, and the variance is the second central moment. We won't be too concerned with these summaries (of data or distributions), but they do come up sometimes.

4.4.5 Median and Mode

The median and mode are two final properties of probability distributions that are not related to moments. The *median* of a distribution is the point that divides the area of the probability density in half, or the point at which the cumulative distribution function is equal to 0.5. It is often useful for describing data, since it is *robust*—outliers change its value less than they change the mean—but for many distributions it's more complicated to compute than the mean. The *mode* is the "most likely value," the maximum of the probability distribution or density function. For symmetric distributions the mean, mode, and median are all equal; for right-skewed distributions, in general mode < median < mean.

4.4.6 The Method of Moments

Suppose you know the theoretical values of the moments (e.g., mean and variance) of a distribution and have calculated the sample values of the moments (by calculating $\bar{x} = \sum x/N$ and $s^2 = \sum(x - \bar{x})^2/N$; don't worry for the moment about whether the denominator in the sample variance should be N or $N - 1$). Then there is a simple way to estimate the parameters of a distribution, called the *method of moments*: just match the sample values up with the theoretical values. For the normal distribution, where the parameters of the distribution are just the mean and the variance, this is trivially simple: $\mu = \bar{x}$, $\sigma^2 = s^2$. For a distribution like the negative binomial, however (p. 124), it involves a little bit of algebra. The negative binomial has parameters μ (equal to the mean, so that's easy) and k; the theoretical variance is $\sigma^2 = \mu(1 + \mu/k)$. Therefore, setting $\mu \approx \bar{x}$, $s^2 \approx \mu(1 + \mu/k)$, and solving for k, we calculate the method-of-moments estimate of k:

$$\sigma^2 = \mu(1 + \mu/k)$$
$$s^2 \approx \bar{x}(1 + \bar{x}/k)$$
$$\frac{s^2}{\bar{x}} - 1 \approx \frac{\bar{x}}{k} \tag{4.4.4}$$
$$k \approx \frac{\bar{x}}{s^2/\bar{x} - 1}.$$

The method of moments is very simple but is often biased; it's a good way to get a first estimate of the parameters of a distribution, but for serious work you should follow it up with a maximum likelihood estimator (Chapter 6).

TABLE 4.1
Summary of Probability Distributions

Distribution	Type	Range	Skew	Examples
Binomial	Discrete	$0, N$	Any	Number surviving, number killed
Poisson	Discrete	$0, \infty$	Right	Seeds per quadrat, settlers (variance/mean ≈ 1)
Negative binomial	Discrete	$0, \infty$	Right	Seeds per quadrat, settlers (variance/mean > 1)
Geometric	Discrete	$0, \infty$	Right	Discrete lifetimes
Beta-binomial	Discrete	$0, N$	Any	Similar to binomial
Uniform	Continous	$0, 1$	None	Cover proportion
Normal	Continuous	$-\infty, \infty$	None	Mass
Gamma	Continuous	$0, \infty$	Right	Survival time, distance to nearest edge
Beta	Continuous	$0, 1$	Any	Cover proportion
Exponential	Continuous	$0, \infty$	Right	Survival time, distance to nearest edge
Lognormal	Continuous	$0, \infty$	Right	Size, mass (exponential growth)

4.5 Bestiary of Distributions

The rest of the chapter presents brief introductions to a variety of useful probability distribution (Table 4.1), including the mechanisms behind them and some of their basic properties. Like the bestiary in Chapter 3, you can skim this bestiary on the first reading. The appendix of Gelman et al. (1996) contains a useful table, more abbreviated than these descriptions but covering a wider range of functions. The book by Evans et al. (2000) is also useful.

4.5.1 Discrete Models

4.5.1.1 BINOMIAL

The binomial (Figure 4.6) is probably the easiest distribution to understand. It applies when you have samples with a fixed number of subsamples or "trials" in each one, and each trial can have one of two values (black/white, heads/tails, alive/dead, species A/species B), and the probability of "success" (black, heads, alive, species A) is the same in every trial. If you flip a coin 10 times ($N = 10$) and the probability of a head

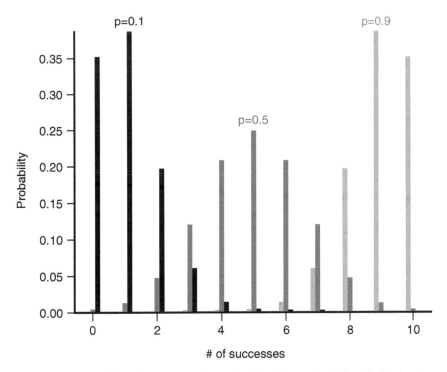

Figure 4.6 Binomial distribution. Number of trials (N) equals 10 for all distributions.

in each coin flip is $p = 0.7$, then the probability of getting 7 heads ($k = 7$) will have a binomial distribution with parameters $N = 10$ and $p = 0.7$.* Don't confuse the trials (subsamples), and the probability of success in each trial, with the number of samples and the probabilities of the number of successful trials in each sample. In the seed predation example, a trial is an individual seed and the trial probability is the probability that an individual seed is taken, while a sample is the observation of a particular station at a particular time and the binomial probabilities are the probabilities that a certain total number of seeds disappears from the station. You can derive the part of the distribution that depends on x, $p^x(1-p)^{N-x}$, by multiplying the probabilities of x independent successes with probability p and $N - x$ independent failures with probability $1 - p$. The rest of the distribution function, $\binom{N}{x} = N!/(x!(N-x)!)$, is a normalization constant that we can justify either with a combinatorial argument about the number of different ways of sampling x objects out of a set of N, or simply by saying that we need a factor in front of the formula to make sure the probabilities add up to 1.

The mean of the binomial is Np and its variance is $Np(1-p)$. Like most discrete sampling distributions (e.g., the binomial, Poisson, negative binomial), this variance is proportional to the number of trials per sample N. When the number of trials per sample increases the variance also increases, but the coefficient of variation

---

* Gelman and Nolan (2002) point out that it is not physically possible to construct a coin that is biased when flipped—although a spinning coin can be biased. Diaconis et al. (2004) even tested a coin made of balsa wood on one side and lead on the other to establish that it was unbiased.

($\sqrt{Np(1-p)}/(Np) = \sqrt{(1-p)/(Np)}$) decreases. The dependence on $p(1-p)$ means the binomial variance is small when p is close to 0 or 1 (and therefore the values are scrunched up near 0 or N) and largest when $p = 0.5$. The coefficient of variation, on the other hand, is largest for small p.

When N is large and p isn't too close to 0 or 1 (i.e., when Np is large), then the binomial distribution is approximately normal (Figure 4.17).

A binomial distribution with only one trial ($N = 1$) is called a *Bernoulli* trial.

You should use the binomial in fitting data only when the number of possible successes has an upper limit. When N is large and p is small, so that the probability of getting N successes is small, the binomial approaches the Poisson distribution, which is covered in the next section (Figure 4.17).

Examples: number of surviving individuals/nests out of an initial sample; number of infested/infected animals, fruits, etc. in a sample; number of a particular class (haplotype, subspecies, etc.) in a larger population.

Summary:

range	discrete, $0 \leq x \leq N$
distribution	$\binom{N}{x}p^x(1-p)^{N-x}$
R	dbinom, pbinom, qbinom, rbinom
parameters	p [real, 0–1], probability of success [prob]
	N [positive integer], number of trials [size]
mean	Np
variance	$Np(1-p)$
CV	$\sqrt{(1-p)/(Np)}$
conjugate prior	Beta

4.5.1.2 POISSON

The Poisson distribution (Figure 4.7) gives the distribution of the number of individuals, arrivals, events, counts, etc., in a given time/space/unit of counting effort if each event is independent of all the others. The most common definition of the Poisson has only one parameter, the average density or arrival rate, λ, which equals the expected number of counts in a sampling unit. An alternative parameterization gives a density *per unit sampling effort* and then specifies the mean as the product of the density per sampling effort r times the sampling effort t, $\lambda = rt$. This parameterization emphasizes that even when the population density is constant, you can change the Poisson distribution of counts by sampling more extensively—for longer times or over larger quadrats.

The Poisson distribution has no upper limit, although values much larger than the mean value are highly improbable. This characteristic provides a rule for choosing between the binomial and Poisson. If you expect to observe a "ceiling" on the number of counts, you should use the binomial; if you expect the number of counts to be effectively unlimited, even if it is theoretically bounded (e.g., there can't really be an infinite number of plants in your sampling quadrat), use the Poisson.

The variance of the Poisson is equal to its mean. However, the coefficient of variation decreases as the mean increases, so in that sense the Poisson distribution becomes

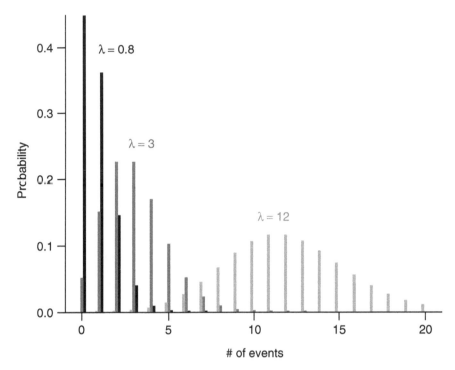

Figure 4.7 Poisson distribution.

more regular as the expected number of counts increases. The Poisson distribution *makes sense only for count data*. Since the CV is unitless, it should not depend on the units we use to express the data; since the CV of the Poisson is $1/\sqrt{\text{mean}}$, if we used a Poisson distribution to describe data on measured lengths, we could reduce the CV by a factor of 10 by changing units from meters to centimeters (which would be silly).

For $\lambda < 1$ the Poisson's mode is at zero. When the expected number of counts gets large (e.g., $\lambda > 10$) the Poisson becomes approximately normal (Figure 4.17).

Examples: number of seeds/seedlings falling in a gap; number of offspring produced in a season (although this might be better fit by a binomial if the number of breeding attempts is fixed); number of prey caught per unit time.

Summary:

range	discrete $(0 \le x)$
distribution	$\frac{e^{-\lambda}\lambda^n}{n!}$
	or $\frac{e^{-rt}(rt)^n}{n!}$
R	dpois, ppois, qpois, rpois
parameters	λ (real, positive), expected number per sample [lambda] or r (real, positive), expected number per unit effort, area, time, etc. (*arrival rate*)

mean	λ (or rt)
variance	λ (or rt)
CV	$1/\sqrt{\lambda}$ (or $1/\sqrt{rt}$)
conjugate prior	Gamma

4.5.1.3 NEGATIVE BINOMIAL

Most probability books derive the negative binomial distribution (Figure 4.8) from a series of independent binary (heads/tails, black/white, male/female, yes/no) trials that all have the same probability of success, like the binomial distribution. However, rather than counting the number of successes obtained in a fixed number of trials as in a binomial distribution, the negative binomial counts the number of *failures* before a predetermined number of successes occurs.

This failure-process parameterization is only occasionally useful in ecological modeling. Ecologists use the negative binomial because it is discrete, like the Poisson, but its variance can be larger than its mean (i.e., it can be *overdispersed*). Thus, it's a good phenomenological description of a patchy or clustered distribution with no intrinsic upper limit that has more variance than the Poisson.

The "ecological" parameterization of the negative binomial replaces the parameters p (probability of success per trial: prob in R) and n (number of successes before you stop counting failures: size in R) with $\mu = n(1-p)/p$, the mean number of failures expected (or of counts in a sample: mu in R), and k, which is typically called an *overdispersion parameter*. Confusingly, k is also called size in R, because it is mathematically equivalent to n in the failure-process parameterization.

The overdispersion parameter measures the amount of clustering, or aggregation, or heterogeneity, in the data: a smaller k means more heterogeneity. The variance of the negative binomial distribution is $\mu + \mu^2/k$, so as k becomes large the variance approaches the mean and the distribution approaches the Poisson distribution. For $k > 10\,\mu$, the negative binomial is hard to tell from a Poisson distribution, but k is often less than μ in ecological applications.[*]

Specifically, you can get a negative binomial distribution as the result of a Poisson sampling process where the rate λ itself varies. If λ is Gamma-distributed (p. 131) with shape parameter k and mean μ, and x is Poisson-distributed with mean λ, then the distribution of x will be a negative binomial distribution with mean μ and overdispersion parameter k (May, 1978; Hilborn and Mangel, 1997). In this case, the negative binomial reflects unmeasured ("random") variability in the population.

Negative binomial distributions can also result from a homogeneous birth-death process, births and deaths (and immigrations) occurring at random in continuous time. Samples from a population that starts from 0 at time $t = 0$, with immigration rate i, birthrate b, and death rate d will be negative binomially distributed with parameters $\mu = i/(b-d)(e^{(b-d)t} - 1)$ and $k = i/b$ (Bailey, 1964, p. 99).

---

[*] Beware of the word "overdispersion," which is sometimes used with an opposite meaning in spatial statistics, where it can mean "more regular than expected from a random distribution of points." If you took quadrat samples from such an "overdispersed" population, the distribution of counts would have variance less than the mean and would be "underdispersed" in the probability distribution sense (Brown and Bolker, 2004)!

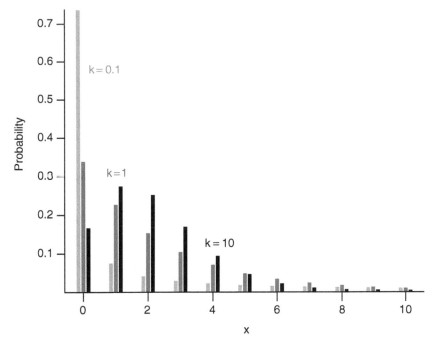

Figure 4.8 Negative binomial distribution. Mean $\mu = 2$ in all cases.

Several different ecological processes can often generate the same probability distribution. We can usually reason forward from knowledge of probable mechanisms operating in the field to plausible distributions for modeling data, but this many-to-one relationship suggests that it is unsafe to reason backwards from probability distributions to particular mechanisms that generate them.

Examples: essentially the same as the Poisson distribution, but allowing for heterogeneity. Numbers of individuals per patch; distributions of numbers of parasites within individual hosts; number of seedlings in a gap, or per unit area, or per seed trap.

Summary:

range	discrete, $x \geq 0$
distribution	$\frac{(n+x-1)!}{(n-1)!x!}p^n(1-p)^x$
	or $\frac{\Gamma(k+x)}{\Gamma(k)x!}(k/(k+\mu))^k(\mu/(k+\mu))^{x*}$
R	dnbinom, pnbinom, qnbinom, rnbinom
parameters	p $(0 < p < 1)$ probability per trial [prob]
	or μ (real, positive) expected number of counts [mu]
	n (positive integer) number of successes awaited [size]
	or k (real, positive), overdispersion parameter [size]
	(= shape parameter of underlying heterogeneity)

* See p. 131 and the appendix for a description of the Γ ("gamma") function.

mean	$\mu = n(1-p)/p$
variance	$\mu + \mu^2/k = n(1-p)/p^2$
CV	$\sqrt{\frac{(1+\mu/k)}{\mu}} = 1/\sqrt{n(1-p)}$
conjugate prior	No simple conjugate prior (Bradlow et al., 2002)

R's default is the coin-flipping (n=size, p=prob) parameterization. In order to use the "ecological" (μ=mu, k=size) parameterization, you *must* name the mu parameter explicitly (e.g., dnbinom(5,size=0.6,mu=1)).

4.5.1.4 GEOMETRIC

The geometric distribution (Figure 4.9) is the number of trials (with a constant probability of failure) until you get a single failure: it's a special case of the negative binomial, with k or $n = 1$.

Examples: number of successful/survived breeding seasons for a seasonally reproducing organism. Lifespans measured in discrete units.

Summary:

range	discrete, $x \geq 0$
distribution	$p(1-p)^x$
R	dgeom, pgeom, qgeom, rgeom
parameters	p ($0 < p < 1$) probability of "success" (death) [prob]
mean	$1/p - 1$
variance	$(1-p)/p^2$
CV	$1/\sqrt{(1-p)}$

4.5.1.5 BETA-BINOMIAL

Just as one can compound the Poisson distribution with a Gamma distribution to allow for heterogeneity in rates, producing a negative binomial, one can compound the binomial distribution with a Beta distribution (p. 133) to allow for heterogeneity in per-trial probability, producing a *beta-binomial* distribution (Figure 4.10) (Crowder, 1978; Reeve and Murdoch, 1985; Hatfield et al., 1996). The most common parameterization of the beta-binomial distribution uses the binomial parameter N (trials per sample), plus two additional parameters a and b that describe the Beta distribution of the per-trial probability. When $a = b = 1$ the per-trial probability is equally likely to be any value between 0 and 1 (the mean is 0.5), and the beta-binomial gives a uniform (discrete) distribution between 0 and N. As $a + b$ increases, the variance of the underlying heterogeneity decreases and the beta-binomial converges to the binomial distribution. Morris (1997) suggests a different parameterization that uses an overdispersion parameter θ, like the k parameter of the negative binomial distribution. In this case the parameters are N, the per-trial probability p ($= a/(a+b)$), and θ ($= a+b$). When θ is large (small overdispersion), the beta-binomial becomes

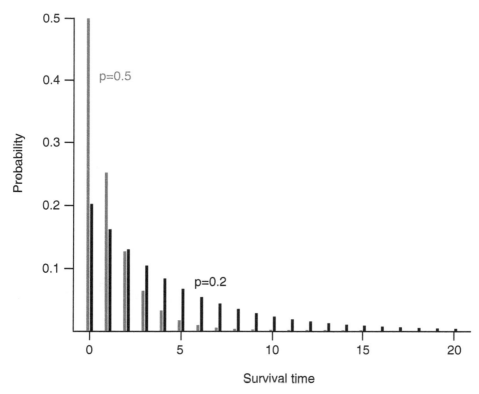

Figure 4.9 Geometric distribution.

binomial. When θ is near zero (large overdispersion), the beta-binomial becomes U-shaped (Figure 4.10).

Examples: as for the binomial.

Summary:

range	discrete, $0 \leq x \leq N$
R	dbetabinom, rbetabinom [emdbook package]
	(pbetabinom and qbetabinom are missing)
density	$\frac{\Gamma(\theta)}{\Gamma(p\theta)\Gamma((1-p)\theta)} \cdot \frac{N!}{x!(N-x)!} \cdot \frac{\Gamma(x+p\theta)\Gamma(N-x+(1-p)\theta)}{\Gamma(N+\theta)}$
parameters	p (real, positive), probability: average per-trial probability [prob]
	θ (real, positive), overdispersion parameter [theta]
	or a and b (shape parameters of Beta distribution for per-trial probability)
	[shape1 and shape2]
	$a = \theta p, b = \theta(1-p)$
mean	Np

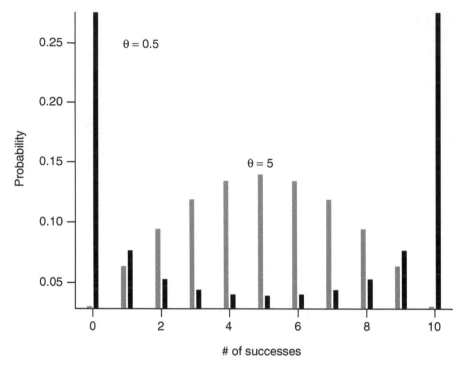

Figure 4.10 Beta-binomial distribution. Number of trials (N) equals 10, average per-trial probability (p) equals 0.5 for all distributions.

variance $\qquad Np(1-p)\left(1+\frac{N-1}{\theta+1}\right)$

CV $\qquad \sqrt{\frac{(1-p)}{Np}\left(1+\frac{N-1}{\theta+1}\right)}$

4.5.2 Continuous Distributions

4.5.2.1 UNIFORM DISTRIBUTION

The uniform distribution (Figure 4.11) with limits a and b, denoted $U(a, b)$, has a constant probability density of $1/(b-a)$ for $a \leq x \leq b$ and zero probability elsewhere. The standard uniform, $U(0, 1)$, is frequently used as a building block for other distributions but the uniform distribution is surprisingly rarely used in ecology otherwise.

Summary:

range $\qquad a \leq x \leq b$
distribution $\qquad 1/(b-a)$

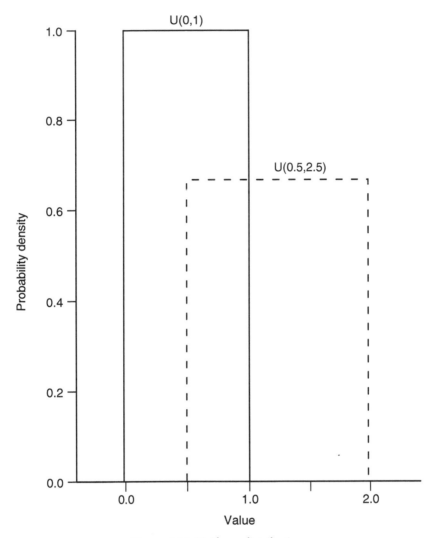

Figure 4.11 Uniform distribution.

R	dunif, punif, qunif, runif
parameters	minimum (a) and maximum (b) limits (real) [min, max]
mean	$(a+b)/2$
variance	$(b-a)^2/12$
CV	$(b-a)/((a+b)\sqrt{3})$

4.5.2.2 NORMAL DISTRIBUTION

Normally distributed variables (Figure 4.12) are everywhere, and most classical statistical methods use this distribution. The explanation for the normal distribution's ubiquity is the *Central Limit Theorem*, which says that if you add a large number

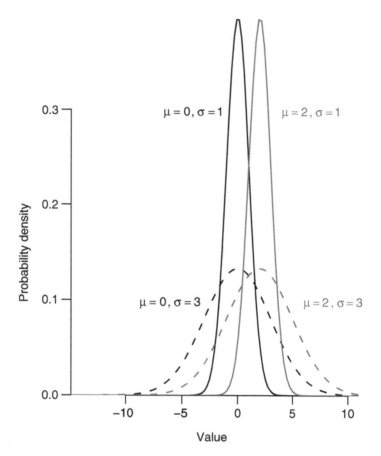

Figure 4.12 Normal distribution.

of independent samples from the same distribution, the distribution of the sum will be approximately normal. "Large," for practical purposes, can mean as few as 5. The Central Limit Theorem does *not* mean that "all samples with large numbers are normal." One obvious counterexample is two different populations with different means that are lumped together, leading to a distribution with two peaks (p. 138). Also, adding isn't the only way to combine samples: if you multiply many independent samples from the same distribution, you get a lognormal distribution instead (p. 135).

Many distributions (binomial, Poisson, negative binomial, Gamma) become approximately normal in some limit (Figure 4.17). You can usually think about this as some form of "adding lots of things together."

The normal distribution specifies the mean and variance separately, with two parameters, which means that one often assumes constant variance (as the mean changes), in contrast to the Poisson and binomial distribution where the variance is a fixed function of the mean.

Examples: many continuous, symmetrically distributed measurements— temperature, pH, nitrogen concentration.

Summary:

range	all real values
distribution	$\frac{1}{\sqrt{2\pi}\sigma} \exp\left(-\frac{(x-\mu)^2}{2\sigma^2}\right)$
R	dnorm, pnorm, qnorm, rnorm
parameters	μ (real), mean [mean]
	σ (real, positive), standard deviation [sd]
mean	μ
variance	σ^2
CV	σ/μ
conjugate prior	Normal (μ); Gamma ($1/\sigma^2$)

4.5.2.3 GAMMA

The *Gamma* distribution (Figure 4.13) is the distribution of *waiting times* until a certain number of events take place. For example, Gamma(shape = 3, scale = 2) is the distribution of the length of time (in days) you'd expect to have to wait for 3 deaths in a population, given that the average survival time is 2 days (or mortality rate is 1/2 per day). The mean waiting time is 6 days = (3 deaths/(1/2 death per day)). (While the gamma *function* (gamma in R; see the appendix) is usually written with a capital Greek gamma, Γ, the Gamma *distribution* (dgamma in R) is written out as Gamma.) Gamma distributions with integer shape parameters are also called *Erlang* distributions. The Gamma distribution is still defined, and useful, for noninteger (positive) shape parameters, but the simple description given above breaks down: how can you define the waiting time until 3.2 events take place?

For shape parameters ≤ 1, the Gamma has its mode at zero; for shape parameter = 1, the Gamma is equivalent to the exponential (see p. 133). For shape parameters greater than 1, the Gamma has a peak (mode) at a value greater than zero; as the shape parameter increases, the Gamma distribution becomes more symmetrical and approaches the normal distribution. This behavior makes sense if you think of the Gamma as the distribution of the sum of independent, identically distributed waiting times, in which case it is governed by the Central Limit Theorem.

The scale parameter (sometimes defined in terms of a rate parameter instead, 1/scale) adjusts the mean of the Gamma by adjusting the waiting time per event; however, multiplying the waiting time by a constant to adjust its mean also changes the variance, so both the variance and the mean depend on the scale parameter.

The Gamma distribution is less familiar than the normal, and new users of the Gamma often find it annoying that in the standard parameterization you can't adjust the mean independently of the variance. You could define a new set of parameters m (mean) and v (variance), with scale = v/m and shape = m^2/v—but then you would find (unlike the normal distribution) the shape changing as you changed the variance. Nevertheless, the Gamma is extremely useful; it solves the problem that many researchers face when they have a continuous, positive variable with "too much variance," whose coefficient of variation is greater than about 0.5. Modeling such data with a normal distribution leads to unrealistic negative values, which then have to

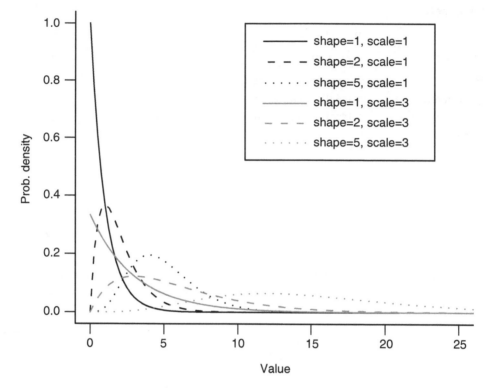

Figure 4.13 Gamma distribution.

be dealt with in some ad hoc way like truncating them or otherwise trying to ignore them. The Gamma is often a more convenient and realistic alternative.

The Gamma is the continuous counterpart of the negative binomial, which is the discrete distribution of a number of trials (rather than length of time) until a certain number of events occur. Both the negative binomial and Gamma distributions are often generalized, however, in ways that are useful phenomenologically but that don't necessarily make sense according to their simple mechanistic descriptions.

The Gamma and negative binomial are both frequently used as phenomenologicaly, skewed, or overdispersed versions of the normal or Poisson distributions. The Gamma is less widely used than the negative binomial because the negative binomial replaces the Poisson, which is restricted to a particular variance, while the Gamma replaces the normal, which can have any variance. Thus you might use the negative binomial for any discrete distribution with a variance greater than its mean, whereas you wouldn't need a Gamma distribution unless the distribution you were trying to match was skewed to the right.

Examples: almost any variable with a large coefficient of variation where negative values don't make sense: nitrogen concentrations, light intensity, growth rates.

Summary:

range   positive real values

R   dgamma, pgamma, qgamma, rgamma

distribution	$\frac{1}{s^a \Gamma(a)} x^{a-1} e^{-x/s}$
parameters	s (real, positive), scale: length per event [scale]
	or r (real, positive), rate = $1/s$; rate at which events occur [rate]
	a (real, positive), shape: number of events [shape]
mean	as or a/r
variance	as^2 or a/r^2
CV	$1/\sqrt{a}$

4.5.2.4 EXPONENTIAL

The exponential distribution (Figure 4.14) describes the distribution of waiting times for a single event to happen, given that there is a constant probability per unit time that it will happen. It is the continuous counterpart of the geometric distribution and a special case (for shape parameter = 1) of the Gamma distribution. It can be useful both mechanistically, as a distribution of interevent times or lifetimes, and phenomenologically, for any continuous distribution that has highest probability for zero or small values.

Examples: times between events (bird sightings, rainfall, etc.); lifespans/survival times; random samples of anything that decreases exponentially with time or distance (e.g., dispersal distances, light levels in a forest canopy).

Summary:

range	positive real values
R	dexp, pexp, qexp, rexp
density	$\lambda e^{-\lambda x}$
parameters	λ (real, positive), rate: death/disappearance rate [rate]
mean	$1/\lambda$
variance	$1/\lambda^2$
CV	1

4.5.2.5 BETA

The Beta distribution (Figure 4.15), a continuous distribution closely related to the binomial distribution, completes our basic family of continuous distributions (Figure 4.17). The Beta distribution is the only standard continuous distribution besides the uniform distribution with a finite range, from 0 to 1. The Beta distribution is the inferred distribution of the *probability* of success in a binomial trial with $a - 1$ observed successes and $b - 1$ observed failures. When $a = b$ the distribution is symmetric around $x = 0.5$, when $a < b$ the peak shifts toward zero, and when $a > b$ it shifts toward 1. With $a = b = 1$, the distribution is $U(0, 1)$. As $a + b$ (equivalent to the total number of trials $+2$) gets larger, the distribution becomes more peaked. For a or b less than 1, the mechanistic description stops making sense (how can you have fewer than zero trials?), but the distribution is still well-defined, and when a and b are both between 0 and 1 it becomes U-shaped—it has peaks at $p = 0$ and $p = 1$.

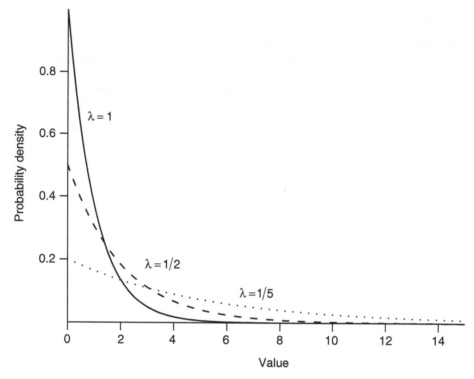

Figure 4.14 Exponential distribution.

The Beta distribution is obviously good for modeling probabilities or proportions. It can also be useful for modeling continuous distributions with peaks at both ends, although in some cases a finite mixture model (p. 138) may be more appropriate. The Beta distribution is also useful whenever you have to define a continuous distribution on a finite range, as it is the only such standard continuous distribution. It's easy to rescale the distribution so that it applies over some other finite range instead of from 0 to 1; for example, Tiwari et al. (2005) used the Beta distribution to describe the distribution of turtles on a beach, so the range would extend from 0 to the length of the beach.

Summary:

range	real, 0 to 1
R	dbeta, pbeta, qbeta, rbeta
density	$\frac{\Gamma(a+b)}{\Gamma(a)\Gamma(b)} x^{a-1}(1-x)^{b-1}$
parameters	a (real, positive), shape 1: number of successes +1 [shape1]
	b (real, positive), shape 2: number of failures +1 [shape2]
mean	$a/(a+b)$
mode	$(a-1)/(a+b-2)$
variance	$ab/((a+b)^2(a+b+1))$
CV	$\sqrt{(b/a)/(a+b+1)}$

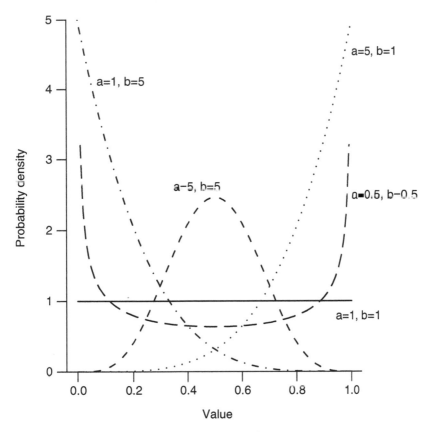

Figure 4.15 Beta distribution.

4.5.2.6 LOGNORMAL

The lognormal (Figure 4.16) falls outside the neat classification scheme we've been building so far; it is not the continuous analogue or limit of some discrete sampling distribution (Figure 4.17).* Its mechanistic justification is like the normal distribution (the Central Limit Theorem), but for the *product* of many independent, identical variates rather than their sum. Just as taking logarithms converts products into sums, taking the logarithm of a lognormally distributed variable—which might result from the product of independent variables—converts it into a normally distributed variable resulting from the sum of the logarithms of those independent variables. The best example of this mechanism is the distribution of the sizes of individuals or populations that grow exponentially, with a per capita growth rate that varies randomly over time. At each time step (daily, yearly, etc.), the current size is *multiplied* by the randomly

---

*The lognormal extends our table in another direction—exponential transformation of a known distribution. Other distributions have this property, most notably the *extreme-value distribution*, which is the log-exponential: if Y is exponentially distributed, then log Y is extreme-value distributed. As its name suggests, the extreme-value distribution occurs mechanistically as the distribution of extreme values (e.g., maxima) of samples of other distributions (Katz et al., 2005).

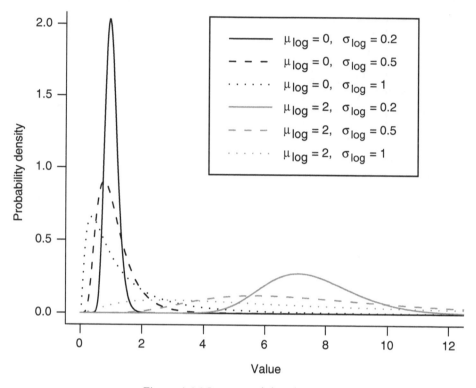

Figure 4.16 Lognormal distribution.

chosen growth increment, so the final size (when measured) is the product of the initial size and all of the random growth increments.

One potentially puzzling aspect of the lognormal distribution is that its mean is not what you might naively expect if you exponentiate a normal distribution with mean μ (i.e., e^μ). Because the exponential function is an accelerating function, the mean of the lognormal, $e^{\mu+\sigma^2/2}$, is greater than e^μ (Jensen's inequality). When the variance is small relative to the mean, the mean is approximately equal to e^μ, and the lognormal itself looks approximately normal (e.g., solid lines in Figure 4.16, with $\sigma(\log) = 0.2$). As with the Gamma distribution, the distribution also changes shape as the variance increases, becoming more skewed.

The lognormal is used phenomenologically in some of the same situations where a Gamma distribution also fits: continuous, positive distributions with long tails or variance much greater than the mean (McGill et al., 2006). Like the distinction between a Michaelis-Menten and a saturating exponential function, you may not be able to tell the difference between a lognormal and a Gamma without large amounts of data. Use the one that is more convenient, or that corresponds to a more plausible mechanism for your data.

Examples: sizes or masses of individuals, especially rapidly growing individuals; abundance vs. frequency curves for plant communities.

Summary:

range   positive real values
R   dlnorm, plnorm, qlnorm, rlnorm

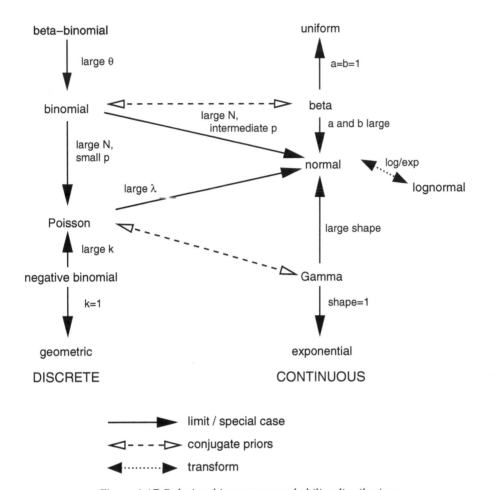

Figure 4.17 Relationships among probability distributions.

density	$\frac{1}{\sqrt{2\pi}\sigma x}e^{-(\log x-\mu)^2/(2\sigma^2)}$
parameters	μ (real): mean of the logarithm [meanlog]
	σ (real): standard deviation of the logarithm [sdlog]
mean	$\exp(\mu+\sigma^2/2)$
variance	$\exp(2\mu+\sigma^2)(\exp(\sigma^2)-1)$
CV	$\sqrt{\exp(\sigma^2)-1}$ ($\approx \sigma$ when $\sigma < 1/2$)

4.6 Extending Simple Distributions: Compounding and Generalizing

What do you do when none of these simple distributions fits your data? You could always explore other distributions. For example, the Weibull distribution (similar to the Gamma distribution in shape: `dweibull` in R) generalizes the exponential to allow for survival probabilities that increase or decrease with age (p. 251). The Cauchy distribution (`dcauchy` in R), described as *fat-tailed* because the probability of extreme events (in the tails of the distribution) is very large—larger than for the exponential or normal distributions—can be useful for modeling distributions with many outliers. You can often find useful distributions for your data in modeling papers from your subfield of ecology.

However, in addition to simply learning more distributions, learning some strategies for generalizing more familiar distributions can also be useful.

4.6.1 Adding Covariates

One obvious strategy is to look for systematic differences within your data that explain the nonstandard shape of the distribution. For example, a *bimodal* or *multimodal* distribution (one with two or more peaks, in contrast to most of the distributions discussed above that have a single peak) may make perfect sense once you realize that your data are a collection of objects from different populations with different means. For example, the sizes or masses of sexually dimorphic animals or animals from several different cryptic species would follow bi- or multimodal distributions, respectively. A distribution that isn't multimodal but is more fat-tailed than a normal distribution might indicate systematic variation in a continuous covariate such as nutrient availability, or maternal size, of environmental temperature, of different individuals. If you can measure these covariates, then you may be able to add them to the deterministic part of the model and use standard distributions to describe the variability *conditioned on* the covariates.

4.6.2 Mixture Models

But what if you can't identify, or measure, systematic differences? You can still extend standard distributions by supposing that your data are really a mixture of observations from different types of individuals, but that you can't observe the (finite) types or (continuous) covariates of individuals. These distributions are called *mixture distributions* or *mixture models*. Fitting them to data can be challenging, but they are very flexible.

4.6.2.1 FINITE MIXTURES

Finite mixture models suppose that your observations are drawn from a discrete set of unobserved categories, each of which has its own distribution. Typically all categories have the same type of distribution, such as normal, but with different mean or variance parameters. Finite mixture distributions often fit multimodal data. Finite mixtures are typically parameterized by the parameters of each component of the mixture, plus a set of probabilities or percentages describing the amount of each component. For example, 30% of the organisms ($p = 0.3$) could be in group 1, normally distributed with mean 1 and standard deviation 2, while 70% ($1 - p = 0.7$) are in group 2, normally distributed with mean 5 and standard deviation 1

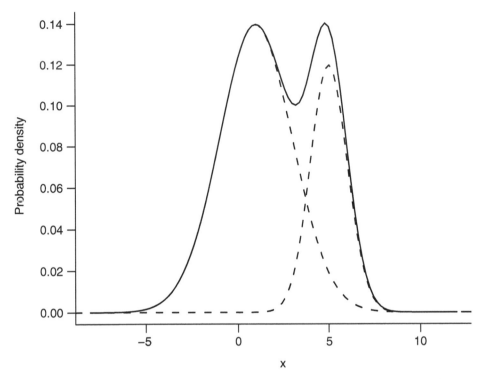

Figure 4.18 Finite mixture distribution: 70% Normal($\mu = 1, \sigma = 2$), 30% Normal($\mu = 5$, $\sigma = 1$).

(Figure 4.18). If the peaks of the distributions are closer together, or their standard deviations are larger so that the distributions overlap, you'll see a broad (and perhaps lumpy) peak rather than two distinct peaks.

Zero-inflated models are a common type of finite mixture model (Inouye, 1999; Martin et al., 2005); we saw a simple example of a zero-inflated binomial at the beginning of this chapter. Zero-inflated models (Figure 4.1) combine a standard discrete probability distribution (e.g., binomial, Poisson, or negative binomial), which typically includes some probability of sampling zero counts even when some individuals are present, with an additional process that can also lead to a zero count (e.g., complete absence of the species or trap failure).

4.6.3 Continuous Mixtures

Continuous mixture distributions, also known as *compounded distributions*, allow the parameters themselves to vary randomly, drawn from their own distribution. They are a sensible choice for overdispersed data, or for data where you suspect that continuous unobserved covariates may be important. Technically, compounded distributions are the distribution of a sampling distribution $S(x, p)$ with parameter(s) p that vary according to another (typically continuous) distribution $P(p)$. The distribution of the compounded distribution C is $C(x) = \int S(x, p)P(p)dp$. For example, compounding a Poisson distribution by drawing the rate parameter λ from a Gamma distribution with shape parameter k (and scale parameter λ/k, to make the mean equal to λ) results in a negative binomial distribution (p. 124). Continuous mixture distributions are growing ever more popular in ecology as ecologists try to account for heterogeneity in their data.

The negative binomial, which could also be called the Gamma-Poisson distribution to highlight its compound origin, is the most common compounded distribution. The beta-binomial is also fairly common: like the negative binomial, it compounds a common discrete distribution (binomial) with its conjugate prior (Beta), resulting in a mathematically simple form that allows for more variability. The *lognormal-Poisson* is very similar to the negative binomial, except that (as its name suggests) it uses the lognormal instead of the Gamma as a compounding distribution. One technical reason to use the less common lognormal-Poisson is that on the log scale the rate parameter is normally distributed, which simplifies some numerical procedures (Elston et al., 2001).

Clark et al. (1999) used the *Student t* distribution to model seed dispersal curves. Seeds often disperse fairly uniformly near parental trees but also have a high probability of long dispersal. These two characteristics are incompatible with standard seed dispersal models like the exponential and normal distributions. Clark et al. assumed that the seed dispersal curve represents a compounding of a normal distribution for the dispersal of any one seed with a Gamma distribution of the inverse variance of the distribution of any particular seed (i.e., $1/\sigma^2 \sim$ Gamma).* This variation in variance accounts for the different distances that different seeds may travel as a function of factors like their size, shape, height on the tree, and the wind speed at the time they are released. Clark et al. used compounding to model these factors as random, unobserved covariates since they are practically impossible to measure for all the individual seeds on a tree or in a forest.

The inverse Gamma-normal model is equivalent to the Student t distribution, which you may recognize from t tests in classical statistics and which statisticians often use as a phenomenological model for fat-tailed distributions. Clark et al. extended the usual one-dimensional t distribution (dt in R) to the two-dimensional distribution of seeds around a parent and called it the 2Dt distribution. The 2Dt distribution has a scale parameter that determines the mean dispersal distance and a shape parameter p. When p is large the underlying Gamma distribution has a small coefficient of variation and the 2Dt distribution is close to normal; when $p = 1$ the 2Dt becomes a Cauchy distribution.

Generalized distributions are an alternative class of mixture distribution that arises when there is a sampling distribution $S(x)$ for the number of individuals within a cluster and another sampling distribution $C(x)$ for number of clusters in a sampling unit. For example, the distribution of number of eggs per quadrat might be generalized from the distribution of clutches per quadrat and of eggs per clutch. A standard example is the "Poisson-Poisson" or "Neyman Type A" distribution (Pielou, 1977), which assumes a Poisson distribution of clusters with a Poisson distribution of individuals in each cluster.

---

* This choice of a compounding distribution, which may seem arbitrary, turns out to be mathematically convenient.

Figuring out the probability distribution or density formulas for compounded distributions analytically is mathematically challenging (see Bailey (1964) or Pielou (1977) for the gory details), but R can easily generate random numbers from these distributions.

The key is that R's functions for generating random distributions (`rpois`, `rbinom`, etc.) can take vectors for their parameters. Rather than generate (say) 20 deviates from a binomial distribution with N trials and a fixed per-trial probability p, you can choose 20 deviates with N trials and a vector of 20 different per-trial probabilities p_1 to p_{20}. Furthermore, you can generate this vector of parameters from another randomizing function! For example, to generate 20 beta-binomial deviates with $N = 10$ and the per-trial probabilities drawn from a Beta distribution with $a = 2$ and $b = 1$, you could use `rbinom(20,prob=rbeta(20,2,1),size=10)`. (See the R supplement for more detail.)

Compounding and generalizing are powerful ways to extend the range of stochastic ecological models. A good fit to a compounded distribution also suggests that environmental variation is shaping the variation in the population. But be careful: Pielou (1977) demonstrates that for Poisson distributions, every generalized distribution (corresponding to variation in the underlying density) can also be generated by a compound distribution (corresponding to individuals occurring in clusters). She concludes that "the fitting of theoretical frequency distributions to observational data can never by itself suffice to 'explain' the pattern of a natural population" (p. 123).

4.7 R Supplement

For all of the probability distributions discussed in this chapter (and many more: try `help.search("distribution")`), R can generate random numbers (*deviates*) drawn from the distribution; compute the cumulative distribution function and the probability distribution function; and compute the quantile function, which gives the x value such that $\int_{-\infty}^{x} P(x)\, dx$ (the probability that $X \leq x$) is a specified value. For example, you can obtain the critical values of the standard normal distribution, ± 1.96, with `qnorm(0.025)` and `qnorm(0.975)` (Figure 4.19).

4.7.1 Discrete Distribution

For example, let's explore the (discrete) negative binomial distribution.

First set the random-number seed for consistency so that you get exactly the same results shown here:

```
> set.seed(1001)
```

Arbitrarily choose parameters $\mu = 10$ and $k = 0.9$—since $k < 1$, this represents a strongly overdispersed population. Remember that R uses `size` to denote k, because k is mathematically equivalent to the number of failures in the failure-process parameterization.

```
> z <- rnbinom(1000, mu = 10, size = 0.9)
```

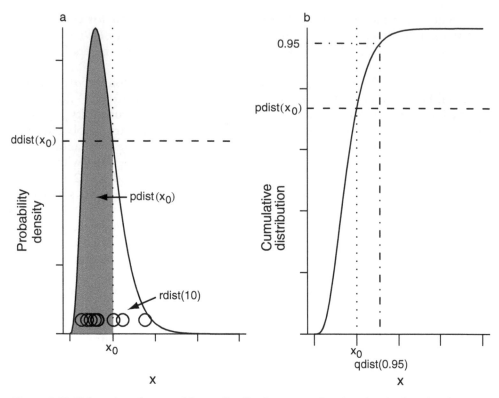

Figure 4.19 R functions for an arbitrary distribution dist, showing density function (ddist), cumulative distribution function (pdist), quantile function (qdist), and random-deviate function (rdist).

Check the first few values:

```
> head(z)
```

```
[1] 41   3   3   0 11 14
```

Since the negative binomial has no set upper limit, we will just plot the results up to the maximum value sampled:

```
> maxz = max(z)
```

The easiest way to plot the results is

```
> f = factor(z, levels = 0:maxz)
> plot(f)
```

using the levels specification to make sure that all values up to the maximum are included in the plot even when none were sampled in this particular experiment.

If we want the observed probabilities (freq/N) rather than the frequencies:

```
> obsprobs = table(f)/1000
> plot(obsprobs)
```

Add theoretical values:

```
> tvals = dnbinom(0:maxz, size = 0.9, mu = 10)
> points(0:maxz, tvals)
```

You could plot the deviations with plot(0:maxz,obsprobs-tvals); this gives you some idea how the variability changes with the mean.

Find the probability that $x > 30$:

```
> pnbinom(30, size = 0.9, mu = 10, lower.tail = FALSE)
```

```
[1] 0.05725252
```

By default R's distribution functions will give you the *lower tail* of the distribution—the probability that x is less than or equal to some particular value. You could use 1-pnbinom(30,size=0.9,mu=10) to get the upper tail since Prob($x > 30$) = 1 − Prob($x \le 30$), but using lower.tail=FALSE to get the upper tail is more accurate.

What is the upper 95th percentile of the distribution?

```
> qnbinom(0.95, size = 0.9, mu = 10)
```

```
[1] 32
```

To get the lower and upper 95% confidence limits, you need

```
> qnbinom(c(0.025, 0.975), size = 0.9, mu = 10)
```

```
[1]  0 40
```

You can also use the random sample z to check that the mean and variance, and the 95th quantile of the sample, agree reasonably well with the theoretical expectations:

```
> mu = 10
> k = 0.9
> c(mu, mean(z))
```

```
[1] 10.000  9.654
```

```
> c(mu * (1 + mu/k), var(z))
```

```
[1] 121.1111 113.6539
```

```
> c(qnbinom(0.95, size = k, mu = mu), quantile(z, 0.95))
```

```
      95%
 32   31
```

4.7.2 Continuous Distribution: Lognormal

Going through the same exercise for the lognormal, a continuous distribution:

```
> z = rlnorm(1000, meanlog = 2, sdlog = 1)
```

Plot the results:

```
> hist(z, breaks = 100, freq = FALSE)
> lines(density(z, from = 0), lwd = 2)
```

Add theoretical values:

```
> curve(dlnorm(x, meanlog = 2, sdlog = 1), add = TRUE,
+      lwd = 2, from = 0, col = "darkgray")
```

The probability of $x > 30$:

```
> plnorm(30, meanlog = 2, sdlog = 1, lower.tail = FALSE)
```

```
[1] 0.08057753
```

95% confidence limits:

```
> qlnorm(c(0.025, 0.975), meanlog = 2, sdlog = 1)
```

```
[1]   1.040848 52.455437
```

Comparing the theoretical values (p. 137) with the observed values for this random sample:

```
> meanlog = 2
> sdlog = 1
> c(exp(meanlog + sdlog^2/2), mean(z))
```

```
[1] 12.18249 12.12708
```

```
> c(exp(2 * meanlog + sdlog^2) * (exp(sdlog^2) - 1),
+      var(z))
```

```
[1] 255.0156 184.7721
```

```
> c(qlnorm(0.95, meanlog = meanlog, sdlog = sdlog),
+      quantile(z, 0.95))
```

```
              95%
38.27717 39.65172
```

The difference between the expected and observed variance is fairly large. This is typical: variances of random samples have larger variances, or absolute differences from their theoretical expected values, than means of random samples.

Sometimes it's easier to deal with lognormal data by taking logarithms of the data and comparing them to the normal distribution:

```
> hist(log(z), freq = FALSE, breaks = 100)
> curve(dnorm(x, mean = meanlog, sd = sdlog), add = TRUE,
+      lwd = 2)
```

4.7.3 *Mixing and Compounding Distributions*

4.7.3.1 FINITE MIXTURE DISTRIBUTIONS

The general recipe for generating samples from finite mixtures is to use a uniform distribution to choose different components of the mixture, then use `ifelse` to pick values from one distribution or the other. To pick 1000 values from a mixture of normal distributions with the parameters shown in Figure 4.18 ($p = 0.3$, $\mu_1 = 1$, $\sigma_1 = 2$, $\mu_2 = 5$, $\sigma_2 = 1$):

```
> u1 = runif(1000)
> z = ifelse(u1 < 0.3, rnorm(1000, mean = 1, sd = 2),
+      rnorm(1000, mean = 5, sd = 1))
> hist(z, breaks = 100, freq = FALSE)
```

The probability density of a finite mixture composed of two distributions D_1 and D_2 in proportions p_1 and $1 - p_1$ is $p_1 D_1 + (1 - p_1)D_2$. We can superimpose the theoretical probability density for the finite mixture above on the histogram:

```
> curve(0.3 * dnorm(x, mean = 1, sd = 2) + 0.7 * dnorm(x,
+      mean = 5, sd = 1), add = TRUE, lwd = 2)
```

The general formula for the probability distribution of a zero-inflated distribution, with an underlying distribution $P(x)$ and a zero-inflation probability of p_z, is

$$\text{Prob}(0) = p_z + (1 - p_z)P(0)$$

$$\text{Prob}(x > 0) = (1 - p_z)P(x)$$

So, for example, we could define a probability distribution for a zero-inflated negative binomial as follows:

```
> dzinbinom = function(x, mu, size, zprob) {
+      ifelse(x == 0, zprob + (1 - zprob) * dnbinom(0,
+          mu = mu, size = size), (1 - zprob) * dnbinom(x,
+          mu = mu, size = size))
+ }
```

(The name, `dzinbinom`, follows the R convention for a probability distribution function: a d followed by the abbreviated name of the distribution, in this case `zinbinom` for "zero-inflated negative **binom**ial.")

The `ifelse` command checks every element of x to see whether it is zero or not and fills in the appropriate value depending on the answer.

Here's a random deviate generator:

```
> rzinbinom = function(n, mu, size, zprob) {
+      ifelse(runif(n) < zprob, 0, rnbinom(n, mu = mu,
+          size = size))
+ }
```

The command `runif(n)` picks n random values between 0 and 1; the `ifelse` command compares them with the value of `zprob`. If an individual value is less than `zprob` (which happens with probability $zprob = p_z$), then the corresponding random number is zero; otherwise it is a value picked out of the appropriate negative binomial distribution (which may also be zero).

4.7.3.2 COMPOUNDED DISTRIBUTIONS

Start by confirming numerically that a negative binomial distribution is really a compounded Poisson-Gamma distribution. Pick 1000 values out of a Gamma distribution, then use those values as the rate (λ) parameters in random draws from a Poisson distribution:

```
> k = 3
> mu = 10
> lambda = rgamma(1000, shape = k, scale = mu/k)
> z = rpois(1000, lambda)
> P1 = table(factor(z, levels = 0:max(z)))/1000
> plot(P1)
> P2 = dnbinom(0:max(z), mu = 10, size = 3)
> points(0:max(z), P2)
```

Establish that a Poisson-lognormal and a Poisson-Gamma (negative binomial) are not very different: pick the Poisson-lognormal with approximately the same mean and variance as the negative binomial just shown.

```
> mlog = mean(log(lambda))
> sdlog = sd(log(lambda))
> lambda2 = rlnorm(1000, meanlog = mlog, sdlog = sdlog)
> z2 = rpois(1000, lambda2)
> P3 = table(factor(z2, levels = 0:max(z)))/1000
> matplot(0:max(z), cbind(P1, P3), pch = 1:2)
> lines(0:max(z), P2)
```

5 Stochastic Simulation and Power Analysis

This chapter introduces techniques and ideas related to simulating ecological patterns. Its main goals are: (1) to show you how to generate patterns you can use to sharpen your intuition and test your estimation tools; and (2) to introduce statistical power and related concepts, and show you how to estimate statistical power by simulation. This chapter and the supplements will also give you more practice working with R.

5.1 Introduction

Chapters 3 and 4 gave a basic overview of functions to describe deterministic patterns and probability distributions to describe stochastic patterns. This chapter will show you how to use stochastic simulation to understand and test your data. Simulation is sometimes called *forward* modeling, to emphasize that you pick a model and parameters and work forward to predict patterns in the data. Parameter estimation, or *inverse* modeling (the main focus of this book), starts from the data and works backward to choose a model and estimate parameters.

Ecologists often use simulation to explore the patterns that emerge from ecological models. Often they use theoretical models without accompanying data, in order to understand qualitative patterns and plan future studies. But even if you have data, you might want to start by simulating your system. You can use simulations to explore the functions and distributions you chose to quantify your data. If you can choose parameters that make the simulated output from those functions and distributions look like your data, you can confirm that the models are reasonable—and simultaneously find a rough estimate of the parameters.

You can also use simulated "data" from your system to test your estimation procedures. Chapters 6–8 will show you how to estimate parameters; in this chapter I'll work with more "canned" procedures like nonlinear regression. Since you never know the true answer to an ecological question—you only have imperfect measurements with which you're trying to get as close to the answer as possible—simulation is the only way to test whether you can correctly estimate the parameters of an ecological system. It's always good to test such a best-case scenario, where you know that the functions and distributions you're using are correct, before you proceed to real data.

Power analysis is a specific kind of simulation testing where you explore how large a sample size you would need to get a reasonably precise estimate of your parameters. You can also also use power analysis to explore how variations in experimental design would change your ability to answer ecological questions.

5.2 Stochastic Simulation

Static ecological processes, where the data represent a snapshot of some ecological system, are easy to simulate.* For static data, we can use a single function to simulate the deterministic process and then add heterogeneity. Often, however, we will chain together several different mathematical functions and probability distributions representing different stages in an ecological process to produce surprisingly complex and rich descriptions of ecological systems.

I'll start with three simple examples that illustrate the general procedure, and then move on to two slightly more in-depth examples.

5.2.1 Simple Examples

5.2.1.1 SINGLE GROUPS

Figure 5.1 shows the results of two simple simulations, each with a single group and single continuous covariate.

The first simulation (Figure 5.1a) is a linear model with normally distributed errors. It might represent productivity as a function of nitrogen concentration, or predation risk as a function of predator density. The mathematical formula is $Y \sim \text{Normal}(a + bx, \sigma^2)$, specifying that Y is a random variable drawn from a normal distribution with mean $a + bx$ and variance σ^2. The symbol $\sim$ means "is distributed according to." This model can also be written as $y_i = a + bx_i + \varepsilon_i, \ \varepsilon_i \sim N(0, \sigma^2)$, specifying that the ith value of Y, y_i, is equal to $a + bx_i$ plus a normally distributed error term with mean zero. I will always use the first form because it is more general: normally distributed error is one of the few kinds that can simply be added onto the deterministic model in this way. The two lines on the plot show the theoretical relationship between y and x and the best-fit line by linear regression, `lm(y~x)` (Section 9.2.1). The lines differ slightly because of the randomness incorporated in the simulation.

A few lines of R code will run this simulation. Set up the values of x, and specify values for the parameters a and b:

```
> x = 1:20
> a = 2
> b = 1
```

---

* Dynamic processes are more challenging. See Chapter 11.

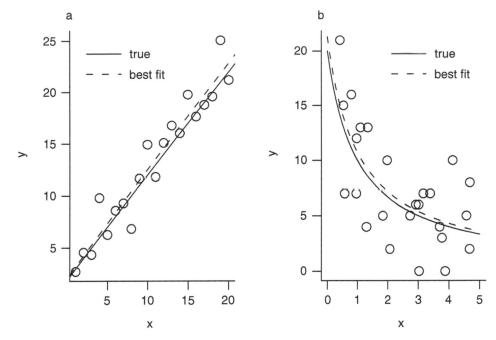

Figure 5.1 Two simple simulations: (a) a linear function with normal errors ($Y \sim$ Normal $(a + bx, \sigma^2)$) and (b) a hyperbolic function with negative binomial errors ($Y \sim$ NegBin($\mu = ab/(b+x), k$)).

Calculate the deterministic part of the model:

```
> y_det = a + b * x
```

Pick 20 random normal deviates with the mean equal to the deterministic equation and $\sigma = 2$:

```
> y = rnorm(20, mean = y_det, sd = 2)
```

(You could also specify this as y = y_det+rnorm(20,sd=2), corresponding to the additive model $y_i = a + bx_i + \varepsilon_i$, $\varepsilon_i \sim N(0, \sigma^2)$ (the mean parameter is zero by default). However, the additive form works only for the normal, and not for most of the other distributions we will be using.)

The second simulation uses hyperbolic functions ($y = ab/(b+x)$) with negative binomial error, or, in symbols, $Y \sim$ NegBin($\mu = ab/(b+x), k$). The function is parameterized so that a is the intercept term (when $x = 0$, $y = ab/b = a$). This simulation might represent the decreasing fecundity of a species as a function of increasing population density; the hyperbolic function is a natural expression of the decreasing quantity of a limiting resource per individual.

In this case, we cannot express the model as the deterministic function "plus error." Instead, we have to incorporate the deterministic model as a control on one of the parameters of the error distribution—in this case, the mean μ. (Although the negative binomial is a discrete distribution, its parameters μ and k are continuous.) Ecological models typically describe the differences in the mean among groups or as covariates change, but we could also allow the variance or the shape of the distribution to change.

The R code for this simulation is easy, too. Define parameters

```
> a = 20
> b = 1
> k = 5
```

How you simulate the x values depends on the experimental design you are trying to simulate. In this case, we choose 50 x values randomly distributed between 0 and 5 to simulate a study where the samples are chosen from natural varying sites, in contrast to the previous simulation where x varied systematically (x=1:20), simulating an experimental or observational study that samples from a gradient in the predictor variable x.

```
> x = runif(50, min = 0, max = 5)
```

Now we calculate the deterministic mean y_det, and then sample negative binomial values with the appropriate mean and overdispersion:

```
> y_det = a * b/(b + x)
> y = rnbinom(50, mu = y_det, size = k)
```

5.2.1.2 MULTIPLE GROUPS

Ecological studies typically compare the properties of organisms in different groups (e.g., control and treatment, parasitized and unparasitized, high and low altitude).

Figure 5.2 shows a simulation that extends the hyperbolic simulation above to compare the effects of a continuous covariate in two different groups (species in this case). Both groups have the same overdispersion parameter k, but the hyperbolic parameters a and b differ:

$$Y \sim \text{NegBin}(\mu = a_i b_i/(b_i + x), k) \qquad (5.2.1)$$

where i is 1 or 2 depending on the species of an individual.

Suppose we still have 50 individuals, but the first 25 are species 1 and the second 25 are species 2. We use rep to set up a factor that describes the group structure (the R commands gl and expand.grid are useful for more complicated group assignments):

```
> g = factor(rep(1:2, each = 25))
```

Defining vectors of parameters, each with one element per species, or a single parameter for k since the species have the same degree of overdispersion:

```
> a = c(20, 10)
> b = c(1, 2)
> k = 5
```

R's vectorization makes it easy to incorporate different parameters for different species into the formula, by using the group vector g to specify which element of the

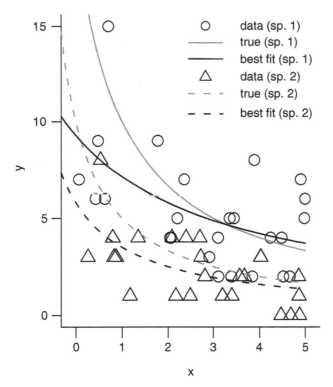

Figure 5.2 Simulation results from a hyperbolic/negative binomial model with groups differing in both intercept and slope: $Y \sim \text{NegBin}(\mu = a_i b_i / (b_i + x), k)$. Parameters: $a = \{20, 10\}$, $b = \{1, 2\}$, $k = 5$.

parameter vectors to use for any particular individual:

```
> y_det = a[g]/(b[g] + x)
> y = rnbinom(50, mu = y_det, size = k)
```

5.2.2 Intermediate Examples

5.2.2.1 REEF FISH SETTLEMENT

The damselfish settlement data from Schmitt et al. (1999; also Section 2.4.3 above) include random variation in settlement density (the density of larvae arriving on a given anemone) and random variation in density-dependent recruitment (number of settlers surviving for 6 months on an anemone).

To simulate the variation in settlement density I took random draws from a zero-inflated negative binomial (p. 145), although a noninflated binomial, or even a geometric distribution (i.e., a negative binomial with $k = 1$) might be sufficient to describe the data.

Schmitt et al. modeled density-dependent recruitment with a Beverton-Holt curve (equivalent to the Michaelis-Menten function). I have simulated this curve with

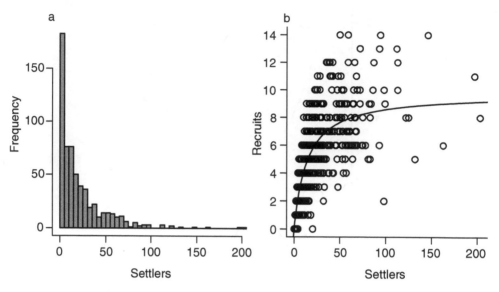

Figure 5.3 Damselfish recruitment: (a) distribution of settlers; (b) recruitment as a function of settlement density.

binomial error (for survival of recruits) superimposed. The model is

$$R \sim \text{Binom}(N = S, p = a/(1 + (a/b)S)). \qquad (5.2.2)$$

With the recruitment probability per settler p given as the hyperbolic function $a/(1 + (a/b)S)$, the mean number of recruits is Beverton-Holt: $Np = aS/(1 + (a/b)S)$. The settlement density S is drawn from the zero-inflated negative binomial distribution shown in Figure 5.3a.

Set up the parameters, including the number of samples (N):

```
> N = 603
> a = 0.696
> b = 9.79
> mu = 25.32
> zprob = 0.123
> k = 0.932
```

Define a function for the recruitment probability:

```
> recrprob = function(S) {
+       a/(1 + (a/b) * S)
+ }
```

Now simulate the number of settlers and the number of recruits, using `rzinbinom` from the `emdbook` package:

```
> settlers = rzinbinom(N, mu = mu, size = k, zprob = zprob)
> recr = rbinom(N, prob = recrprob(settlers),
+          size = settlers)
```

5.2.2.2 PIGWEED DISTRIBUTION AND FECUNDITY

Pacala and Silander (1990) quantified the strength and spatial scale of competition between the annual weeds velvetweed (*Abutilon theophrasti*) and pigweed (*Amaranthus retroflexus*). They were interested in neighborhood competition among nearby plants. Local dispersal of seeds changes the distribution of the number of neighbors per plant. If plants were randomly distributed, we would expect a Poisson distribution of neighbors within a given distance, but if seeds have a limited dispersal range so that plants are spatially aggregated, we expect a distribution with higher variance (and a higher mean number of neighbors for a given overall plant density) such as the negative binomial. Neighbors increase local competition for nutrients, which in turn decreases plants' growth rate, their biomass at the end of the growing season, and their fecundity (seed set). Thus interspecific differences in dispersal and spatial patterning could change competitive outcomes (Bolker et al., 2003), although Pacala and Silander found that spatial structure had little effect in their system.

To explore the patterns of competition driven by local dispersal and crowding, we can simulate this spatial competitive process. Let's start by simulating a spatial distribution of plants in an $L \times L$ plot ($L = 30$ m below). We'll use a *Poisson cluster process*, where mothers are located randomly in space at points $\{x_p, y_p\}$ (called a *Poisson process* in spatial ecology), and their children are distributed nearby (only the children, and not the mothers, are included in the final pattern). The simulation includes $N = 50$ parents, for which we pick 50 x and 50 y values, each uniformly distributed between 0 and L. The distance of each child from its parent is exponentially distributed with rate $= 1/d$ (mean dispersal distance d), and the direction is random—that is, uniformly distributed between 0 and 2π radians.* I use a little bit of trigonometry to calculate the offspring locations (Figure 5.4a).

The formal mathematical definition of the model for offspring location is:

parent locations	$x_p, y_p \sim U(0, L)$
distance from parent	$r \sim \text{Exp}(1/d)$
dispersal angle	$\theta \sim U(0, 2\pi)$
offspring x	$x_c = x_p + r\cos\theta$
offspring y	$y_c = y_p + r\sin\theta$

In R, set up the parameters:

```
> set.seed(1001)
> L = 30
> nparents = 50
> offspr_per_parent = 10
> noffspr = nparents * offspr_per_parent
> dispdist = 2
```

Pick locations for the parents:

```
> parent_x = runif(nparents, min = 0, max = L)
> parent_y = runif(nparents, min = 0, max = L)
```

---

*R, like most computer languages, works in radians rather than degrees; to convert from degrees to radians, multiply by $\pi/180$. Since R doesn't understand Greek letters, use pi to denote π: `radians=degrees*pi/180`.

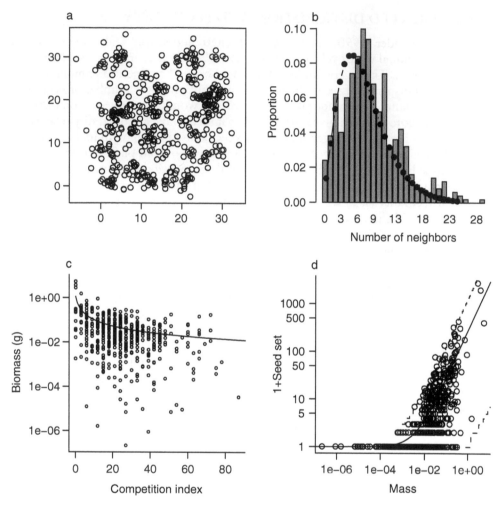

Figure 5.4 Pigweed simulations. (a) Spatial pattern (Poisson cluster process). (b) Distribution of number of neighbors within 2 m. (c) End-of-year biomass, based on a hyperbolic function of crowding index with a Gamma error distribution. (d) Seed set, proportional to biomass with a negative binomial error distribution.

Pick angles and distances for dispersal:

```
> angle = runif(noffspr, min = 0, max = 2 * pi)
> dist = rexp(noffspr, 1/dispdist)
```

Add the offspring displacements to the parent coordinates (using `rep(...,each= offspr_per_parent)`):

```
> offspr_x = rep(parent_x, each = offspr_per_parent) +
+     cos(angle) * dist
> offspr_y = rep(parent_y, each = offspr_per_parent) +
+     sin(angle) * dist
```

If you wanted to allow different numbers of offspring for each parent—for example, drawn from a Poisson distribution—you could use `offspr_per_parent=rpois(nparents,lambda)` and then `rep(...,times=offspr_per_parent)`. Instead of specifying that each parent's coordinates should be repeated the same number of times, you would be telling R to repeat each parent's coordinates according to its number of offspring.

Next we calculate the neighborhood density, or the number of individuals within 2 m of each plant (not counting itself). Figure 5.4b shows this distribution, along with a fitted negative binomial distribution. This calculation reduces the spatial pattern to a simpler nonspatial distribution of crowding.

```
> pos = cbind(offspr_x, offspr_y)
> ndist = as.matrix(dist(pos, upper = TRUE, diag = TRUE))
> nbrcrowd = rowSums(ndist < 2) - 1
```

The `dist` command calculates the distances among plant positions, while `rowSums` counts the number of distances in each row that satisfy the condition `ndist < 2`; we subtract 1 to ignore self-crowding.

Next we use a relationship that Pacala and Silander found between end-of-year mass (M) and competition index (C) (Figure 5.4c). They fitted this relationship based on a competition index estimated as a function of the neighborhood density of conspecific (pigweed) and heterospecific (velvetleaf) competitors, $C = 1 + c_{pp}n_p + c_{vp}n_v$. For this example, I simplified the crowding index to $C = 1 + 3n_p$. Pacala and Silander found that biomass $M \sim \text{Gamma}(\text{shape} = m/(1 + C), \text{scale} = \alpha)$, with $m = 2.3$ and $\alpha = 0.49$.

```
> ci = nbrcrowd * 3
> m = 2.3
> alpha = 0.49
> Mass_det = m/(1 + ci)
> Mass = rgamma(length(Mass_det), scale = Mass_det,
+       shape = alpha)
```

Finally, we simulate seed set as a function of biomass, again using a relationship estimated by Pacala and Silander. Seed set is proportional to mass, with negative binomial errors: $S \sim \text{NegBin}(\mu = bM, k)$, with $b = 271.6$, $k = 0.569$.

```
> b = 271.6
> k = 0.569
> seed_det = b * Mass
> seed = rnbinom(length(seed_det), mu = seed_det, size = k)
```

Figure 5.4d shows both mass and (1 + seed set) on a logarithmic scale, along with dashed lines showing the 95% confidence limits of the theoretical distribution.

The idea behind realistic static models is that they can link together simple deterministic and stochastic models of each process in a chain of ecological processes—in this case from spatial distribution to neighborhood crowding to biomass to seed set. (Pacala and Silander actually went a step further and computed the density-dependent survival probability. We could simulate this using a standard model like survival $\sim$ Binom$(N = 1, p = $ logistic$(a + bC))$, where the logistic function allows the survival probability to be an increasing function of competition index that cannot exceed 1.)

Thus, although writing down a single function that describes the relationship between competition index and the number surviving would be extremely difficult, as shown here we can break the relationship down into stages in the ecological process and use a simple model for each stage.

5.3 Power Analysis

Power analysis in the narrow sense means figuring out the (frequentist) statistical power, the probability of correctly rejecting the null hypothesis when it is false (Figure 5.5). Power analysis is important, but the narrow frequentist definition suffers from some of the problems that we are trying to move beyond by learning new statistical methods, such as a focus on p values and on the "truth" of a particular null hypothesis. Thinking about power analysis even in this narrow sense is already a vast improvement on the naive and erroneous "the null hypothesis is false if $p < 0.05$ and true if $p > 0.05$" approach. However, we should really be considering a much broader question: *How do the quality and quantity of my data and the true properties (parameters) of my ecological system affect the quality of the answers to my questions about ecological systems?*

For any real experiment or observation situation, we don't know what is really going on (the "true" model or parameters), so we don't have the information required to answer these questions from the data alone. But we can approach them by analysis or simulation. Historically, questions about statistical power could be answered only by sophisticated analyses, and only for standard statistical models and experimental designs such as one-way ANOVA or linear regression. Increases in computing power have extended power analyses to many new areas, and R's capability to run repeated stochastic simulations is a great help. Paradoxically, the mathematical difficulty of deriving power formulas is a great equalizer: since even research statisticians typically use simulations to estimate power, it's now possible (by learning simulation, which is easier than learning advanced mathematical statistics) to work on an equal footing with even cutting-edge researchers.

The first part of the rather vague (but commonsense) question above is about "quantity and quality of data and the true properties of the ecological system." These properties include:

- Number of data points (number of observations/sampling intensity).
- Distribution of data (experimental design):
  - Number of observations per site, number of sites.
  - Temporal and spatial extent (distance between the farthest samples, controlling the largest scale you can measure) and grain (distance between the closest samples, controlling the smallest scale you can measure).
  - Even or clustered distribution in space and/or time. Blocking. Balance (i.e., equal or similar numbers of observations in each treatment).

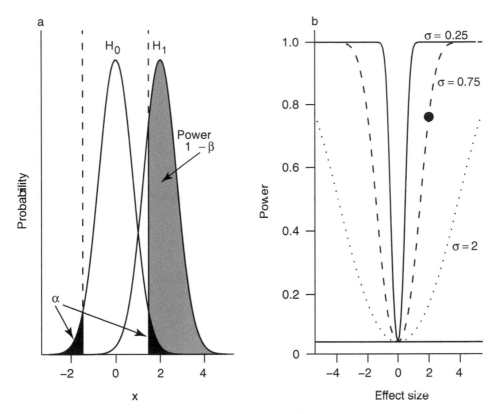

Figure 5.5 The frequentist definition of power. In the left-hand plot, the type I (false positive) rate α is the area under the tails of the null hypothesis H_0; the type II error rate β is the area under the sampling distribution of the alternative hypothesis (H_1) between the tails of the null hypothesis; thus the power $1 - \beta$ is the gray area shown that lies above the upper critical value of the null hypothesis curve. (There is also a tiny area, too small to see, where H_1 overlaps the lower tail of H_0.) The right-hand plot shows power as a function of effect size (distance between the means) and standard deviation; the point shows the scenario (effect size = 2, $\sigma = 0.75$) illustrated in the left figure.

- Distribution of continuous covariates—mimicking the natural distribution, or *stratified* to sample evenly across the natural range of values, or artificially extended to a wider range.
- Amount of variation (measurement/sampling error, demographic stochasticity, environmental variation). Experimental control or quantification of variation.
- Effect size (small or large), or the distance of the true parameter from the null-hypothesis value.

These properties will determine how much information you can extract from your data. Large data sets are better than smaller ones; balanced data sets with wide ranges are better than unbalanced data sets with narrow ranges; data sets with large extent (maximum spatial and/or temporal range) and small grain (minimum distance between samples) are best; and larger effects are obviously easier to detect and characterize. There are obvious trade-offs between effort (measured in person-hours or dollars) and the number of samples, and in how you allocate that effort. Would you prefer more information about fewer samples, or less information about more? More observations at fewer sites, or fewer at more sites? Should you spend your effort increasing extent or decreasing grain?

Subtler trade-offs also affect the value of an experiment. For example, controlling extraneous variation allows a more powerful answer to a statistical question—but how do we know what is "extraneous"? Variation actually affects the function of ecological systems (Jensen's inequality; Ruel and Ayres, 1999). Measuring a plant in a constant laboratory environment may answer the wrong question: we ultimately want to know how the plant performs in the natural environment, not in the lab, and variability is an important part of most environments. In contrast, performing "unrealistic" manipulations like pushing population densities beyond their natural limits may help to identify density-dependent processes that are real and important but undetectable at ambient densities (Osenberg et al., 2002). These questions have no simple answer, but they're important to consider.

The quality of the answers we get from our analyses is as multifaceted as the quality of the data. *Precision* specifies how finely you can estimate a parameter—the number of significant digits, or the narrowness of the confidence interval—while *accuracy* specifies how likely your answer is to be correct. Accurate but imprecise answers are better than precise but inaccurate ones: at least in this case you know that your answer is imprecise, rather than having misleadingly precise but inaccurate answers. But you need both precision and accuracy to understand and predict ecological systems.

More specifically, I will show how to estimate the following aspects of precision and accuracy for the damselfish system:

- *Bias* (accuracy): Bias is the expected difference between the estimate and the true value of the parameter. If you run a large number of simulations with a true value of d and estimate a value of $\hat{d}$ for each one, then the bias is $E[\hat{d} - d]$. Most simple statistical estimators are unbiased, and so most of us have come to expect (wrongly) that statistical estimates are generally unbiased. Most statistical estimators are indeed *asymptotically* unbiased, which means that in the limit of a large amount of data they will give the right answer on average, but surprisingly many common estimators are biased (Poulin, 1996; Doak et al., 2005).

- *Variance* (precision): Variance, or $E[(\hat{d} - E[\hat{d}])^2]$, measures the variability of the point estimates $(\hat{d})$ around their mean value. Just as an accurate but imprecise answer is worthless, unbiased answers are worthless if they have high variance. With low bias we know that we get the right answer *on average*, but high variability means that any particular estimate could be way off. With real data, we never know which estimates are right and which are wrong.

- *Confidence interval width* (precision): The width of the confidence intervals, either in absolute terms or as a proportion of the estimated value, provides useful information on the precision of your estimate. If the confidence interval is estimated correctly (see coverage, below), then low variance will give rise to narrow confidence intervals and high variance will give rise to broad confidence intervals.

- *Mean squared error* (MSE: accuracy and precision): MSE combines bias and variance as (bias^2+ variance). It represents the total variation around the true value, rather than around the average estimated value $(E[d - \hat{d}])^2 + E[(\hat{d} - E[\hat{d}])^2] = E[(\hat{d} - d)^2]$. MSE gives an overall measure of the quality of the estimator.
- *Coverage* (accuracy): When we sample data and estimate parameters, we try to estimate the uncertainty in those parameters. Coverage describes how accurate those confidence intervals are and (once again) can be estimated only via simulation. If the confidence intervals (for a given confidence level $1 - \alpha$) are d_{low} and d_{high}, then the coverage describes the proportion or percentage of simulations in which the confidence intervals actually include the true value: coverage $= \text{Prob}(d_{\text{low}} < d < d_{\text{high}})$. Ideally, the observed coverage should equal the nominal coverage of $1 - \alpha$; values that are larger than $1 - \alpha$ are pessimistic or conservative, overstating the level of uncertainty, while values that are smaller than $1 - \alpha$ are optimistic or "anticonservative." (It often takes several hundred simulations to get a reasonably precise estimate of the coverage, especially when estimating the coverage for 95% confidence intervals.)
- *Power* (precision): Finally, the narrow-sense power gives the probability of correctly rejecting the null hypothesis, or in other words the fraction of the times that the null-hypothesis value d_0 will be outside of the confidence limits: Power $= \text{Prob}(d_0 < d_{\text{low}} \text{ or } d_0 > d_{\text{high}})$. In frequentist language, it is $1 - \beta$, where β is the probability of making a type II error.

	H_0 true	H_0 false
fail to reject H_0	$1 - \alpha$	β
reject H_0	α	$1 - \beta$

Typically you specify an alternative hypothesis H_1, a desired type I error rate α, and a desired power $(1 - \beta)$ and then calculate the required sample size, or alternatively calculate $(1 - \beta)$ as a function of sample size, *for some particular H_1*. When the effect size is zero (the difference between the null and the alternate hypotheses is zero—i.e., the null hypothesis is true), the power is undefined, but it approaches α as the effect size gets small $(H_1 \rightarrow H_0)$.*

R has built-in functions for several standard cases (power of tests of difference between means of two normal populations [power.t.test], tests of difference in proportions, [power.prop.test], and one-way, balanced ANOVA [power.anova.test]).[†] For more discussion of these cases, or for other straightforward examples, you can look in any relatively advanced biometry book (e.g., Sokal and Rohlf (1995) or Quinn and Keough (2002)), or even find a calculator on the Web (search for "statistical power calculator"). For more complicated and ecologically realistic examples, however, you'll probably have to find the answer through simulation, as demonstrated below.

---

* The power does not approach zero! Even when the null hypothesis is true, we reject it a proportion α of the time. Thus we can expect to correctly reject the null hypothesis, even for very small effects, with probability at least α.

[†] The Hmisc package, available on CRAN, has a few more power calculators.

5.3.1 Simple Examples

5.3.1.1 LINEAR REGRESSION

Let's start by estimating the statistical power of detecting the linear trend in Figure 5.1a, as a function of sample size. To find out whether we can reject the null hypothesis in a single "experiment," we simulate a data set with a given slope, intercept, and number of data points; run a linear regression; extract the p-value; and see whether it is less than our specified α criterion (usually 0.05). For example:

```
> x = 1:20; a = 2; b = 1; sd = 8; N = 20; set.seed(1)
> y_det = a + b * x
> y = rnorm(N, mean = y_det, sd = sd)
> m = lm(y ~ x)
> coef(summary(m))["x", "Pr(>|t|)"]
```

```
[1] 0.0007208296
```

Extracting p-values from R analyses can be tricky. In this case, the coefficients of the *summary* of the linear fit are a matrix including the standard error, t statistic, and p-value for each parameter; I used matrix indexing based on the row and column names to pull out the specific value I wanted. More generally, you will have to use the names and str commands to pick through the results of a test to find the p-value.

To estimate the *probability* of successfully rejecting the null hypothesis when it is false (the power), we have to repeat this procedure many times and calculate the proportion of the time that we reject the null hypothesis.

Specify the number of simulations to run (400 is a reasonable number if we want to calculate a percentage—even 100 would do to get a crude estimate):

```
> nsim = 400
```

Set up a vector to hold the p-value for each simulation:

```
> pval = numeric(nsim)
```

Now repeat what we did above 400 times, each time saving the p-value in the storage vector:

```
> for (i in 1:nsim) {
+     y_det = a + b * x
+     y = rnorm(N, mean = y_det, sd = sd)
+     m = lm(y ~ x)
+     pval[i] = coef(summary(m))["x", "Pr(>|t|)"]
+ }
```

Calculate the power:

```
> sum(pval < 0.05)/nsim
```

```
[1] 0.87
```

However, we don't just want to know the power for a single experimental design. Rather, we want to know how the power changes as we change some aspect of the design such as the sample size or the variance. Thus we have to repeat the entire procedure multiple times, each time changing some parameter of the simulation such as the slope, or the error variance, or the distribution of the x values. Coding this in R usually involves nested for loops. For example:

```
> bvec = seq(-2, 2, by = 0.1)
> power.b = numeric(length(bvec))
> for (j in 1:length(bvec)) {
+       b = bvec[j]
+       for (i in 1:nsim) {
+           y_det = a + b * x
+           y = rnorm(N, mean = y_det, sd = sd)
+           m = lm(y ~ x)
+           pval[i] - coef(summary(m))["x", "Pr(>|t|)"]
+       }
+       power.b[j] = sum(pval < 0.05)/nsim
+ }
```

The results would resemble a noisy version of Figure 5.5b. The power equals $\alpha = 0.05$ when the slope is zero, rising to 0.8 for slope $\approx \pm 1$.

You could repeat these calculations for a different set of parameters (e.g., changing the sample size or the number of parameters). If you were feeling ambitious, you could calculate the power for many combinations of (e.g.) slope and sample size, using yet another for loop, saving the results in a matrix, and using contour or persp to plot the results.

5.3.1.2 HYPERBOLIC/NEGATIVE BINOMIAL DATA

What about the power to detect the difference between the two groups shown in Figure 5.1b with hyperbolic dependence on x, negative binomial errors, and different intercepts and hyperbolic slopes?

To estimate the power of the analysis, we have to know how to test statistically for a difference between the two groups. Jumping the gun a little bit (this topic will be covered in much greater detail in Chapter 6), we can define *negative log-likelihood functions* for a null model that assumes the intercept is the same for both groups as well as for a more complex model that allows for differences in the intercept.

The mle2 command in the bbmle package lets us fit the parameters of these models, and the anova command gives us a p-value for the difference between the models (p. 206):

```
> m0 = mle2(y ~ dnbinom(mu = a * b * x/(b + x), size = k),
+       start = list(a = 15, b = 1, k = 5))
> m1 = mle2(y ~ dnbinom(mu = a * b * x/(b + x), size = k),
+       parameters = list(a ~ g, b ~ g), start = list(a = 15,
+           b = 1, k = 5))
> anova(m0, m1)[2, "Pr(>Chisq)"]
```

Without showing the details, we now run a for loop that simulates the system 200 times each for a range of sample sizes, uses anova to calculate the p-values, and calculates the proportion of p-values that are smaller than 0.05 for each sample size (Figure 5.6). For small sample sizes (< 20), the power is abysmal (≈ 0.2–0.4).

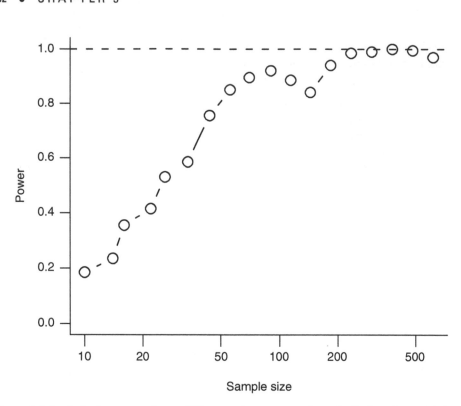

Figure 5.6 Statistical power to detect differences between two hyperbolic functions with intercepts $a = \{10, 20\}$, slopes $b = \{2, 1\}$, and negative binomial $k = 5$, as a function of sample size. Sample size is plotted on a logarithmic scale.

Power then rises approximately linearly, reaching acceptable levels (0.8 and up) at sample sizes of 50–100 and greater. The variation in Figure 5.6 is due to stochastic variation in the simulations. We could run more simulations per sample size to reduce the variation, but it's probably unnecessary since all power analysis is approximate anyway.

5.3.1.3 BIAS AND VARIANCE IN ESTIMATES OF THE NEGATIVE BINOMIAL K PARAMETER

For another simple example, one that demonstrates that there's more to life than p-values, consider the problem of estimating the k parameter of a negative binomial distribution. Are standard estimators biased? How large a sample do you need for a reasonably accurate estimate of aggregation?

Statisticians have long been aware that maximum likelihood estimates of the negative binomial k and similar aggregation indices, while better than simpler method of moments estimates (p. 119), are biased for small sample sizes (Pieters et al., 1977; Piegorsch, 1990; Poulin, 1996; Lloyd-Smith, 2007). Although you could delve into the statistical literature on this topic and even find special-purpose estimators that reduce the bias (Saha and Paul, 2005), being able to explore the problem yourself through simulation is empowering.

We can generate negative binomial samples with `rnbinom`, and the `fitdistr` command from the MASS package is a convenient way to estimate the parameters. `fitdistr` finds maximum likelihood estimates, which generally have good properties—but are not infallible, as we will see shortly. For a single sample:

```
> x = rnbinom(100, mu = 1, size = 0.5)
> f = fitdistr(x, "negative binomial")
> f
```

```
            size             mu
       0.21908756      1.05996103
      (0.05712932)    (0.24875054)
```

(The standard deviations of the parameter estimates are given in parentheses.) You can see that for this example the value of k (`size`) is underestimated relative to the true value of 0.5—but how do the estimates behave in general?

To dig the particular values we want (estimated k and standard deviation of the estimate) out of the object that `fitdistr` returns, we have to use `str(f)` to examine its internal structure. It turns out that `f$estimate["size"]` and `f$sd["size"]` give us the numbers we want.

Set up a vector of sample sizes (`lseq` is a function from the `emdbook` package that generates a logarithmically spaced sequence) and set aside space for the estimate of k and its standard deviation:

```
> Nvec = round(lseq(20, 500, length = 100))
> estk = numeric(100)
> estksd = numeric(100)
```

Now pick samples and estimate the parameters:

```
> set.seed(1001)
> for (i in 1:100) {
+       N = Nvec[i]
+       x = rnbinom(N, mu = 1, size = 0.5)
+       f = fitdistr(x, "negative binomial")
+       estk[i] = f$estimate["size"]
+       estksd[i] = f$sd["size"]
+ }
```

The estimate is indeed biased, and highly variable, for small sample sizes (Figure 5.7). For sample sizes below about 100, the estimate k is biased upward by about 20% on average. The coefficient of variation (standard deviation divided by the mean) is similarly greater than 0.2 for sample sizes less than 100.

5.3.2 Detecting Under- and Overcompensation in Fish Data

Finally, we will explore a more extended and complex example—the difficulty of estimating the exponent d in the Shepherd function, $R = aS/(1 + (a/b)S^d)$ (Figure 3.10). This parameter controls whether the Shepherd function is undercompensating ($d < 1$: recruitment increases indefinitely as the number of settlers grows), saturating

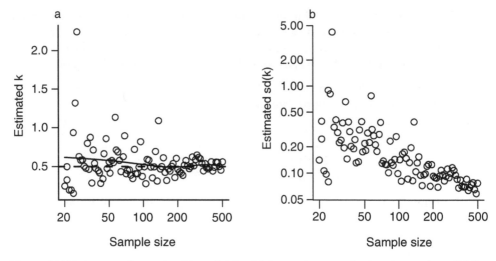

Figure 5.7 Estimates of negative binomial k with increasing sample size. In (a) the solid line is a loess fit; horizontal dashed line is the true value. The y axis in (b) is logarithmic.

($d = 1$: recruitment reaches an asymptote), or overcompensating ($d > 1$: recruitment decreases at high settlement). Schmitt et al. (1999) set $d = 1$ in part because d is very hard to estimate reliably—we are about to see just how hard.

You can use the simulation approach described above to generate simulated "data sets" of different sizes whose characteristics match Schmitt et al.'s data: a zero-inflated negative binomial distribution of numbers of settlers and a Shepherd function relationship (with a specified value of d) between the number of settlers and the number of recruits. For each simulated data set, use R's nls function to estimate the values of the parameters by nonlinear least squares.* Then calculate the confidence limits on d (using the confint function) and record the estimated value of the parameter and the lower and upper confidence limits.

Figure 5.8 shows the point estimates ($\hat{d}$) and 95% confidence limits (d_{low}, d_{high}) for the first 20 out of 400 simulations each with 1000 simulated observations and a true value of $d = 1.2$. The figure also illustrates several of the summary statistics discussed above: bias, variance, power, and coverage (see the caption for details).

For this particular case ($n = 1000$, $d = 1.2$) I can compute the bias (0.0039), variance (0.003, or $\sigma_{\hat{d}} = 0.059$), mean-squared error (0.003), coverage (0.921), and power (0.986). With 1000 observations, things look great, but 1000 observations is a lot and $d = 1.2$ represents strong overcompensation. The real value of power analyses comes when we compare the quality of estimates across a range of sample sizes and effect sizes.

Figure 5.9 gives a gloomier picture, showing the bias, precision, coverage, and power for a range of d values from 0.7 to 1.3 and a range of sample sizes from 50 to 2000. Sample sizes of at least 500 are needed to obtain reasonably unbiased estimates with adequate precision, and even then the coverage may be low if $d < 1.0$

---

* Nonlinear least-squares fitting assumes constant, normally distributed error, ignoring the fact that the data are really binomially distributed. Chapter 5, 6 and 7 will present more sophisticated maximum likelihood approaches to this problem.

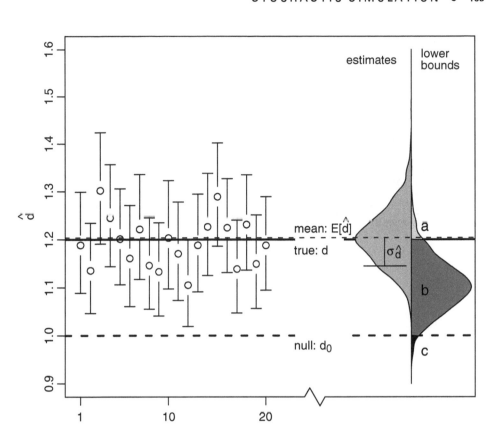

Figure 5.8 Simulations and power/coverage. Points and error bars show point estimates ($\hat{d}$) and 95% confidence limits (d_{low}, d_{high}) for the first 20 out of 400 simulations with a true value of $d = 1.2$ and 1000 samples. Horizontal lines show the mean value of $\hat{d}$, $E[\hat{d}] = 1.204$; the true value for this set of simulations, $d = 1.2$; and the null value, $d_0 = 1$. The left-hand vertical density plot shows the distribution of $\hat{d}$ for all 400 simulations. The right-hand density shows the distribution of the lower confidence limit, d_{low}. The distance between d (solid horizontal line) and $E[\hat{d}]$ (short-dashed horizontal line) shows the bias. The error bar showing the standard deviation of $\hat{d}$, $\sigma_{\hat{d}}$, shows the square root of the variance of $\hat{d}$. The coverage is the proportion of lower confidence limits that fall below the true value, area $b + c$ in the lower-bound density. The power is the proportion of lower confidence limits that fall above the null value, area $a + b$ in the lower-bound density. For simplicity, I have omitted the distribution of the upper bounds d_{high}.

and the power low if d is close to 1 ($0.9 \leq d \leq 1.1$). Because of the upward bias in d at low sample sizes, the calculated power is actually *higher* at very low sample sizes, but this is not particularly comforting. The power of the analysis is slightly greater for overcompensation than undercompensation. The relatively low power values are as expected from Figure 5.9b, which shows wide confidence intervals. Low power would also be predictable from the high variance of the estimates, which I didn't even bother to show in Figure 5.9a because they obscured the figure too much.

Another use for our simulations is to take a first look at the trade-offs involved in adding complexity to models. Figure 5.10 shows estimates of b, the asymptote if

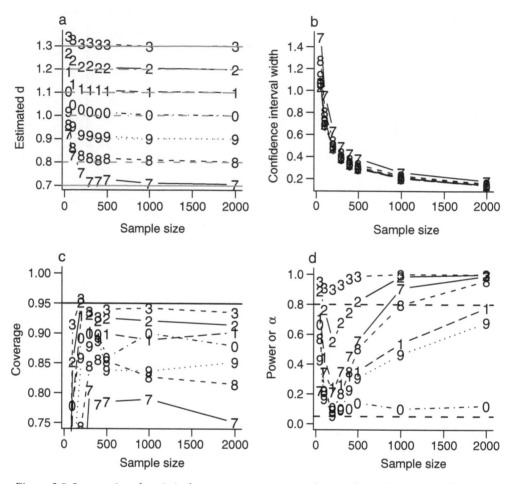

Figure 5.9 Summaries of statistical accuracy, precision, and power for estimating the Shepherd exponent d for a range of d values from undercompensation, $d = 0.7$ (line marked "7"), to overcompensation, $d = 1.3$ (line marked "3"). (a) Estimated d: the estimates are strongly biased upward for sample sizes less than 500, especially for undercompensation ($d < 1$). (b) Confidence interval width: the confidence intervals are large (> 0.4) for sample sizes smaller than about 500, for any value of d. (c) Coverage of the nominal 95% confidence intervals is adequate for large sample size (> 250) and overcompensation ($d > 1$), but poor even for large sample sizes when $d < 1$. (d) For statistical power ($1 - \beta$) of at least 0.8, sample sizes of 500–1000 are required if $d \leq 0.7$ or $d \geq 1.2$; sample sizes of 1000 if $d = 0.8$; and sample sizes of at least 2000 if $d = 0.9$ or $d = 1.1$. When $d = 1.0$ ("0" line), the probability of rejecting the null hypothesis is a little above the nominal value of $\alpha = 0.05$.

$d = 1$, for different sample sizes and values of d. If $d = 1$, then the Shepherd model reduces to the Beverton-Holt model. In this case, you might think that it wouldn't matter whether you used the Shepherd or the Beverton-Holt model to estimate the b parameter, but the Shepherd function presents serious disadvantages. First, even when $d = 1$, the Shepherd estimate of d is biased upward for low sample sizes, leading to a severe upward bias in the estimate of b. Second, not shown on the graph because it would have obscured everything else, the variance of the Shepherd estimate is far

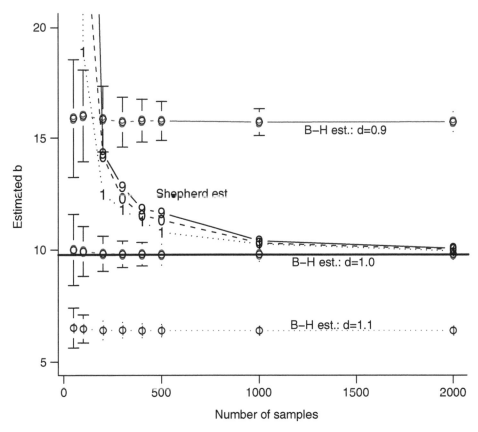

Figure 5.10 Estimates of b, using Beverton-Holt or Shepherd functions, for different values of d and sample sizes. True value of $b = 10.0$. Shepherd estimate lines labeled "9", "0", and "1", correspond to estimates of b when $d = 0.9, 1.0$, or 1.1.

higher than the variance of the Beverton-Holt estimate (e.g., for a sample size of 200, the Beverton-Holt estimate is 9.83 ± 0.78 (s.d.), while the Shepherd estimate is 14.16 ± 13.94 (s.d.)).

On the other hand, if d is *not* equal to 1, the Beverton-Holt estimate of b (horizontal lines in Figure 5.10) is strongly biased, independent of the sample size. For reasonable sample sizes, if $d = 0.9$, the Beverton-Holt estimate is biased upward by 6, to 16; if $d = 1.1$, it is biased downward by about 4, to 6. Since the Beverton-Holt model isn't flexible enough to account for the changes in shape caused by d, it has to modify b in order to compensate.

This general phenomenon is called the *bias–variance trade-off* (see p. 204): more complex models reduce bias at the price of increased variance. (The small-sample bias of the Shepherd is a separate, and slightly less general, phenomenon.)

Because estimating parameters or testing hypotheses with noisy data is fundamentally difficult, and most ecological data sets are noisy, power analyses are often depressing. On the other hand, even if things are bad, it's better to know how bad they are than just to guess; knowing how much you really know is important. In addition, you can make some design decisions (e.g., number of treatments vs. number of replicates per treatment) that will optimize power given the constraints of time and money.

Remember that the overall quality of your experimental design—including good technique; proper randomization, replication, and controls; and common sense—is often far more important than the fussy details of your statistical design (Hurlbert, 1984). While you should quantify the power of your experiment to make sure it has a reasonable chance of success, thoughtful experimental design (e.g., measuring and statistically accounting for covariates such as mass and rainfall; pairing control and treatment samples; or expanding the range of covariates tested) will make a much bigger difference than tweaking experimental details to squeeze out a little more statistical power.

6 Likelihood and All That

This chapter presents the basic concepts and methods you need in order to estimate parameters, establish confidence limits, and choose among competing hypotheses and models. It defines likelihood and discusses frequentist, Bayesian, and information-theoretic inference based on likelihood.

6.1 Introduction

Previous chapters introduced all the ingredients you need to define a model—mathematical functions to describe the deterministic patterns and probability distributions to describe the stochastic patterns—and showed how to use these ingredients to simulate simple ecological systems. The final steps of the modeling process are estimating parameters from data and testing models against each other. You may be wondering by now how you would actually do this.

Estimating the parameters of a model means finding the parameters that make that model fit the data best. To compare among models we have to figure out which one fits the data best, and decide if one or more models fit sufficiently better than the rest that we can declare them the winners. Our goodness-of-fit metrics will be based on the *likelihood*, the probability of seeing the data we actually collected given a particular model. Depending on the context, "model" could mean either the general form of the model or a specific set of parameter values.

6.2 Parameter Estimation: Single Distributions

Parameter estimation is simplest when we have a a collection of independent data that are drawn from a distribution (e.g., Poisson, binomial, normal), with the same parameters for all observations.* As an example with discrete data, we will select one particular case out of Vonesh's tadpole predation data (p. 47)—small tadpoles at a density of 10—and estimate the per-trial probability parameter of a binomial distribution (i.e., each individual's probability of being eaten by a predator). As an

---

* In statistical jargon, such data are called *independent and identically distributed* (iid).

example with continuous data, we will introduce a new data set on myxomatosis virus concentration (titer) in experimentally infected rabbits (Myxo in the emdbook package; Fenner et al., 1956; Dwyer et al., 1990). Although the titer actually changes systematically over time, we will gloss over that problem for now and pretend that all the measurements are drawn from the same distribution so that we can estimate the parameters of a Gamma distribution that describes the variation in titer among different rabbits.

6.2.1 Maximum Likelihood

We want the *maximum likelihood estimates* of the parameters—those parameter values that make the observed data most likely to have happened. Since the observations are independent, the joint likelihood of the whole data set is the product of the likelihoods of each individual observation. Since the observations are identically distributed, we can write the likelihood as a product of similar terms. For mathematical convenience, we almost always maximize the logarithm of the likelihood (log-likelihood) instead of the likelihood itself. Since the logarithm is a monotonically increasing function, the maximum log-likelihood estimate is the same as the maximum likelihood estimate. Actually, it is conventional to *minimize* the negative log-likelihood rather than maximizing the log-likelihood. For continuous probability distributions, we compute the probability *density* of observing the data rather than the probability itself. Since we are interested in relative (log-)likelihoods, not the absolute probability of observing the data, we can ignore the distinction between the density $(P(x))$ and the probability (which includes a term for the measurement precision: $P(x)\,dx$).

6.2.1.1 TADPOLE PREDATION DATA: BINOMIAL LIKELIHOOD

For a single observation from the binomial distribution (e.g., the number of small tadpoles killed by predators in a single tank at a density of 10), the likelihood that k out of N individuals are eaten as a function of the per capita predation probability p is $\mathrm{Prob}(k|p, N) = \binom{N}{k}p^k(1-p)^{N-k}$. If we have n observations, each with the same total number of tadpoles N, and the number of tadpoles killed in the ith observation is k_i, then the likelihood is

The log-likelihood is
$$\mathcal{L} = \prod_{i=1}^{n} \binom{N}{k_i}p^{k_i}(1-p)^{N-k_i}.^* \tag{6.2.1}$$

$$L = \sum_{i=1}^{n} \left(\log \binom{N}{k_i} + k_i \log p + (N - k_i) \log (1-p) \right). \tag{6.2.2}$$

In R, this would be sum(dbinom(k,size=N,prob=p,log=TRUE)).

---

* The symbol $\prod$ denotes a product, like $\sum$ but for multiplication.

Analytical Approach

In this simple case, we can actually solve the problem analytically, by differentiating with respect to p and setting the derivative to zero. Let $\hat{p}$ be the maximum likelihood estimate, the value of p that satisfies

$$\frac{dL}{dp} = \frac{d \sum_{i=1}^{n} \left(\log \binom{N}{k_i} + k_i \log p + (N - k_i) \log (1 - p) \right)}{dp} = 0. \tag{6.2.3}$$

Since the derivative of a sum equals the sum of the derivatives,

$$\sum_{i=1}^{n} \frac{d \log \binom{N}{k_i}}{dp} + \sum_{i=1}^{n} \frac{d \, k_i \log p}{dp} + \sum_{i=1}^{n} \frac{d \, (N - k_i) \log (1 - p)}{dp} = 0. \tag{6.2.4}$$

The term $\log \binom{N}{k_i}$ is a constant with respect to p, so its derivative is zero and the first term disappears. Since k_i and $(N - k_i)$ are constant factors, they come out of the derivatives and the equation becomes

$$\sum_{i=1}^{n} k_i \frac{d \log p}{dp} + \sum_{i=1}^{n} (N - k_i) \frac{d \log (1 - p)}{dp} = 0. \tag{6.2.5}$$

The derivative of $\log p$ is $1/p$, so the chain rule says the derivative of $\log (1 - p)$ is $d(\log (1 - p))/d(1 - p) \cdot d(1 - p)/dp = -1/(1 - p)$. Remembering that $\hat{p}$ is the value of p that satisfies this equation:

$$\frac{1}{\hat{p}} \sum_{i=1}^{n} k_i - \frac{1}{1 - \hat{p}} \sum_{i=1}^{n} (N - k_i) = 0$$

$$\frac{1}{\hat{p}} \sum_{i=1}^{n} k_i = \frac{1}{1 - \hat{p}} \sum_{i=1}^{n} (N - k_i)$$

$$(1 - \hat{p}) \sum_{i=1}^{n} k_i = \hat{p} \sum_{i=1}^{n} (N - k_i)$$

$$\sum_{i=1}^{n} k_i = \hat{p} \left(\sum_{i=1}^{n} k_i + \sum_{i=1}^{n} (N - k_i) \right) = \hat{p} \sum_{i=1}^{n} N$$

$$\sum_{i=1}^{n} k_i = \hat{p} n N$$

$$\hat{p} = \frac{\sum_{i=1}^{n} k_i}{nN}. \tag{6.2.6}$$

So the maximum likelihood estimate, $\hat{p}$, is just the overall fraction of tadpoles eaten, lumping all the observations together: a total of $\sum k_i$ tadpoles were eaten out of a total of nN tadpoles exposed in all of the observations.

We seem to have gone to a lot of effort to prove the obvious, that the best estimate of the per capita predation probability is the observed frequency of predation.

Other simple distributions like the Poisson behave similarly. If we differentiate the likelihood, or the log-likelihood, and solve for the maximum likelihood estimate, we get a sensible answer. For the Poisson, the estimate of the rate parameter $\hat{\lambda}$ is equal to the mean number of counts observed per sample. For the normal distribution, with two parameters μ and σ^2, we have to compute the *partial derivatives* (see the appendix) of the likelihood with respect to both parameters and solve the two equations simultaneously $(\partial L/\partial \mu = \partial L/\partial \sigma^2 = 0)$. The answer is again obvious in hindsight: $\hat{\mu} = \bar{x}$ (the estimate of the mean is the observed mean) and $\hat{\sigma}^2 = \sum (x_i - \bar{x})^2/n$ (the estimate of the variance is the variance of the sample).*

Some simple distributions like the negative binomial, and all the complex problems we will be dealing with hereafter, have no easy analytical solution, so we will have to find the maximum likelihood estimates of the parameters numerically. The point of the algebra here is just to convince you that maximum likelihood estimation makes sense in simple cases.

Numerics

This chapter presents the basic process of computing and maximizing likelihoods (or minimizing negative log-likelihoods) in R; Chapter 7 will go into much more technical detail. First, you need to define a function that calculates the negative log-likelihood for a particular set of parameters. Here's the R code for a binomial negative log-likelihood function:

```
> binomNLL1 = function(p, k, N) {
+      -sum(dbinom(k, prob = p, size = N, log = TRUE))
+ }
```

The dbinom function calculates the binomial likelihood for a specified data set (vector of number of successes) k, probability p, and number of trials N; the log=TRUE option gives the log-probability instead of the probability (more accurately than taking the log of the product of the probabilities); -sum adds the log-likelihoods and changes the sign to compute an overall negative log-likelihood for the data set.

Load the data and extract the subset we plan to work with:

```
> data(ReedfrogPred)
> x = subset(ReedfrogPred, pred == "pred" & density == 10
+   & size == "small")
> k = x$surv
```

The total number of tadpoles exposed in this subset of the data is 40 (10 in each of 4 trials), 30 of which were eaten by predators, so the maximum likelihood estimate will be $\hat{p} = 0.75$.

We can use the optim function to numerically **optim**ize (by default, minimizing rather than maximizing) this function. You need to give optim the *objective function*—the function you want to minimize (binomNLL1 in this case)—and a vector of starting parameters. You can also give it other information, such as a data set, to

---

* Maximum likelihood estimation gives a biased estimate of the variance, dividing the sum of squares $\sum (x_i - \bar{x})^2$ by n instead of $n - 1$.

be passed on to the objective function. The starting parameters don't have to be very accurate (if we had accurate estimates already we wouldn't need optim), but they do have to be reasonable. That's why we spent so much time in Chapters 3 and 4 on eyeballing curves and the method of moments.

```
> opt1 = optim(fn = binomNLL1, par = c(p = 0.5), N = 10,
+     k = k, method = "BFGS")
```

fn is the argument that specifies the objective function and par specifies the vector of starting parameters. Using c(p=0.5) names the parameter p—probably not necessary here but very useful for keeping track when you start fitting models with more parameters. The rest of the command specifies other parameters and data and optimization details; Chapter 7 explains why you should use method="BFGS" for a single-parameter fit.

Check the estimated parameter value and the maximum likelihood—we need to change sign and exponentiate the minimum negative log-likelihood that optim returns to get the maximum log-likelihood:

```
> opt1$par
```

```
p
0.7499998
```

Because it was computed numerically the answer is almost, but not exactly, equal to the theoretical answer of 0.75.

```
> exp(-opt1$value)
```

```
[1] 0.0005150149
```

The mle2 function in the bbmle package provides a "wrapper" for optim that gives prettier output and makes standard tasks easier.[*] Unlike optim, which is designed for general-purpose optimization, mle2 assumes that the objective function is a negative log-likelihood function. The names of the arguments are easier to understand: minuslogl instead of fn for the negative log-likelihood function, start instead of par for the starting parameters, and data for additional parameters and data.

```
> library(bbmle)
> m1 = mle2(minuslogl = binomNLL1, start = list(p = 0.5),
+     data = list(N = 10, k = k))
> m1
```

```
Call:
mle2(minuslogl = binomNLL1, start = list(p = 0.5), data =
    list(N = 10, k = k))
```

---

[*] Why mle2? There is an mle function in the stats4 package that comes with R, but I added some features—and then renamed it to avoid confusion with the original R function.

```
Coefficients:
        p
0.7499998

Log-likelihood: -7.57
```

The `mle2` package has a shortcut for simple likelihood functions. Instead of writing an R function to compute the negative log-likehood, you can specify a formula:

```
> mle2(k ~ dbinom(prob = p, size = 10),
+      start = list(p = 0.5))
```

gives exactly the same answer as the previous commands. R assumes that the variable on the left-hand side of the formula is the response variable (k in this case) and that you want to sum the negative log-likelihood of the expression on the right-hand side for all values of the response variable.

Another way to find maximum likelihood estimates for data drawn from most simple distributions—although not for the binomial distribution—is the `fitdistr` command in the MASS package, which will even guess reasonable starting values for you. However, it works only in the very simple case where none of the parameters of the distribution depend on other covariates.

The estimated value of the per capita predation probability, 0.7499..., is very close to the analytic solution of 0.75. The estimated value of the maximum likelihood (Figure 6.1) is quite small ($\mathcal{L} = 5.15 \times 10^{-4}$). That is, the probability of *this particular outcome*—5, 7, 9 and 9 out of 10 tadpoles eaten in four replicates—is low.* In general, however, we will be interested only in the relative likelihoods (or log-likelihoods) of different parameters and models rather than their absolute likelihoods.

Having fitted a model to the data (even a very simple one), it's worth plotting the predictions of the model. In this case the data set is so small (four points) that sampling variability dominates the plot (Figure 6.1b).

6.2.1.2 MYXOMATOSIS DATA: GAMMA LIKELIHOOD

As part of the effort to use myxomatosis as a biocontrol agent against introduced European rabbits in Australia, Fenner and co-workers (1956) studied the virus concentrations (*titer*) in the skin of rabbits that had been infected with different virus strains. We'll choose a Gamma distribution to model these continuously distributed, positive data.[†] For the sake of illustration, we'll use just the data for one viral strain (grade 1).

```
> data(MyxoTiter_sum)
> myxdat = subset(MyxoTiter_sum, grade == 1)
```

---

*I randomly simulated 1000 samples of four values drawn from the binomial distribution with $p = 0.75$, $N = 10$. The maximum likelihood was smaller than the observed value given in the text 12.4% of the time (or 54% of the time if we keep only the samples where 30/40 tadpoles were eaten). Thus this likelihood, although small, is not significantly lower than would be expected by chance.

[†] We could also use a log-normal distribution or (since the minimum values are far from zero and the distributions are reasonably symmetric) a normal distribution.

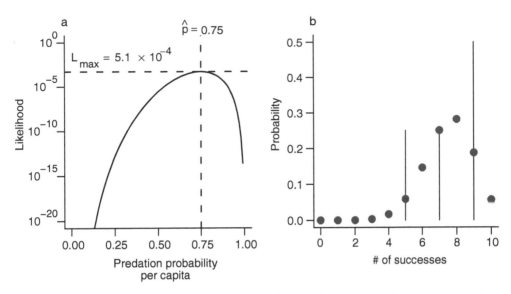

Figure 6.1 Binomial-distributed predation. (a) Likelihood curve, on a logarithmic y scale. (b) Best-fit model prediction compared with the data.

The likelihood equation for Gamma-distributed data is hard to maximize analytically, so we'll go straight to a numerical solution. The negative log-likelihood function looks just like the one for binomial data.*

```
> gammaNLL1 = function(shape, scale) {
+      -sum(dgamma(myxdat$titer, shape = shape, scale = scale,
+           log = TRUE))
+ }
```

It's harder to find starting parameters for the Gamma distribution. We can use the method of moments (Chapter 4) to determine reasonable starting values for the scale (= variance/mean) and shape (= mean^2/variance = 1/(coefficient of variation)^2) parameters.†

```
> gm = mean(myxdat$titer)
> gvm = var(myxdat$titer)/mean(myxdat$titer)
```

Now fit the data:

```
> m3 = mle2(gammaNLL1, start = list(shape = gm/gvm,
+      scale = gvm))

> m3
```

---

* optim insists that you specify all of the parameters packed into a single numeric vector in your negative log-likelihood function. mle prefers the parameters as a list. mle2 will accept either a list, or, if you use parnames to specify the parameter names, a numeric vector (p. 183).

† Because the estimates of the shape and scale are very strongly correlated in this case, I ended up having to tweak the starting conditions slightly away from the method of moments estimates, to {45.8,0.151}.

```
Call:
mle2(minuslogl = gammaNLL1, start = list(shape = 45.8,
    scale = 0.151))

Coefficients:
  shape        scale
49.3421124    0.1403326

Log-likelihood: -37.67
```

I could also use the formula interface,

```
> m3 = mle2(myxdat$titer ~ dgamma(shape, scale = scale),
+      start = list(shape = gm/gvm, scale = gvm))
```

Since the default parameterization of the Gamma distribution in R uses a rate parameter instead of a scale parameter, I have to make sure to specify the scale parameter explicitly. Or I could use `fitdistr` from the MASS package:

```
> f1 = fitdistr(myxdat$titer, "gamma")
```

`fitdistr` gives slightly different values for the parameters and the likelihood, but not different enough to worry about. A greater possibility for confusion is that `fitdistr` reports the rate (= 1/scale) instead of the scale parameter.

Figure 6.2 shows the negative log-likelihood (now a negative log-likelihood *surface* as a function of two parameters, the shape and scale) and the fit of the model to the data (virus titer for grade 1). Since the "true" distribution of the data is hard to visualize (all of the distinct values of virus titer are displayed as jittered values along the bottom axis), I've plotted the nonparametric (kernel) estimate of the probability density in gray for comparison. The Gamma fit is very similar, although it takes account of the lowest point (a virus titer of 4.2) by spreading out slightly rather than allowing the bump in the left-hand tail that the nonparametric density estimate shows. The large shape parameter of the best-fit Gamma distribution (shape = 49.34) indicates that the distribution is nearly symmetrical and approaching normality (Chapter 4). Ironically, in this case the plain old normal distribution actually fits slightly better than the Gamma distribution, despite the fact that we would have said the Gamma was a better model on biological grounds (it doesn't allow virus titer to be negative). However, according to criteria we will discuss later in the chapter, the models are not significantly different and you could choose either on the basis of convenience and appropriateness for the rest of the story you were telling. If we fitted a more skewed distribution, like the damselfish settlement distribution, the Gamma would certainly win over the normal.

6.2.2 Bayesian Analysis

Bayesian estimation also uses the likelihood, but it differs in two ways from maximum likelihood analysis. First, we combine the likelihood with a prior probability distribution in order to determine a posterior probability distribution. Second, we often report the mean of the posterior distribution rather than its mode (which would

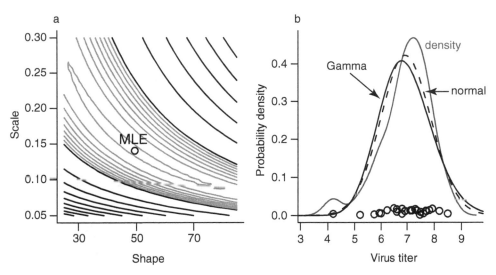

Figure 6.2 Likelihood curves for a simple distribution: Gamma-distributed virus titer. Black contours are spaced 200 log-likelihood units apart; gray contours are spaced 20 log-likelihood units apart. In the right-hand plot, the gray line is a kernel density estimate; solid line is the Gamma fit; and dashed line is the normal fit.

equal the MLE if we were using a completely uninformative, or "flat," prior). Unlike the mode, which reflects only local information about the peak of the distribution, the mean incorporates the entire pattern of the distribution, so it can be harder to compute.

6.2.2.1 BINOMIAL DISTRIBUTION: CONJUGATE PRIORS

In the particular case when we have so-called *conjugate priors* for the distribution of interest, Bayesian estimation is easy. As introduced in Chapter 4, a conjugate prior is a choice of the prior distribution that matches the likelihood model so that the posterior distribution has the same form as the prior distribution. Conjugate priors also allow us to interpret the strength of the prior in simple ways.

For example, the conjugate prior of the binomial likelihood that we used for the tadpole predation data is the Beta distribution. If we pick a Beta prior with shape parameters a and b, and if our data include a total of $\sum k$ "successes" (predation events) and $nN - \sum k$ "failures" (surviving tadpoles) out of a total of nN "trials" (exposed tadpoles), the posterior distribution is a Beta distribution with shape parameters $a + \sum k$ and $b + (nN - \sum k)$. If we interpret $a - 1$ as the total number of previously observed successes and $b - 1$ as the number of previously observed failures, then the new distribution just combines the total number of successes and failures in the complete (prior plus current) data set. When $a = b = 1$, the Beta distribution is flat, corresponding to no prior information ($a - 1 = b - 1 = 0$). As a and b increase, the prior distribution gains more information and becomes peaked. We can also see that, as far as a Bayesian is concerned, how we divide our experiments up doesn't matter. Many small experiments, aggregated with successive uses of Bayes'

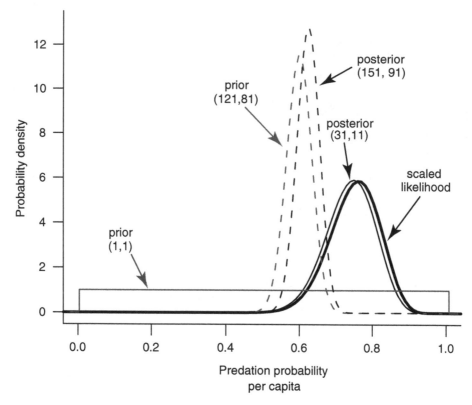

Figure 6.3 Bayesian priors and posteriors for the tadpole predation data. The scaled likelihood is the normalized likelihood curve, corresponding to the weakest prior possible. Prior(1,1) is weak, corresponding to zero prior samples and leading to a posterior (31,11) that is almost identical to the scaled likelihood curve. Prior(121,81) is strong, corresponding to a previous sample size of 200 trials and leading to a posterior (151,111) that is much closer to the prior than to the scaled likelihood.

Rule, give the same information as one big experiment (provided of course that there is no variation in per-trial probability among sets of observations, which we have assumed in our statistical model for both the likelihood and the Bayesian analysis).

We can also examine the effect of different priors on our estimate of the mean (Figure 6.3). If we have no prior information and choose a flat prior with $a_{\mathrm{prior}} = b_{\mathrm{prior}} = 1$, then our final answer is that the per capita predation probability is distributed as a Beta distribution with shape parameters $a = \sum k + 1 = 31, b = nN - \sum k + 1 = 11$. The mode of this Beta distribution occurs at $(a - 1)/(a + b - 2) = \sum k/(nN) = 0.75$—exactly the same as the maximum likelihood estimate of the per capita predation probability. Its mean is $a/(a + b) = 0.738$—very slightly shifted toward 0.5 (the mean of our prior distribution) from the MLE. If we wanted a distribution whose *mean* was equal to the maximum likelihood estimate, we could generate a *scaled likelihood* by normalizing the likelihood so that it integrated to 1. However, to create the Beta prior that would lead to this posterior distribution we would have to take the limit as a and b go to zero, implying a very peculiar prior distribution with infinite spikes at 0 and 1.

If we had much more prior data—say a set of experiments with a total of $(nN)_{\text{prior}} = 200$ tadpoles, of which $\sum k_{\text{prior}} = 120$ were eaten—then the parameters of the prior distribution would be $a_{\text{prior}} = 121$ and $b_{\text{prior}} = 81$, the posterior mode would be 0.625, and the posterior mean would be 0.624. In this case both the posterior mode and mean are much closer to the prior values than to the maximum likelihood estimate because the prior information is much stronger than the information we can obtain from the data.

If our data were Poisson, we could use a conjugate prior Gamma distribution with shape α and scale s and interpret the parameters as α = total counts in previous observations and $1/s$ = number of previous observations. Then if we observed C counts in our data, the posterior would be a Gamma distribution with $\alpha' = \alpha + C$, $1/s' = 1/s + 1$.

The conjugate prior for the mean of a normal distribution, if we know the variance, is also a normal distribution. The posterior mean is an average of the prior mean and the observed mean, weighted by the *precisions*—the reciprocals of the prior and observed variances. The conjugate prior for the precision, if we know the mean, is the Gamma distribution.

6.2.2.2 GAMMA DISTRIBUTION: MULTIPARAMETER DISTRIBUTIONS AND NONCONJUGATE PRIORS

Unfortunately simple conjugate priors aren't always available, and we often have to resort to numerical integration to evaluate Bayes' Rule. Just plotting the numerator of Bayes' Rule $(\text{prior}(p) \times \mathcal{L}(p))$ is easy; for anything else, we need to integrate (or use summation to approximate an integral).

In the absence of much prior information for the myxomatosis parameters a (shape) and s (scale), I chose a weak, independent prior distribution:

$$\text{Prior}(a) \sim \text{Gamma}(\text{shape} = 0.01, \text{scale} = 100)$$

$$\text{Prior}(s) \sim \text{Gamma}(\text{shape} = 0.1, \text{scale} = 10)$$

$$\text{Prior}(a, s) = \text{Prior}(a) \cdot \text{Prior}(s).$$

Bayesians often use the Gamma as a prior distribution for parameters that must be positive (although Gelman (2006) has other suggestions). Using a small shape parameter gives the distribution a large variance (corresponding to little prior information) and means that the distribution will be peaked at small values but is likely to be flat over the range of interest. Finally, the scale is usually set large enough to make the mean of the parameter (= shape · scale) reasonable. Finally, I made the probabilities of a and s independent, which keeps the form of the prior simple.

As introduced in Chapter 4, the posterior probability is proportional to the prior times the likelihood. To compute the actual posterior probability, we need to divide the numerator $\text{Prior}(p) \times L(p)$ by its integral to make sure the total area (or volume) under the probability distribution is 1:

$$\text{Posterior}(a, s) = \frac{\text{Prior}(a, s) \times \mathcal{L}(a, s)}{\iint \text{Prior}(a, s)\mathcal{L}(a, s) \, da \, ds}$$

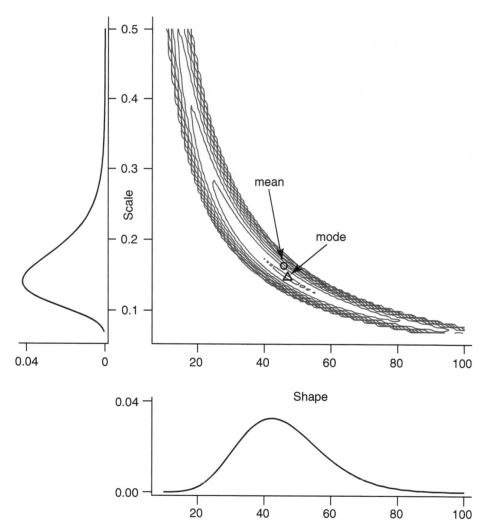

Figure 6.4 Bivariate and marginal posterior distributions for the myxomatosis titer data. Contours are drawn, logarithmically spaced, at probability levels from 0.01 to 10^{-10}. Posterior distributions are weak and independent, Gamma(shape = 0.1, scale = 10) for scale and Gamma(shape = 0.01, scale = 100) for shape.

Figure 6.4 shows the two-dimensional posterior distribution for the myxomatosis data. As is typical for reasonably large data sets, the probability density is very sharp. The contours shown on the plot illustrate a rapid decrease from a probability density of 0.01 at the mode down to a probability density of 10^{-10}, and most of the posterior density is even lower than this minimum contour line.

If we want to know the distribution of each parameter individually, we have to calculate its *marginal* distribution: that is, what is the probability that *a* or *s* falls within a particular range, independent of the value of the other variable? To calculate the marginal distribution, we have to integrate (take the expectation) over all possible values of the other parameter:

$$\text{Posterior}(a) = \int \text{Posterior}(a, s)\, ds$$

$$\text{Posterior}(s) = \int \text{Posterior}(a, s)\, da$$

(6.2.7)

Figure 6.4 also shows the marginal distributions of a and s.

What if we want to summarize the results still further and give a single value for each parameter (a *point estimate*) representing our conclusions about the virus titer? Bayesians generally prefer to quote the mean of a parameter (its expected value) rather than the mode (its most probable value). Neither summary statistic is more correct than the other—they give different information about the distribution—but they can lead to radically different inferences about ecological systems (Ludwig, 1996). The differences will be largest when the posterior distribution is asymmetric (the only time the mean can differ from the mode) and when uncertainty is large. In Figure 6.4, the mean and the mode are close together.

To compute mean values for the parameters, we need to compute some more integrals, finding the weighted average of the parameters over the posterior distribution:

$$\bar{a} = \int \text{Posterior}(a) \cdot a\, da$$

$$\bar{s} = \int \text{Posterior}(s) \cdot s\, ds.$$

(We can also compute these means from the full rather than the marginal distributions: e.g., $\bar{a} = \iint \text{Posterior}(a, s)a\, da\, ds$.)[*]

R can compute all of these integrals numerically. We can define functions

```
> prior.as = function(a, s) {
+       dgamma(a, shape = 0.01, scale = 100) * dgamma(s,
+           shape = 0.1, scale = 10)
+ }
> unscaled.posterior = function(a, s) {
+       prior.as(a, s) * exp(-gammaNLL1(shape = a, scale = s))
+ }
```

and use `integrate` (for one-dimensional integrals) or `adapt` (in the `adapt` package; for multidimensional integrals) to do the integration. More crudely, we can approximate the integral by a sum, calculating values of the integrand for discrete values (e.g., $s = 0, 0.01, \ldots, 10$) and then calculating $\sum P(s)\Delta s$—this is how I created Figure 6.4.

However, integrating probabilities is tricky for two reasons. (1) Prior probabilities and likelihoods are often tiny for some parameter values, leading to roundoff error; tricks like calculating log probabilities for the prior and likelihood, adding, and then exponentiating can help. (2) You must pick the number and range of points

---

[*] The means of the marginal distributions are the same as the mean of the full distribution. Confusingly, the modes of the marginal distributions are *not* the same as the mode of the full distribution.

at which to evaluate the integral carefully. Too coarse a grid leads to approximation error, which may be severe if the function has sharp peaks. Too small a range, or the wrong range, can miss important parts of the surface. Large, fine grids are very slow. The numerical integration functions built into R help—you give them a range and they try to evaluate the number of points at which to evaluate the integral—but they can still miss peaks in the function if the initial range is set too large so that their initial grid fails to pick up the peaks. Integrals over more than two dimensions make these problem even worse, since you have to compute a huge number of points to cover a reasonably fine grid. This problem is the first appearance of the *curse of dimensionality* (Chapter 7).

In practice, brute-force numerical integration is no longer feasible with models with more than about two parameters. The only practical alternatives are *Markov chain Monte Carlo* approaches, introduced later in this chapter and in more detail in Chapter 7.

For the myxomatosis data, the posterior mode is $(a = 47, s = 0.15)$, close to the maximum likelihood estimate of $(a = 49.34, s = 0.14)$ (the differences are probably caused as much by round-off error as by the effects of the prior). The posterior mean is $(a = 45.84, s = 0.16)$.

6.3 Estimation for More Complex Functions

So far we've estimated the parameters of a single distribution (e.g., $X \sim \text{Binom}(p)$ or $X \sim \text{Gamma}(a, s)$). We can easily extend these techniques to more interesting ecological models like the ones simulated in Chapter 5, where the mean or variance parameters of the model vary among groups or depend on covariates.

6.3.1 Maximum Likelihood

6.3.1.1 TADPOLE PREDATION

We can combine deterministic and stochastic functions to calculate likelihoods, just as we did to simulate ecological processes in Chapter 5. For example, suppose tadpole predators have a Holling type II functional response (predation rate $= aN/(1 + abN)$), meaning that the per capita predation rate of tadpoles decreases hyperbolically with density $(= a/(1 + abN))$. The distribution of the actual number eaten is likely to be binomial with this probability. If N is the number of tadpoles in a tank,

$$p = \frac{a}{1 + abN}$$

$$k \sim \text{Binom}(p, N).$$

(6.3.1)

Since the distribution and density functions in R (such as dbinom) operate on vectors just as do the random-deviate functions (such as rbinom) used in Chapter 5,

I can translate this model definition directly into R, using a numeric vector p={*a*, *h*} for the parameters:

```
> binomNLL2 = function(params, N, k) {
+       a = params[1]
+       h = params[2]
+       predprob = a/(1 + a * h * N)
+       -sum(dbinom(k, prob = predprob, size = N, log = TRUE))
+ }
```

Now we can dig out the data from the functional response experiment of Vonesh and Bolker (2005), which contains the variables Initial (*N*) and Killed (*k*). Plotting the data (Figure 2.8) and eyeballing the initial slope and asymptote gives us crude starting estimates of *a* (initial slope) around 0.5 and *h* (1/asymptote) around 1/80 = 0.0125.

```
> data(ReedfrogFuncresp)
> attach(ReedfrogFuncresp)
> opt2 = optim(fn = binomNLL2, par = c(a = 0.5, h = 0.0125),
+       N = Initial, k = Killed)
> detach (ReedfrogFuncresp)
```

This optimization gives us parameters (*a* = 0.526, *h* = 0.017)—so our starting guesses were pretty good.

To use mle2 for this purpose, you would normally have to rewrite the negative log-likelihood function with the parameters a and h as separate arguments (i.e., function(a,h,p,N,k)). However, mle2 will let you pass the parameters inside a vector as long as you use parnames to attach the names of the parameters to the function.

```
> parnames(binomNLL2) = c("a", "h")
> m2 = mle2(binomNLL2, start = c(a = 0.5, h = 0.0125),
+       data = list(N = Initial, k = Killed))
> m2
```

```
Call:
mle2(minuslogl = binomNLL2, start = c(a = 0.5, h = 0.0125),
    data = list(N = Initial, k = Killed))

Coefficients:
     a              h
0.52630319     0.01664362

Log-likelihood: -46.72
```

The answers are very slightly different from the optim results (mle2 uses a different numerical optimizer by default).

As always, we should plot the fit to the data to make sure it is sensible. Figure 6.5a shows the expected number killed (a Holling type II function) and uses the qbinom

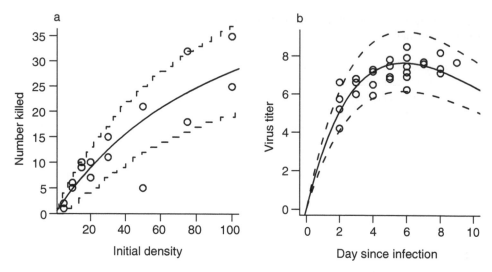

Figure 6.5 Maximum-likelihood fits to (a) tadpole predation (Holling type II/binomial) and (b) myxomatosis (Ricker/Gamma) models.

function to plot the 95% confidence intervals of the binomial distribution.* One point falls outside of the confidence limits; for 16 points, this isn't surprising (we would expect 1 point out of 20 to fall outside the limits on average), although this point is quite low (5/50, compared to an expectation of 18.3/50—the probability of getting this extreme an outlier is only 2.11×10^{-5}).

6.3.1.2 MYXOMATOSIS VIRUS

When we looked at the myxomatosis titer data earlier, we treated it as though it all came from a single distribution. In reality, titers typically change considerably as a function of the time since infection. Following Dwyer et al. (1990), we will fit a Ricker model to the mean titer level. Figure 6.5b shows the data for the grade 1 virus. The Ricker is a good function for fitting data that start from zero, grow to a peak, and then decline, although for the grade 1 virus we have only biological common sense, and the evidence from the other virus grades, to say that the titer would eventually decrease. Grade 1 is so virulent that rabbits die before titer has a chance to drop off. We'll stick with the Gamma distribution for the distribution of titer T at time t, but parameterize it with shape (s) and mean rather than shape and scale parameters (i.e., scale = mean/shape):

$$m = ate^{-bt}$$
$$T \sim \text{Gamma}(\text{shape} = s, \text{scale} = m/s).$$

(6.3.2)

Translating this into R is straightforward:

```
> gammaNLL2 = function(a, b, shape) {
+     meantiter = a * myxdat$day * exp(-b * myxdat$day)
```

---

* These confidence limits, sometimes called *plug-in estimates*, ignore the uncertainty in the parameters.

```
+        -sum(dgamma(myxdat$titer, shape = shape,
+           scale = meantiter/shape, log = TRUE))
+ }
```

We need initial values, which we can guess knowing from Chapter 3 that a is the initial slope of the Ricker function and $1/b$ is the x-location of the peak. Figure 6.5 suggests that $a \approx 1$, $1/b \approx 5$. I knew from the previous fit that the shape parameter is large, so I started with shape = 50. When I tried to fit the model with the default optimization method I got a warning that the optimization had not converged, so I used an alternative optimization method, the Nelder-Mead simplex (p. 229).

```
> m4 = mle2(gammaNLL2, start = list(a = 1, b = 0.2,
+       shape = 50), method = "Nelder-Mead")
> m4
```

```
Call:
mle2(minuslogl = gammaNLL2, start = list(a = 1, b = 0.2, shape = 50),
    method = "Nelder-Mead")
```

```
Coefficients:
    a             b            shape
 3.5614933     0.1713346     90.6790545
```

```
Log-likelihood: -29.51
```

We could run the same analysis a bit more compactly, without explicitly defining a negative log-likelihood function, using mle2's formula interface:

```
> mle2(titer ~ dgamma(shape, scale = a * day * exp(-b *
+       day)/shape), start = list(a = 1, b = 0.2, shape = 50),
+       data = myxdat, method = "Nelder-Mead")
```

Specifying data=myxdat lets us use day and titer in the formula instead of myxdat$day and myxdat$titer.

6.3.2 Bayesian Analysis

Extending the tools to use a Bayesian approach is straightforward, although the details are more complicated than maximum likelihood estimation. We can use the same likelihood models (e.g., (6.3.1) for the tadpole predation data or (6.3.2) for myxomatosis). All we have to do to complete the model definition for Bayesian analysis is specify prior probability distributions for the parameters. However, defining the model is not the end of the story. For the binomial model, which has only two parameters, we could proceed more or less as in the Gamma distribution example above (Figure 6.4), calculating the posterior density for many combinations of the parameters and computing integrals to calculate marginal distributions and means. To evaluate integrals for the three-parameter myxomatosis model we would have to integrate the posterior distribution over a three-dimensional grid, which would quickly become impractical.

Markov chain Monte Carlo (MCMC) is a numerical technique that makes Bayesian analysis of more complicated models feasible. BUGS is a program that allows you to run MCMC analyses without doing lots of programming. Here is the BUGS code for the myxomatosis example:

```
1.  model {
2.    for (i in 1:n) {
3.        mean[i] <- a*day[i]*exp(-b*day[i])
4.        rate[i] <- shape/mean[i]
5.        titer[i] ~ dgamma(shape,rate[i])
6.    }
7.  ## priors
8.  a ~ dgamma(0.1,0.1)
9.  b ~ dgamma(0.1,0.1)
10. shape ~ dgamma(0.1,0.01)
11. }
```

BUGS's modeling language is similar but not identical to R. For example, BUGS requires you to use <- instead of = for assignments.

As you can see, the BUGS model also looks a lot like the likelihood model (6.3.2). Lines 3–5 specify the model (BUGS uses shape and rate parameters to define the Gamma distribution rather than shape and scale parameters: differences in parameterization are some of the most important differences between the BUGS and R languages). Lines 8–10 give the prior distributions for the parameters, all Gamma in this case. The BUGS model is more explicit than (6.3.2)—in particular, you have to put in an explicit for loop to calculate the expected values for each data point—but the broad outlines are the same, even up to using a tilde (~) to mean "is distributed as."

You can run BUGS either as a standalone program or from within R, using the R2WinBUGS package as an interface to the WinBUGS program for running BUGS on Windows.*

```
> library(R2WinBUGS)
```

You have to specify the names of the data exactly as they are listed in the BUGS model (given above, but stored in a separate text file myxo1.bug):

```
> titer = myxdat$titer
> day = myxdat$day
> n = length(titer)
```

You also have to specify starting points for multiple chains, which should vary among reasonable values (p. 237), as a list of lists:

```
> inits = list(list(a = 4, b = 0.2, shape = 90), list(a = 1,
+     b = 0.1, shape = 50), list(a = 8, b = 0, shape = 150))
```

(I originally started b at 1.0 for the third chain, but WinBUGS kept giving me an error saying "cannot bracket slice for node a." By trial and error—eliminating chains and changing parameters—I established that the value of b in chain 3 was the problem.)

---

* WinBUGS runs on Windows and on Intel machines under Linux or MacOS (using Wine or Crossover Office). Chapter 7 gives more details.

Now you can run the model through WinBUGS:

```
> myxo1.bugs = bugs(data = list("titer", "day", "n"),
+     inits, parameters.to.save = c("a", "b", "shape"),
+     model.file = "myxo1.bug", n.chains = length(inits),
+     n.iter = 3000)
```

As we will see shortly, you can recover lots of information for a Bayesian analysis from a WinBUGS run—for now, you can use `print(myxo1.bugs,digits=4)` to see that the estimates of the means, $\{a = 3.55, b = 0.17, s = 79.9\}$, are reassuringly close to the maximum likelihood estimates (p. 185).

6.4 Likelihood Surfaces, Profiles, and Confidence Intervals

So far, we've used R and WinBUGS to find point estimates (maximum likelihood estimates or posterior means) automatically, without looking very carefully at the curves and surfaces that describe how the likelihood varies with the parameters. This approach gives little insight when things go wrong with the fitting (as happens all too often). Furthermore, point estimates are useless without measures of uncertainty. We really want to know the uncertainty associated with the parameter estimates, both individually (univariate confidence intervals) and together (bi- or multivariate confidence regions). This section will show how to draw and interpret goodness-of-fit curves (likelihood curves and profiles, Bayesian posterior joint and marginal distributions) and their connections to confidence intervals.

6.4.1 Frequentist Analysis: Likelihood Curves and Profiles

The most basic tool for understanding how likelihood depends on one or more parameters is the *likelihood curve* or *likelihood surface*, which is just the likelihood plotted as a function of parameter values (e.g., Figure 6.1). By convention, we plot the negative log-likelihood rather than log-likelihood, so the best estimate is a minimum rather than a maximum. (I sometimes call negative log-likelihood curves *badness-of-fit* curves, since higher points indicate a poorer fit to the data.) Figure 6.6a shows the negative log-likelihood curve (like Figure 6.1 but upside-down and with a different y axis), indicating the minimum negative log-likelihood (=maximum likelihood) point, and lines showing the upper and lower 95% confidence limits (we'll soon see how these are defined). Every point on a likelihood curve or surface represents a different fit to the data: Figure 6.6b shows the observed distribution of the binomial data along with three separate curves corresponding to the lower estimate ($p = 0.6$), best fit ($p = 0.75$), and upper estimate ($p = 0.87$) of the per capita predation probability.

For models with more than one parameter, we draw likelihood surfaces instead of curves. Figure 6.7 shows the negative log-likelihood surface of the tadpole predation data as a function of attack rate a and handling time h. The minimum is where we found it before, at ($a = 0.526, h = 0.017$). The likelihood contours are roughly elliptical and are tilted near a 45 degree angle, which means (as we will see) that the estimates of the parameters are correlated. Remember that each point on the

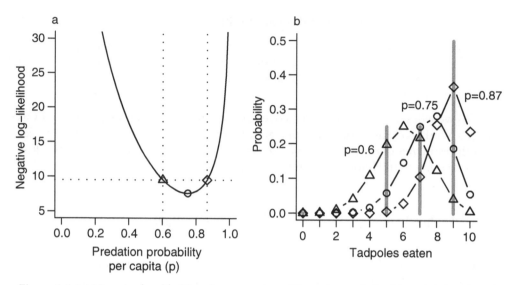

Figure 6.6 (a) Negative log-likelihood curve and confidence intervals for binomial-distributed tadpole predation. (b) Comparison of fits to data. Gray vertical bars show proportion of trials with different outcomes; lines and symbols show fits corresponding to different parameters indicated on the curve in (a).

likelihood surface corresponds to a fit to the data, which we can (and should) look at in terms of a curve through the actual data values: Figure 6.8 shows the fit of several sets of parameters (the ML estimates, and two other less well-fitting a-h pairs) on the scale of the original data.

If we want to deal with models with more than two parameters, or if we want to analyze a single parameter at a time, we have to find a way to isolate the effects of one or more parameters while still accounting for the rest. A simple, but usually wrong, way of doing this is to calculate a likelihood *slice*, fixing the values of all but one parameter (usually at their maximum likelihood estimates) and then calculating the likelihood for a range of values of the focal parameter. The horizontal line in the middle of Figure 6.7 shows a likelihood slice for a, with h held constant at its MLE. Figure 6.9 shows an elevational view, the negative log-likelihood for each value of a. Slices can be useful for visualizing the geometry of a many-parameter likelihood surface near its minimum, but they are statistically misleading because they don't allow the other parameters to vary and thus they don't show the minimum negative log-likelihood achievable for a particular value of the focal parameter.

Instead, we calculate likelihood *profiles*, which represent "ridgelines" in parameter space showing the minimum negative log-likelihood for particular values of a single parameter. To calculate a likelihood profile for a focal parameter, we have to set the focal parameter in turn to a range of values, and for each value optimize the likelihood with respect to all of the other parameters. The likelihood profile for a in Figure 6.7 runs through the contour lines (such as the confidence regions shown) at the points where the contours run exactly vertical. Think about looking for the minimum along a fixed-a transect (varying h—vertical lines in Figure 6.7); the minimum will occur at a point where the transect is just touching (tangent to) a contour line. Slices are always steeper than profiles (e.g., Figure 6.8), because they don't allow the

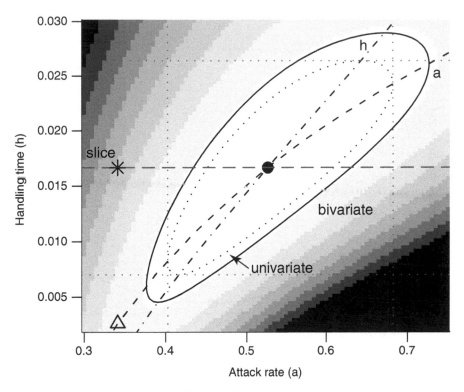

Figure 6.7 Likelihood surface for tadpole predation data, showing univariate and bivariate 95% confidence limits and likelihood profiles for a and h. Darker shades of gray represent higher (i.e., worse) negative log-likelihoods. The solid line shows the 95% bivariate confidence region. Dotted black and gray lines indicate 95% univariate confidence regions. The dash-dotted line and dashed line show likelihood profiles for h and a. The long-dash gray line shows the likelihood slice with varying a and constant h. The black dot indicates the maximum likelihood estimate; the star is an alternate fit along the slice with the same handling time; the triangle is an alternate fit along the likelihood profile for a.

other parameters to adjust to changes in the focal parameter. Figure 6.9 shows that the fit corresponding to a point on the profile (triangle/dashed line) has a lower value of h (handling time, corresponding to a higher asymptote) that compensates for its enforced lower value of a (attack rate/initial slope), while the equivalent point from the slice (star/dotted line) has the same handling time as the MLE fit, and hence fits the data worse—corresponding to the higher negative log-likelihood in Figure 6.8.

6.4.1.1 THE LIKELIHOOD RATIO TEST

On a negative log-likelihood curve or surface, higher points represent worse fits. The steeper and narrower the valley (i.e., the faster the fit degrades as we move away from the best fit), the more precisely we can estimate the parameters. Since the negative log-likelihood for a set of independent observations is the sum of the individual negative log-likelihoods, adding more data makes likelihood curves steeper. For example,

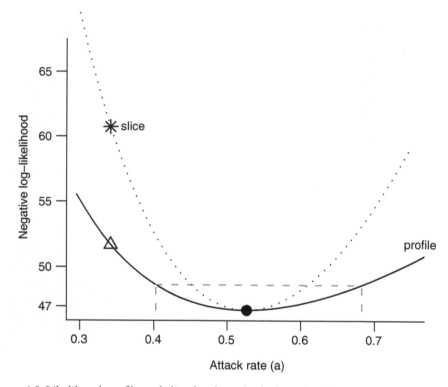

Figure 6.8 Likelihood profile and slice for the tadpole data, for the attack rate parameter *a*. Gray dashed lines show the negative log-likelihood cutoff and 95% confidence limits for *a*. Points correspond to parameter combinations marked in Figure 6.6.

doubling the number of observations will double the negative log-likelihood curve across the board—in particular, doubling the slope of the negative log-likelihood surface.*

It makes sense to determine confidence limits by setting some upper limit on the negative log-likelihood and declaring that any parameters that fit the data at least that well are within the confidence limits. The steeper the likelihood surface, the faster we reach the limit and the narrower are the confidence limits. Since we care only about the relative fit of different models and parameters, the limits should be relative to the maximum log-likelihood (minimum negative log-likelihood).

For example, Edwards (1992) suggested that one could set reasonable confidence regions by including all parameters within 2 log-likelihood units of the maximum log-likelihood, corresponding to all fits that gave likelihoods within a factor of $e^2 \approx 7.4$ of the maximum. However, this approach lacks a frequentist

---

* Doubling the sample size also typically doubles the minimum negative log-likelihood as well, which may seem odd—why would adding more data worsen the fit of the model?—until you remember that we're not really interested in the probability of a *particular* set of data, just the relative likelihood of different models and parameters. The probability of flipping a fair coin ($p = 0.5$) twice and getting one head and one tail is 0.5. The probability of flipping the same coin 1000 times and getting 500 heads and 500 tails is only 0.025; that doesn't mean that we should reject the hypothesis that the coin is fair.

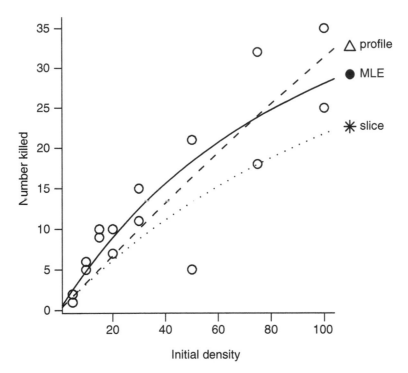

Figure 6.9 Fits to tadpole predation data corresponding to the parameter values marked in Figures 6.7 and 6.8.

probability interpretation—there is no corresponding p-value. This deficiency may be an advantage, since it makes dogmatic null-hypothesis testing impossible.

If you insist on p-values, you can also use differences in log-likelihoods (corresponding to ratios of likelihoods) in a frequentist approach called the *Likelihood Ratio Test* (LRT). Take some likelihood function $\mathcal{L}(p_1, p_2, \ldots, p_n)$, and find the overall best (maximum likelihood) value, $\hat{\mathcal{L}} = \mathcal{L}(\hat{p}_1, \hat{p}_2, \ldots, \hat{p}_n)$. Now fix some of the parameters (say $p_1, \ldots, p_r$) to specific values $(p_1^*, \ldots, p_r^*)$, and maximize with respect to the remaining parameters to get $\mathcal{L}_r = \mathcal{L}(p_1^*, \ldots, p_r^*, \hat{p}_{r+1}, \ldots, \hat{p}_n)$ (r stands for "restricted," sometimes also called a *reduced* or *nested* model). The Likelihood Ratio Test says that twice the negative log of the likelihood ratio, $-2\log(\mathcal{L}_r/\hat{\mathcal{L}})$, called the *deviance*, is approximately χ^2 ("chi-squared") distributed* with r degrees of freedom.[†]

---

* You may associate the χ^2 distribution with contingency table analysis, chisq.test in R, but it is a distribution that appears much more broadly in statistics.

† Here's a heuristic explanation: you can prove that the distribution of the maximum likelihood estimate is asymptotically normally distributed (i.e., with sufficiently large sample sizes). You can also show, by Taylor expanding, that the log-likelihood surface is quadratic, with curvature determined by the variances of the parameters. If we are restricting r parameters, then we are moving away from the maximum likelihood of the more complex model in r directions, by a normally distributed amount θ_i in each direction. Since the log-likelihood surface is quadratic, the drop in the negative log-likelihood is $\sum_{i=1}^{r} \theta_i^2$. Since the θ_i values (likelihood estimates of each parameter) are each normally distributed, the sum of squares of r of them is χ^2 distributed with r degrees of freedom. This explanation is necessarily crude; for the real derivation, see Kendall and Stuart (1979).

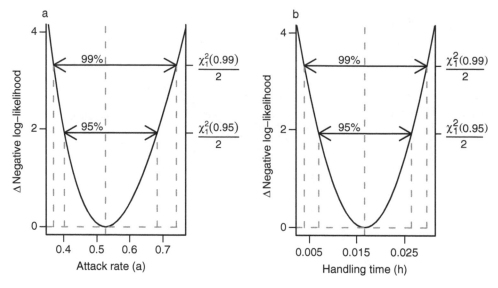

Figure 6.10 Likelihood profiles and LRT confidence intervals for tadpole predation data.

The log of the likelihood ratio is the difference in the log-likelihoods, so

$$2(-\log \mathcal{L}_r - (-\log \hat{\mathcal{L}})) = 2(-L_r - (-\hat{L})) \sim \chi_r^2. \qquad (6.4.1)$$

The definition of the LRT echoes the definition of the likelihood profile, where we fix one parameter and maximize the likelihood/minimize the negative log-likelihood with respect to all the other parameters: $r = 1$ in the definition above. Thus, for *univariate* confidence limits we cut off the likelihood profile at $(-\hat{L} + \chi_1^2(1-\alpha)/2)$, where α is our chosen type I error level (e.g., 0.05 or 0.01). The cutoff is a one-tailed test, since we are interested only in differences in likelihood that are larger than expected under the null hypothesis. Figure 6.10 shows the likelihood profiles for a and h, along with the 95% and 99% confidence intervals; you can see how the confidence intervals on the parameters are drawn as vertical lines through the intersection points of the (horizontal) likelihood cutoff levels with the profile.

The 99% confidence intervals have a higher cutoff than the 95% confidence intervals ($\chi_1^2(0.99)/2 = 3.32 > \chi_1^2(0.95)/2 = 1.92$), and hence the 99% intervals are wider. The numbers are given in Table 6.1.

R can compute profiles and profile confidence limits automatically. Given an mle2 fit m, profile(m) will compute a likelihood profile and confint(m) will compute profile confidence limits. plot(profile(m2)) will plot the profile, square-root transformed so that a quadratic profile will appear V-shaped (or linear if you specify absVal=FALSE). This transformation makes it easier to see whether the profile is quadratic, since it's easier to see whether a line is straight than it is to see whether it's quadratic. Computing the profile can be slow, so if you want to plot the profile and find confidence limits, or find several different confidence limits, you can save the profile and then use confint on the profile:

```
> p2 = profile(m2)
> confint(p2)
```

TABLE 6.1
Likelihood profile confidence limits

α	$\frac{\chi_1^2(\alpha)}{2}$	$-\hat{L} + \frac{\chi_1^2(\alpha)}{2}$	Variable	Lower	Upper
0.95	1.92	48.6	a	0.40200	0.6820
			b	0.00699	0.0264
0.99	3.32	50.0	a	0.37000	0.7390
			b	0.00387	0.0296

It's also useful to know how to calculate profiles and profile confidence limits yourself, both to understand them better and for the not-so-rare times when the automatic procedures break down. Because profiling requires many separate optimizations, it can fail if your likelihood surface has multiple minima (p. 245) or if the optimization is otherwise finicky. You can try to tune your optimization procedures using the techniques discussed in Chapter 7, but in difficult cases you may have to settle for approximate quadratic confidence intervals (Section 6.5).

To compute profiles by hand, you need to write a new negative log-likelihood function that holds one of the parameters fixed while minimizing the likelihood with respect to the rest. For example, to compute the profile for a (minimizing with respect to b for many values of a), you could use the following reduced negative log-likelihood function (compare this with the full function on p. 183):

```
> binomNLL2.a = function(params, N, k, a) {
+     h = params[1]
+     predprob = a/(1 + a * h * N)
+     -sum(dbinom(k, prob = predprob, size = N, log = TRUE))
+ }
```

Compute the profile likelihood for a range of a values:

```
> avec = seq(0.3, 0.8, length = 100)
> aprof = numeric(100)
> for (i in 1:100) {
+     aprof[i] = optim(binomNLL2.a, par = 0.02,
+         k = ReedfrogFuncresp$Killed,
+         N = ReedfrogFuncresp$Initial, a = avec[i],
+         method = "BFGS")$value
+ }
```

The curve drawn by plot(avec,aprof) would look just like the one in Figure 6.10a.

To find the profile confidence limits for a, we have to take one branch of the profile at a time. Starting with the lower branch, the a values below the maximum likelihood estimate:

```
> prof.lower = aprof[1:which.min(aprof)]
> prof.avec = avec[1:which.min(aprof)]
```

Finally, use the approx function to calculate the a value for which $-L = -\hat{L} + \chi_1^2(0.95)/2$:

```
> approx(prof.lower, prof.avec, xout = -logLik(m2) +
+        qchisq(0.95, 1)/2)
```

```
$x
```

```
'log Lik.' 48.64212 (df=2)
```

```
$y
```

```
[1] 0.4024598
```

Now let's go back and look at the *bivariate* confidence region in Figure 6.7. The 95% bivariate confidence region (solid black line) occurs at negative log-likelihood equal to $-\hat{L} + \chi_2^2(0.95)/2 \approx -\hat{L} + 3$. I've also drawn the univariate region ($\hat{L} + \chi_1^2(0.95)/2$ contour). That region is not really appropriate for this figure, because it applies to a single parameter at a time, but it illustrates that univariate intervals are smaller than the bivariate confidence region, and that the confidence intervals, like the profiles, are tangent to the univariate confidence region.

The LRT is correct only *asymptotically*, for large data sets. For small data sets it is an approximation, although one that people use very freely. The other limitation of the LRT that frequently arises, although it is often ignored, is that it applies only when the best estimate of the parameter is away from the edge of its allowable range (Pinheiro and Bates, 2000). For example, if the MLE of the mean parameter of a Poisson distribution λ (which must be ≥ 0) is equal to 0, then the LRT estimate for the upper bound of the confidence limit is not technically correct (see p. 250).

6.4.2 Bayesian Approach: Posterior Distributions and Marginal Distributions

What about the Bayesians? Instead of drawing likelihood curves, Bayesians draw the posterior distribution (proportional to prior $\times L$, e.g., Figure 6.4). Instead of calculating confidence limits using the (frequentist) LRT, they define the *credible interval*, which is the region in the center of the distribution containing 95% (or some other standard proportion) of the probability of the distribution, bounded by values on either side that have the same probability (or probability density). Technically, the credible interval is the interval $[x_1, x_2]$ such that $P(x_1) = P(x_2)$ and $C(x_2) - C(x_1) = 1 - \alpha$, where P is the probability density and C is the cumulative density. The credible interval is slightly different from the frequentist confidence interval, which is defined as $[x_1, x_2]$ such that $C(x_1) = \alpha/2$ and $C(x_2) = 1 - \alpha/2$. For empirical samples, use quantile to compute confidence intervals and HPDinterval ("highest posterior density interval"), in the coda package, to compute credible intervals. For theoretical distributions, use the appropriate "q" function (e.g., qnorm) to compute confidence intervals and tcredint, in the emdbook package, to compute credible intervals.

Figure 6.11 shows the posterior distribution for the tadpole predation (from Figure 6.4), along with the 95% credible interval and the lower and upper 2.5% tails

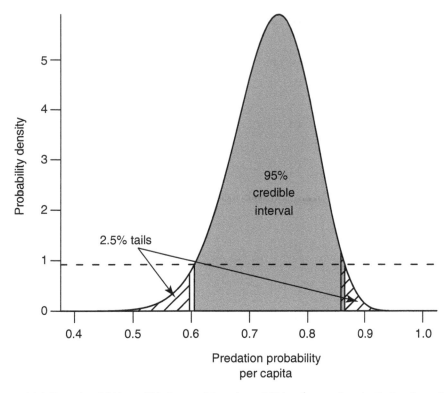

Figure 6.11 Bayesian 95% credible interval (gray), and 5% tail areas (hatched), for the tadpole predation data (weak Beta prior: shape = (1,1)).

for comparison. The credible interval is symmetrical in height; the cutoff value on either end of the distribution has the same posterior probability density. The extreme tails are symmetrical in area; the likelihood of extreme values in either direction is the same. The credible interval's height symmetry leads to a uniform probability cutoff: we never include a less probable value on one boundary than on the other. To a Bayesian, this property makes more sense than insisting (as the frequentists do in defining confidence intervals) that the probabilities of extremes in either direction are the same.

For multiparameter models, the likelihood surface is analogous to a bivariate or multivariate probability distribution (Figure 6.12). The *marginal probability density* is the Bayesian analogue of the likelihood profile. Where frequentists use likelihood profiles to make inferences about a single parameter while taking the effects of the other parameters into account, Bayesians use the marginal posterior probability density, the overall probability for a particular value of a focal parameter integrated over all the other parameters. Figure 6.12 shows the 95% credible intervals for the tadpole predation analysis, both bivariate and marginal (univariate). In this case, when the prior is weak and the posterior distribution is reasonably symmetrical, there is little difference between the bivariate 95% confidence region and the bivariate 95% credible interval (Figure 6.12), but Bayesian and frequentist conclusions are not always so similar.

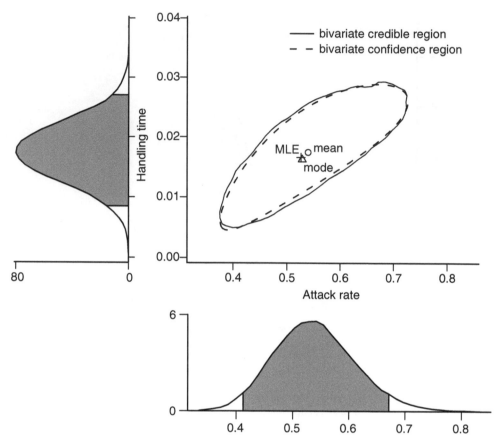

Figure 6.12 Bayesian credible intervals (bivariate and marginal) for tadpole predation analysis.

6.5 Confidence Intervals for Complex Models: Quadratic Approximation

The methods I've discussed so far (calculating likelihood profiles or marginal likelihoods numerically) work fine when you have only two, or maybe three, parameters, but they become impractical for models with many parameters. To calculate a likelihood profile for n parameters, you have to optimize over $n-1$ parameters for every point in a univariate likelihood profile. If you want to look at the bivariate confidence limits of any two parameters, you can't just compute a likelihood surface. To compute a 2D likelihood profile, the analogue of the 1D profiles we calculated previously, you would have to take every combination of the two parameters you're interested in (e.g., a 50×50 grid of parameter values) and maximize with respect to all the other $n-2$ parameters for *every point* on that surface, and then use the values you've calculated to draw contours. Especially when the likelihood function itself is hard to calculate, this procedure can be extremely tedious.

A powerful, general, but approximate shortcut is to examine the second derivatives of the log-likelihood as a function of the parameters. The second derivatives provide information about the curvature of the surface, which tells us how rapidly the log-likelihood gets worse, which in turn allows us to estimate the confidence intervals. This procedure involves a second level of approximation (like the LRT, becoming more accurate as the number of data points increases), but it can be useful when you run into numerical difficulties calculating the profile confidence limits, or when you want to compute bivariate confidence regions for complex models or more generally explore correlations in high-dimensional parameter spaces.

To motivate this procedure, let's briefly go back to a one-dimensional normal distribution and compute an analytical expression for the profile confidence limits. The likelihood of a set of independent samples from a normal distribution is $\mathcal{L} = \prod_{i=1}^{n} \frac{1}{\sqrt{2\pi}\sigma} \exp\left(-(x_i - \mu)^2/(2\sigma^2)\right)$. That means the negative log-likelihood as a function of μ (assuming we know σ) is

$$-L(\mu) = C + n \log \sigma + \sum_i \left(\frac{(x_i - \mu)^2}{2\sigma^2} \right), \qquad (6.5.1)$$

where we've lumped the parameter-independent parts of the likelihood into the constant C. We could differentiate this expression with respect to μ and solve for μ when the derivative is zero to show that $\hat{\mu} = \sum x_i/n$. We could then substitute $\mu = \hat{\mu}$ into (6.5.1) to find the minimum negative log-likelihood. Once we have done this we want to calculate the width of the profile confidence interval c—that is, we want to find the value of c such that

$$-L(\hat{\mu} \pm c) = -L(\hat{\mu}) + \chi_1^2(1 - \alpha)/2. \qquad (6.5.2)$$

Some slightly nasty algebra leads to

$$c = \sqrt{\chi_1^2(1 - \alpha)} \cdot \frac{\sigma}{\sqrt{n}}. \qquad (6.5.3)$$

This expression might look familiar: we've just rederived the expression for the confidence limits of the mean! The term $\sigma/\sqrt{n}$ is the standard error of the mean; it turns out that the term $\sqrt{\chi_1^2(1 - \alpha)}$ is the same as the $(1 - \alpha)/2$ quantile for the normal distribution.* The test uses the quantile of a normal distribution, rather than a Student t distribution, because we have assumed the variance is known.

How does this relate to the second derivative? For the normal distribution, the second derivative of the negative log-likelihood with respect to μ is

$$D_2 = \frac{d^2 \left(\sum (x_i - \mu)^2/(2\sigma^2) \right)}{d\mu^2} = \frac{n}{\sigma^2}. \qquad (6.5.4)$$

---

* Try `sqrt(qchisq(0.95,1))` and `qnorm(0.975)` in R to test this idea; use 0.975 instead of 0.95 in the second expression because this procedure involves a two-tailed test on the normal distribution but a one-tailed test on the χ^2 distribution, because χ^2 is the distribution of a *squared* normal deviate.

So we can rewrite the term $\sigma/\sqrt{n}$ in (6.5.3) as $\sqrt{1/D_2}$; the standard deviation of the parameter, which determines the width of the confidence interval, is proportional to the square root of the reciprocal of the curvature (i.e., the second derivative).

While we have derived these conclusions for the normal distribution, they're true for any model *if* the data set is large enough. In general, for a one-parameter model with parameter p, the width of our confidence region is

$$N(\alpha) \left(\frac{d^2(-\log \mathcal{L})}{dp^2} \right)^{-1/2}, \tag{6.5.5}$$

where $N(\alpha)$ is the appropriate quantile for the standard normal distribution. This equation gives us a general recipe for finding the confidence region without doing any extra computation, if we know the second derivative of the negative log-likelihood at the maximum likelihood estimate. We can find that second derivative either by calculating it analytically or, when this is too difficult, by calculating it numerically by *finite differences*. Extending the general rule that the derivative $df(p)/dp$ is approximately $(f(p+\Delta p) - f(p))/\Delta p$:

$$\left. \frac{d^2 f}{dp^2} \right|_{p=\hat{p}} \approx \frac{f(\hat{p}+\Delta p) - 2f(\hat{p}) + f(\hat{p}-\Delta p)}{(\Delta p)^2}. \tag{6.5.6}$$

The `hessian=TRUE` option in `optim` tells R to calculate the second derivative in this way; this option is set automatically in `mle2`.

The same idea works for multiparameter models, but we have to know a little bit more about second derivatives to understand it. A multiparameter likelihood surface has more than one second partial derivative; in fact, it has a matrix of second partial derivatives, called the *Hessian*. When calculated for a likelihood surface, the negative of the expected value of the Hessian is called the *Fisher information*; when evaluated at the maximum likelihood estimate, it is the *observed information* matrix. The second partial derivatives with respect to the same variable twice (e.g., $\partial^2 L/\partial\mu^2$) represent the curvature of the likelihood surface along a particular axis, while the *cross-derivatives*, e.g., $\partial^2 L/(\partial\mu\partial\sigma)$, describe how the slope in one direction changes as you move along another direction. For example, for the negative log-likelihood $-L$ of the normal distribution with parameters μ and σ, the Hessian is

$$\begin{pmatrix} \frac{\partial^2 -L}{\partial\mu^2} & \frac{\partial^2 -L}{\partial\mu\partial\sigma} \\ \frac{\partial^2 -L}{\partial\mu\partial\sigma} & \frac{\partial^2 -L}{\partial\sigma^2} \end{pmatrix}. \tag{6.5.7}$$

In the simplest case of a one-parameter model, the Hessian reduces to a single number $(d^2 - L/dp^2)$, the curvature of the likelihood curve at the MLE, and the estimated standard deviation of the parameter is just $(\partial^2 - L/\partial p^2)^{-1/2}$, as above.

In some simple two-parameter models such as the normal distribution the parameters are uncorrelated, and the matrix is diagonal:

$$\begin{pmatrix} \frac{\partial^2 -L}{\partial\mu^2} & 0 \\ 0 & \frac{\partial^2 -L}{\partial\sigma^2} \end{pmatrix}. \tag{6.5.8}$$

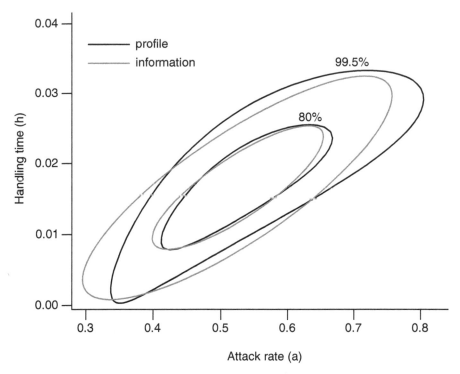

Figure 6.13 Likelihood ratio and information-matrix confidence limits on the tadpole predation model parameters.

The off-diagonal zeros mean that the slope of the surface in one direction doesn't change as you move in the other direction, and hence the shapes of the likelihood surface in the μ direction and the σ direction are unrelated. In this case we can compute the standard deviations of each parameter independently—they're the inverse square roots of the second partial derivative with respect to each parameter (i.e., $(\partial^2 - L/\partial\mu^2)^{-1/2}$ and $(\partial^2 - L/\partial\sigma^2)^{-1/2}$).

In general, when the off-diagonal elements are different from zero, we have to invert the matrix numerically, which we can do with solve. For a two-parameter model with parameters a and b we obtain the variance-covariance matrix

$$\mathbf{V} = \begin{pmatrix} \sigma_a^2 & \sigma_{ab} \\ \sigma_{ab} & \sigma_b^2 \end{pmatrix}, \qquad (6.5.9)$$

where σ_a^2 and σ_b^2 are the variances of a and b and σ_{ab} is the covariance between them; the correlation between the parameters is $\sigma_{ab}/(\sigma_a\sigma_b)$.

The approximate 80% and 99.5% confidence ellipses calculated in this way are reasonably close to the more accurate profile confidence regions for the tadpole predation data set. The profile region is slightly skewed—it includes more points where d and r are both larger than the maximum likelihood estimate, and fewer where both are smaller—while the approximate ellipse is symmetric around the maximum likelihood estimate.

This method extends to more than two parameters, although it is difficult to draw the analogous pictures in multiple dimensions. The information matrix of a p-parameter model is a $p \times p$ matrix. Using `solve` to invert the information matrix gives the variance-covariance matrix

$$\mathbf{V} = \begin{pmatrix} \sigma_1^2 & \sigma_{12} & \cdots & \sigma_{1p} \\ \sigma_{21} & \sigma_2^2 & \cdots & \sigma_{2p} \\ \vdots & \vdots & \ddots & \vdots \\ \sigma_{p1} & \sigma_{p2} & \cdots & \sigma_p^2 \end{pmatrix}, \qquad (6.5.10)$$

where σ_i^2 is the estimated variance of variable i and $\sigma_{ij} = \sigma_{ji}$ is the estimated covariance between variables i and j: the correlation between variables i and j is $\sigma_{ij}/(\sigma_i \sigma_j)$. For an `mle2` fit `m`, `vcov(m)` will give the approximate variance-covariance matrix computed in this way and `cov2cor(vcov(m))` will scale the variance-covariance matrix by the variances to give a correlation matrix with entries of 1 on the diagonal and parameter correlations as the off-diagonal elements.

The shape of the likelihood surface contains essentially all of the information about the model fit and its uncertainty. For example, a large curvature or steep slope in one direction corresponds to high precision for the estimate of that parameter or combination of parameters. If the curvature is different in different directions (leading to ellipses that are longer in one direction than another), then the data provide unequal amounts of precision for the different estimates. If the contours are oriented vertically or horizontally, then the estimates of the parameters are independent, but if they are diagonal, then the parameter estimates are correlated. If the contours are roughly elliptical (at least near the MLE), then the surface can be described by a quadratic function.

These characteristics also help determine which methods and approximations will work well (Figure 6.14). If the parameters are uncorrelated (i.e., the contours are oriented horizontally and vertically), then you can estimate them separately and still get the correct confidence intervals: the likelihood slice is the same as the profile (Figure 6.14a). If they are correlated, on the other hand, you will need to calculate a profile or invert the information matrix to allow for variation in the other parameters (Figure 6.14b). If the likelihood contours are elliptical—which happens when the likelihood surface has a quadratic shape—the information matrix approximation will work well (Figure 6.14a, b); otherwise, you must use a full profile likelihood to calculate the confidence intervals accurately (Figure 6.14c, d).

You should usually handle nonquadratic and correlated surfaces by computing profile confidence limits, but in extreme cases these characteristics may cause problems for fitting (Chapter 7) and you will have to fall back on the less-accurate quadratic approximations. All other things being equal, smaller confidence regions (i.e., for larger and less noisy data sets and for higher α levels) are more elliptical. Reparameterizing functions can sometimes make the likelihood surface closer to quadratic and decrease correlation between the parameters. For example, one might fit the asymptote and half-maximum of a Michaelis-Menten function rather than the asymptote and initial slope, or fit log-transformed parameters.

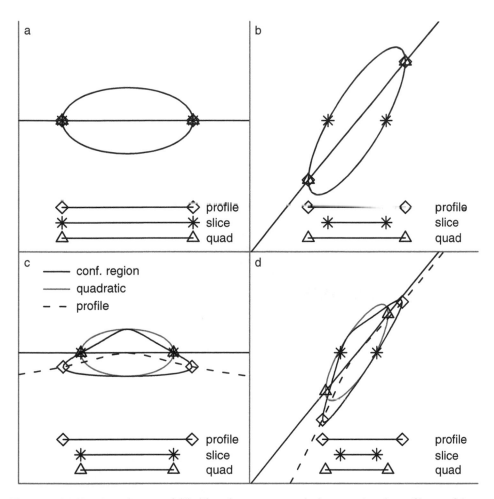

Figure 6.14 Varying shapes of likelihood contours and the associated profile confidence intervals, approximate information matrix (quadratic) confidence intervals, and slice intervals. (a) Quadratic contours, no correlation. (b) Quadratic contours, positive correlation. (c) Nonquadratic contours, no correlation. (d) Nonquadratic contours, positive correlation.

6.6 Comparing Models

The last topic for this chapter, a controversial and important one, is *model comparison* or *model selection*. Model comparison and selection are closely related to the techniques for estimating confidence regions that we just covered.

Dodd and Silvertown did a series of studies on fir (*Abies balsamea*) in New York state, exploring the relationships among growth, size, age, competition, and number of cones produced in a given year (Silvertown and Dodd, 1999; Dodd and Silvertown, 2000); see ?Fir in the emdbook package. Figure 6.15 shows the relationship between size (diameter at breast height, DBH) and the total fecundity over the study period, contrasting populations that have experienced wavelike die-offs ("wave") with those

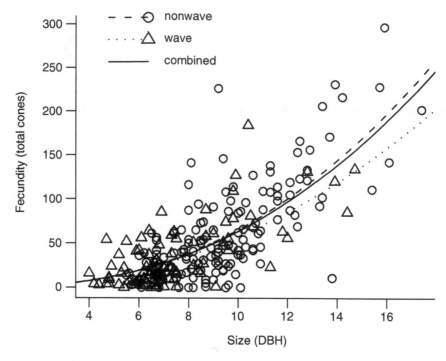

Figure 6.15 Fir fecundity as a function of DBH for wave and nonwave populations. Lines show estimates of the model $y = a \cdot \mathrm{DBH}^b$ fitted to the populations separately and combined.

that have not ("nonwave"). A power-law (*allometric*) dependence of expected fecundity on size allows for increasing fecundity with size while preventing the fecundity from being negative for any parameter values. It also agrees with the general observation in morphometrics that many traits increase as a power function of size. A negative binomial distribution in size around the expected fecundity describes discrete count data with potentially high variance. The resulting model is

$$
\begin{aligned}
\mu &= a \cdot \mathrm{DBH}^b \\
Y &\sim \mathrm{NegBin}(\mu, k).
\end{aligned}
\tag{6.6.1}
$$

We might ask any of these biological/statistical questions:

- Does fir fecundity (total number of cones) change (increase) with size (DBH)?
- Do the confidence intervals (credible intervals) of the allometric parameter b include zero (no change)? Do they include one (*isometry*)?
- Is the allometric parameter b significantly different from (greater than) zero? One?
- Does a model incorporating the allometric parameter fit the data significantly better than a model without an allometric parameter, or equivalently where the allometric parameter is set to zero ($\mu = a$) or one ($\mu = a \cdot \mathrm{DBH}$)?
- What is the best model to explain, or predict, fir fecundity? Does it include DBH?

Figure 6.15 shows very clearly that fecundity does increase with size. While we might want to know *how much* it increases (based on the estimation and confidence-limits procedures discussed above), any statistical test of the null hypothesis $b = 0$ would be *pro forma*. More interesting questions in this case ask whether and how the size-fecundity curve differs in wave and nonwave populations. We can extend the model to allow for differences between the two populations:

$$\mu = a_i \cdot \text{DBH}^{b_i}$$
$$Y_i \sim \text{NegBin}(\mu, k_i)$$

(6.6.2)

where the subscripts i denote different populations—wave ($i = w$) or nonwave ($i = n$).

Now our questions become:

- Is baseline fecundity the same for small trees in both populations? (Can we reject the null hypothesis $a_n = a_w$? Do the confidence intervals of $a_n - a_w$ include zero? Does a model with $a_n \neq a_w$ fit significantly better?)
- Does fecundity increase with DBH at the same rate in both populations? (Can we reject the null hypothesis $b_n = b_w$? Do the confidence intervals of $b_n - b_w$ include zero? Does a model with $b_n \neq b_w$ fit significantly better?)
- Is variability around the mean the same in both populations? (Can we reject the null hypothesis $k_n = k_w$? Do the confidence intervals of $k_n - k_w$ include zero? Does a model with $k_n \neq k_w$ fit significantly better?)

We can boil any of these questions down to the same basic statistical question: for any one of a, b, and k, does a simpler model (with a single parameter for both populations rather than separate parameters for each population) fit adequately? Does adding extra parameters improve the fit sufficiently to justify the additional complexity?

As we will see, these questions one can translate into statistical hypotheses and tests in many ways. While there are stark differences in the assumptions and philosophy behind different statistical approaches, and hot debate over which ones are best, it's worth remembering that in many cases they will all give reasonably consistent answers to the underlying ecological questions. The rest of this introductory section explores some general ideas about model selection. The following sections describe the basics of different approaches, and the final section summarizes the pros and cons of various approaches.

If we ask "does fecundity change with size?" or "do two populations differ?" we know as ecologists that the answer is "yes"—*every* ecological factor has some impact, and all populations differ in some way. The real questions are, given the data we have, whether we can tell what the differences are, and how we decide which model best explains the data or predicts new results.

Parsimony (sometimes called "Occam's razor") is a general argument for choosing simpler models even though we know the world is complex. All other things being equal, we should prefer a simpler model to a more complex one—especially when the data don't tell a clear story. Model selection approaches typically go beyond parsimony to say that a more complex model must be not just better than, but a specified amount better than, a simpler model. If the more complex model doesn't exceed a threshold of improvement in fit (we will see below exactly where this threshold comes from), we typically reject it in favor of the simpler model.

Model complexity also affects our predictive ability. Walters and Ludwig (1981) simulated fish population dynamics using a complex age-structured model and showed that when data were realistically sparse and noisy they could best predict future (simulated) dynamics using a simpler non-age-structured model. In other words, even though they knew for sure that juveniles and adults had different mortality rates (because they simulated the data from a model with mortality differences), a model that ignored this distinction gave more accurate predictions. This apparent paradox is an example of the *bias-variance trade-off* introduced in Chapter 5. As we add more parameters to a model, we necessarily get an increasingly accurate fit to the particular data we have observed (the bias of our predictions decreases), but our precision for predicting future observations decreases as well (the variance of our predictions increases). Data contain a fixed amount of information; as we estimate more and more parameters we spread the data thinner and thinner. Eventually the gain in accuracy from having more details in the model is outweighed by the loss in precision from estimating the effect of each of those details more poorly. In Ludwig and Walters's case, spreading the data out across age classes meant there was not enough data to estimate each age class's dynamics accurately.

Figure 6.16 shows two sets of simulated data generated from a generalized Ricker model, $Y \sim \text{Normal}((a + bx + cx^2)e^{-dx})$. I fitted the first data set with a constant model (y equal to the mean of data), a Ricker model ($y = ae^{-bx}$), and the generalized Ricker model. Despite being the true model that generated the data, the generalized Ricker model is overly flexible and adjusts the fit to go through an unusual point at (1.5,0.24). It fits the first data set better than the Ricker ($R^2 = 0.55$ for the generalized Ricker vs. $R^2 = 0.29$ for the Ricker). However, the generalized Ricker has *overfitted* these data. It does poorly when we try to predict a second data set generated from the same underlying model. In the new set of data shown in Figure 6.16, the generalized Ricker fit misses the point near $x = 1.5$ so badly that it actually fits the data worse than the constant model and has a negative R^2! In 500 new simulations, the Ricker prediction was closest to the data 83% of the time, while the generalized Ricker prediction won only 11% of the time; the other 6% of the time, the constant model was best.

6.6.1 Likelihood Ratio Test: Nested Models

How can we tell when we are overfitting real data? We can use the Likelihood Ratio Test, which we used before to find confidence intervals and regions, to choose models in certain cases. A simpler model (with fewer parameters) is *nested* in another, more complex, model (with more parameters) if the complex model reduces to the simpler model by setting some parameters to particular values (often zero). For example, a constant model, $y = a$, is nested in the linear model, $y = a + bx$ because setting $b = 0$ makes the linear model constant. The linear model is nested in turn in the quadratic model, $y = a + bx + cx^2$. The linear model is also nested in the Beverton-Holt model, $y = ax/(1 + (a/b)x)$, for $b \to \infty$. The Beverton-Holt is in turn nested

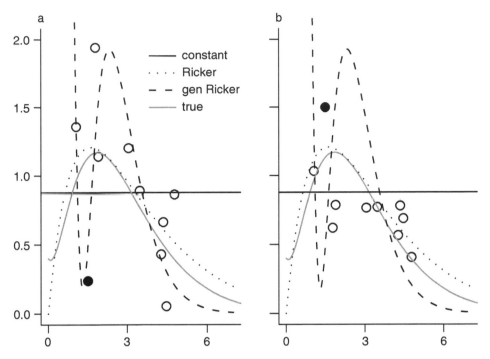

Figure 6.16 Fits to simulated "data" generated with $y = (0.4 + 0.1 \cdot x + 2 \cdot x^2)e^{-x}$, plus normal error with $\sigma = 0.35$. Models fitted: constant ($y = \bar{x}$), Ricker ($y = ae^{-bx}$), and generalized Ricker ($y = (a + bx + cx^2)e^{-dx}$). The highlighted point at $x \approx 1.5$ drives much of the fit to the original data, and much of the failure to fit new data sets. (a) Data set 1. (b) Data set 2.

in the Shepherd model, $y = ax/(1 + (a/b)x^d)$, for $d = 1$. (The nesting of the linear model in the Beverton-Holt model is clearer if we use the parameterization of the Holling type II model, $y = ax/(1 + ahx)$. The handling time h is equivalent to $1/b$ in the Beverton-Holt. When $h = 0$ predators handle prey instantaneously and their per capita consumption rate increases linearly forever as prey densities increase.)

Comparisons among different groups can also be framed as a comparison of nested models. If the more complex model has the mean of group 1 equal to a_1 and the mean of group 2 equal to a_2, then the nested model (both groups equivalent) applies when $a_1 = a_2$. It is also common to parameterize this model as $a_2 = a_1 + \delta_{12}$, where $\delta_{12} = a_2 - a_1$, so that the simpler model applies when $\delta_{12} = 0$. This parameterization works better for model comparisons since testing the hypothesis that the more complex model is better becomes a test of the value of one parameter ($\delta_{12} = 0$?) rather than a test of the relationship between two parameters ($a_1 = a_2$?).*

---

* We can also interpret these parameterizations geometrically. In (a_1, a_2) parameter space, we're testing to see whether the best fit falls on the line through the origin $a_1 = a_2$; in (a_1, δ_{12}) parameter space, we're testing whether the best fit lies on the line $\delta_{12} = 0$. To explore further how different parameterizations relate to testing different hypotheses, look for the topic of *contrasts* in statistical models (Crawley, 2002; Venables and Ripley, 2002).

To prepare to ask these questions with the fir data, we read in the data, drop NAs, and pull out the variables we want. The fecundity data are not always integers, but a negative binomial model requires integer responses so we round the data.

```
> data(FirDBHFec)
> X = na.omit(FirDBHFec[, c("TOTCONES", "DBH", "WAVE_NON")])
> X$TOTCONES = round(X$TOTCONES)
```

Using `mle2`'s formula interface is the easiest way to estimate the nested series of models in R. The reduced model (no variation among populations) is

```
> nbfit.0 = mle2(TOTCONES ~ dnbinom(mu = a * DBH^b,
+     size = k),
+     start = list(a = 1, b = 1, k = 1), data = X)
```

To fit more complex models, use the `parameters` argument to specify which parameters differ among groups. For example, the argument `list(a~WAVE_NON, b~WAVE_NON)` would allow a and b to have different values for wave and nonwave populations, corresponding to the hypothesis that the populations differ in both a and b but not in variability ($a_w \neq a_n$, $b_w \neq b_n$, $k_w = k_n$). The statistical model is $Y_i \sim \text{NegBin}(a_i \cdot \text{DBH}^{b_i}, k)$, and the R code is

```
> start.ab = as.list(coef(nbfit.0))
> nbfit.ab = mle2(TOTCONES ~ dnbinom(mu = a * DBH^b,
+     size = k),
+     start = start.ab, data = X,
+     parameters = list(a  ~ WAVE_NON, b ~ WAVE_NON))
```

Here I have used the best-fit parameters of the simpler model as starting parameters for the complex model. Using the best available starting parameters avoids many optimization problems.

`mle2`'s formula interface automatically expands the starting parameter list (which includes only a single value for each of a and b) to include the appropriate number of parameters. `mle2` uses default starting parameter values corresponding to equality of all groups, which for this parameterization means that all of the additional parameters for groups other than the first are set to zero.

The formula interface is convenient, but as with likelihood profiles you often encounter situations where you have to know how to do things by hand. Here's an explicit negative log-likelihood model for the model with differences in a and b between groups (we attach the data first for simplicity—don't forget to detach it later):

```
> attach(X)
> nbNLL.ab = function(a.w, b.w, a.n, b.n, k) {
+     wcode = as.numeric(WAVE_NON)
+     a = c(a.n, a.w)[wcode]
+     b = c(b.n, b.w)[wcode]
+     predcones = a * DBH^b
+     -sum(dnbinom(TOTCONES, mu = predcones, size = k,
+         log = TRUE))
+ }
```

The first three lines of nbNLL.ab turn the factor WAVE_NON into a numeric code (1 or 2) and use the resulting code as an index to decide which value of a or b to use in predicting the value for each individual. To make k differ by group as well, just change k in the argument list to k.n and k.w and add the line

```
> k = c(k.n, k.w)[wcode]
```

To simplify the model by making a or b homogeneous, reduce the argument list and eliminate the line of code that specifies the value of the parameter by group.

The only difference between this negative log-likelihood function and the one that mle2 constructs when you use the formula interface is that the mle2-constructed function uses the parameterization $\{a_1, a_1 + \delta_{12}\}$, whereas our hand-coded function uses $\{a_1, a_2\}$ (see p. 205). The former is more convenient for statistical tests, while the latter is more convenient if you want to know the parameter values for each group. To tell mle2 to use the latter parameterization, specify parameters=list(a~ WAVE_NON-1, b~WAVE_NON-1). The -1 tells mle2 to fit the model without an intercept, which in this case means that the parameters for each group are specified relative to 0 rather than relative to the parameter value for the first group. When mle2 fills in default starting values for this parameterization, it sets the starting parameters for all groups equal.

The anova function* performs likelihood ratio tests on a series of nested mle2 fits, automatically calculating the difference in numbers of parameters (denoted by Df for "degrees of freedom") and deviance and calculating p values:

```
> anova(nbfit.0, nbfit.a, nbfit.ab)
```

```
Likelihood Ratio Tests
Model 1: nbfit.0, TOTCONES~dnbinom(mu=a*DBH^b,size=k)
Model 2: nbfit.a, TOTCONES~dnbinom(mu=a*DBH^b,size=k):
          a~WAVE_NON
Model 3: nbfit.ab, TOTCONES~dnbinom(mu=a*DBH^b,size=k):
          a~WAVE_NON, b~WAVE_NON
```

Tot	Df	Deviance	Chisq	Df	Pr(>Chisq)
1	3	2272.0			
2	4	2271.6	0.4276	1	0.5132
3	5	2271.3	0.2496	1	0.6173

The Likelihood Ratio test can compare any two nested models, testing whether the nesting parameters of the more complex model differ significantly from their null values. Put another way, the LRT tests whether the extra goodness of fit to the data is worth the added complexity of the additional parameters. To use the LRT to compare models, compare the difference in deviances (the more complex model should always have a smaller deviance—if not, check for problems with the optimization) to the critical value of the χ^2 distribution, with degrees of freedom equal to the additional number of parameters in the more complex model. If the difference in deviances is greater than $\chi^2_{n_2-n_1}(1-\alpha)$, then the more complex model is significantly better at the $p = \alpha$ level. If not, then the additional complexity is not justified.

---

* Why anova? The corresponding series of tests for a simple linear model with categorical predictors is an analysis of variance (Chapter 9).

Choosing among statistical distributions can often be reduced to comparing among nested models. As a reminder, Figure 4.17 (p. 137) shows some of the relationships among common distributions. The most common use of the LRT in this context is to see whether we need to use an overdispersed distribution such as the negative binomial or beta-binomial instead of their lower-variance counterparts (Poisson or binomial). The Poisson distribution is nested in the negative binomial distribution when $k \to \infty$. If we fit a model with a and b varying but using a Poisson distribution instead of a negative binomial, we can then use the LRT to see if adding the overdispersion parameter is justified:

```
> poisfit.ab = mle2(TOTCONES ~ dpois(a * DBH^b),
+      start = list(a = 1, b = 1), data = X,
+      parameters = list(a ~ WAVE_NON, b ~ WAVE_NON))
> anova(poisfit.ab, nbfit.ab)
```

```
Likelihood Ratio Tests
Model 1: poisfit.ab, TOTCONES~dpois(a*DBH^b): a~WAVE_NON,
         b~WAVE_NON
Model 2: nbfit.ab, TOTCONES~dnbinom(mu=a*DBH^b,size=k):
         a~WAVE_NON, b~WAVE_NON
Tot  Df  Deviance  Chisq  Df  Pr(>Chisq)
1    4    6302.7
2    5    2271.4   4031.4  1   < 2.2e-16  ***
---
Signif. codes:  0 '***' 0.001 '**' 0.01 '*' 0.05 '.' 0.1 ' ' 1
```

We conclude that negative binomial is clearly justified: the difference in deviance is greater than 4000, compared to a critical value of 3.84! This analysis ignores the nonapplicability of the LRT on the boundary of the allowable parameter space ($k \to \infty$ or $1/k = 0$; see p. 250), but the evidence is so overwhelming in this case that it doesn't matter.

Models with multiple parameters and multiple groups naturally lead to a web of nested models. Figure 6.17 shows all of the model comparisons for the fir data—even for this relatively simple example there are seven possible models and nine possible series of nested comparisons. In this case the answer is easy, because none of the comparisons is significant according to the LRT (i.e., none of the one-step comparisons differ by more than 3.84). In more complex scenarios deciding which set of comparisons to do first can be quite hard. Two simple options are forward selection (try to add parameters one at a time to the simplest model) and backward selection (try to subtract parameters from the most complex model). Either of these approaches will work, but for comparisons that are close to the edge of statistical significance, or where the effects of the parameters are strongly correlated, you'll often find that you get different answers. Similar problems arise in multiple regression (in fact, in any complex modeling exercise). With too large a set of possibilities, this kind of model selection quickly devolves into data-dredging. You should (1) use common sense and ecological knowledge to isolate the most important comparisons; (2) draw plots of the best candidate fits to try to understand why different models fit the data approximately equally well; and (3) try to rule out differences in variance

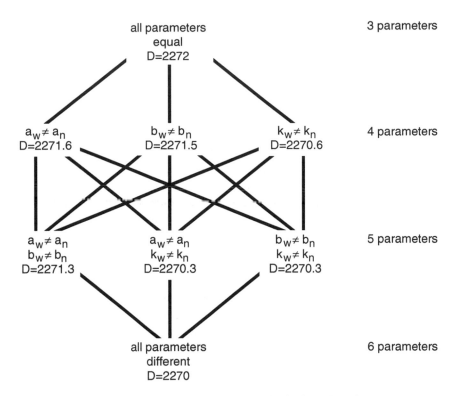

Figure 6.17 Nested hierarchy of models for the fir data. D = deviance.

parameters (k in this case) first. If you can simplify the model in this way it will be more comparable with classical models. If not, something interesting may be happening.

6.6.2 Information Criteria

One way to avoid a plethora of pairwise model comparisons is to select models based on *information criteria*, which compare all candidate models at once and do not require nested alternatives. These relatively recent alternatives to Likelihood Ratio Tests are based on the expected distance (quantified in a way that comes from information theory) between a particular model and the "true" model (Burnham and Anderson, 1998, 2002). In practice, all information-theoretic methods reduce to finding the model that minimizes some criterion that is the sum of a term based on the likelihood (usually twice the negative log-likelihood) and a *penalty term* which is different for different information criteria.

The *Akaike Information Criterion*, or AIC, is the most widespread information criterion, and is defined as

$$AIC = -2L + 2k \qquad (6.6.3)$$

where L is the log-likelihood and k is the number of parameters in the model.* As with all information criteria, small values represent better overall fits; adding a parameter with a negligible improvement in fit penalizes the AIC by 2 log-likelihood units. For small sample sizes (n)—such as when $n/k < 40$ (Burnham and Anderson, 2004, p. 66)—you should use a finite-size correction and apply the AIC_c ("corrected AIC") instead:

$$\text{AIC}_c = \text{AIC} + \frac{2k(k+1)}{n-k-1}. \tag{6.6.4}$$

As n grows large, the correction term in (6.6.4) vanishes and the AIC_c matches the AIC. The AIC_c was originally derived on the basis of linear models with normally distributed errors, so it may apply to a smaller range of models than the AIC—but this is really an open question. Shono (2000) found using simulation studies that the AIC_c gave accurate answers for typical fisheries data sets, but Richards (2005) suggests that AIC_c might not perform as well for other kinds of ecological data sets. (I would recommend using AIC_c for small samples, but being careful with the results if they disagree with the results based on large-sample AIC.)

The second most common information criterion, the *Schwarz* or *Bayesian* information criterion (BIC),[†] uses a penalty term of $(\log n)k$. When n is greater than $e^2 \approx 8$ observations (so that $\log n > 2$), the BIC is more conservative than the AIC, insisting on a greater improvement in fit before it will accept a more complex model.

Information criteria do not allow frequentist significance tests based on the estimated probability of getting more extreme results in repeated experiments (some statisticians would say this is an advantage). With ICs, you cannot say that there is a statistically significant difference between models; a model with a lower IC is better, but there is no p-value associated with how much better it is.[‡] Instead, there are commonly used rules of thumb: models with ICs less than 2 apart ($\Delta \text{IC} < 2$) are more or less equivalent; those with ICs 4–7 apart are clearly distinguishable; and models with ICs more than 10 apart are definitely different. Richards (2005) concurs with these recommendations, but cautions that simply dropping models with $\Delta \text{AIC} > 2$ (as some ecologists do) risks discarding useful models.

---

* Where does the magic penalty term $2k$ come from? AIC is the expected value of the *Kullback-Leibler distance* $\int f(\mathbf{x}) \log (f(\mathbf{x})/g(\mathbf{x})) \, d\mathbf{x}$: in words, the $K - L$ distance is the log ratio between $f(\mathbf{x})$, the likelihood of the true model and parameters, and $g(\mathbf{x}_0)$, the likelihood of a candidate model evaluated at its best parameters, averaged over the true distribution of the data. Separating terms and dropping constants that don't involve $g(\mathbf{x}_0)$, we get $E[-\log g(\mathbf{x}_0)]$. We don't know the true MLE $\mathbf{x}_0$, only the *observed* MLE $\hat{\mathbf{x}}$, so we take another expectation: $E[E[-\log g(\hat{\mathbf{x}})]]$. Taylor expanding $-\log g(\hat{\mathbf{x}})$ around $\mathbf{x}_0$, the expectation of the linear term drops out, leaving the constant and quadratic terms: $E[E[-\log g(\hat{\mathbf{x}}) - \frac{1}{2}(\mathbf{x} - \hat{\mathbf{x}})^T V(\mathbf{x} - \hat{\mathbf{x}})]]$. V is the matrix of second derivatives of the log-likelihood (the information matrix): $-V^{-1} \approx \mathbf{\Sigma}$, the variance-covariance matrix of the parameters. By definition, $E[(\mathbf{x} - \hat{\mathbf{x}})^T(\mathbf{x} - \hat{\mathbf{x}})]$ also equals $\mathbf{\Sigma}$. After more math, we get $-\log g(\hat{\mathbf{x}}) + \text{trace}(\mathbf{\Sigma}^{-1}\mathbf{\Sigma})$, where the *trace* is the sum of the diagonal elements of a matrix. Since a matrix times its inverse is the identity matrix, this becomes $-\log g(\hat{\mathbf{x}}) + k$, where k is the size of the matrix—which equals the number of parameters. Doubling this expression so that the first term is the deviance ($-2L$) gives $-2L + 2k$. For more information, see Ripley (2004) or Chapter 7 of Burnham and Anderson (2002).

† While the BIC is derived from a Bayesian argument, it is not inherently a Bayesian technique. It is also not how most Bayesians would compare models (Section 6.6.3).

‡ Burnham and Anderson (2002, p. 84) recommend avoiding the word "significant" in conjunction with AIC-based model selection; no matter how carefully you phrase your conclusions, some readers will impose a frequentist hypothesis-testing interpretation.

TABLE 6.2
Results of fir model fits

Model	k	ΔAIC	ΔAIC$_c$	ΔBIC
nbfit.0	3	0.00	0.00	0.00
nbfit.a	4	1.57	1.64	5.06
nbfit.b	4	1.48	1.55	4.97
nbfit.k	4	0.62	0.69	4.11
nbfit.ab	5	3.32	3.48	10.30
nbfit.ak	5	2.24	2.39	9.21
nbfit.bk	5	2.24	2.39	9.21
nbfit.abk	6	3.99	4.25	14.46

One big advantage of IC-based approaches is that they do not require nested models.* You can compare all models to each other simultaneously, rather than stepping through a sometimes confusing sequence of pairwise tests. In IC-based approaches, you simply compute the likelihood and IC for all of the candidate models and rank them in order of increasing IC. The model with the lowest IC is the best fit to the data; those models with ICs within 10 units of the minimum IC are worth considering. As with the LRT, the absolute size of the ICs is unimportant—only the differences in ICs matter.

The `AICtab`, `AICctab`, and `BICtab` commands in the bbmle package will compute IC tables from lists of `mle` fits. Use the options `delta=TRUE` to get a list of the ΔIC values, `weights=TRUE` to get AIC weights (see below), `sort=TRUE` to sort models in order of increasing IC, and `nobs` to specify the number of observations for BIC or AIC$_c$. Table 6.2 gives the results for the fir models. All three approaches pick the simplest model as the best model (minimum IC). AIC would keep all models under consideration (ΔAIC < 4 for all models), while AIC$_c$ might rule out the most complex model (ΔAIC$_c$ = 4.25), and BIC would definitely rule out complex models where a and b both change (ΔBIC > 10).

ICs can also be useful to choose among stochastic models, which are often not nested. For example, the Gamma, log-normal, and negative binomial models can all describe skewed data, and they all converge to the normal distribution in some limit (Figure 4.17), but there is no easy way to nest them. We can fit the same deterministic model as before (fecundity $= a_i \cdot \text{DBH}_i^b$) with different probability distributions and then use AIC to compare the results.

For each distribution I have to modify the parameters slightly. The lognormal's parameters are the mean and standard deviation of the distribution on the log scale, so I set $\mu_{\log} = \log(a \cdot \text{DBH}^b) = \log a + b \log \text{DBH}$. The Gamma's are shape and scale, with the mean equal to shape · scale, so I set scale $= (a \cdot \text{DBH}^b)/\text{shape}$. I also added 0.001 to `TOTCONES` for the lognormal and Gamma fits because zero values are impossible for the lognormal distribution and for the Gamma distribution with shape > 1,

---

* Although some, such as Ripley (2004), disagree.

TABLE 6.3
Comparison of stochastic models for fir data

	AIC	df	∆AIC
Negative binomial	2281.4	5	0.0
Gamma	2288.7	5	7.4
Lognormal	2556.3	5	274.9
Poisson	6310.7	4	4029.4

leading to infinite negative log-likelihoods. This problem warns us that a discrete distribution like the negative binomial might make more sense, but a better fit to a continuous distribution might override this concern.

```
> lnormfit.ab = mle2(TOTCONES + 0.001 ~ dlnorm(meanlog = b *
+     log(DBH) + log(a), sdlog = sdlog), start = list(a = 1,
+     b = 1, sdlog = 0.1), data = X, parameters = list(a ~
+     WAVE_NON, b ~ WAVE_NON), method = "Nelder-Mead")
> gammafit.ab = mle2(TOTCONES + 0.001 ~ dgamma(scale = a *
+     DBH^b/shape, shape = shape), start = list(a = 1,
+     b = 1, shape = 2), data = X, parameters = list(a ~
+     WAVE_NON, b ~ WAVE_NON))
```

Table 6.3 shows that the negative binomial is best after all.

6.6.3 Bayesian Analyses

Bayesians generally have little interest in formal methods of model selection. Dropping a parameter from a model is often equivalent to testing a null hypothesis that the parameter is exactly zero, and Bayesians consider such *point* null hypotheses silly. They would describe a parameter's distribution as being concentrated near zero rather than saying its value is exactly zero.[*]

Nevertheless, Bayesians do compute the relative probability of different models, in a way that implicitly recognizes the bias-variance trade-off and penalizes more complex models (Kass and Raftery, 1995). Bayesians prefer to make inferences based on averages rather than on most-likely values; for example, they generally use the posterior mean values of parameters rather than the posterior mode. This preference extends to model selection. The *marginal likelihood* of a model is the probability of observing the data (likelihood), averaged over the *prior* distribution of the parameters:

$$\bar{\mathcal{L}} = \int \mathcal{L}(x) \cdot \text{Prior}(x) \, dx, \qquad (6.6.5)$$

where x represents a parameter or set of parameters (if a set, then the integral would be a multiple integral). The marginal likelihood (the average probability of observing a particular data set *exactly*) is often very small, and we are really interested in the

---

[*] Although they might consider testing a hypothesis about whether a parameter is small (i.e., whether its absolute value is below some threshold (Gelman and Tuerlinckx, 2000)).

TABLE 6.4
Rules of thumb for Bayes factors

$2 \log B_{12}$	Evidence in Favor of Model 1
0–2	Weak
2–6	Positive
6–10	Strong
> 10	Very strong

From Jeffreys (1961, p. 432).

relative probability of different models. If two models have marginal likelihoods $\bar{\mathcal{L}}_1$ and $\bar{\mathcal{L}}_2$, the *Bayes factor* is the ratio of the marginal likelihoods, $B_{12} = \bar{\mathcal{L}}_1/\bar{\mathcal{L}}_2$, or the odds in favor of model 1.* If we want to compare several different (not necessarily nested) models, we can look at the pairwise Bayes factors or compute a set of posterior probabilities—assuming that all the models have the same prior probability—by computing the relative values of the marginal likelihoods:

$$\text{Prob}(M_i) = \frac{\bar{\mathcal{L}}_i}{\sum_{j=1}^{N} \bar{\mathcal{L}}_j}. \tag{6.6.6}$$

Marginal likelihoods and Bayes factors incorporate an implicit penalty for over-parameterization. When you add more parameters to a model, it can fit better—the maximum likelihood and the maximum posterior probability increase—but at the same time the posterior probability distribution spreads out to cover more less-well-fitting possibilities. Since marginal likelihoods express the mean and not the maximum posterior probability, they will actually decrease when the model becomes too complex.

In principle, using Bayes factors to select the better of two models is simple. If we compare twice the logarithm of the Bayes factors (thus putting them on the deviance scale), the generally accepted rules of thumb for Bayes factors are seen in Table 6.4. That these rules of thumb are similar to those quoted for the AIC is no coincidence. With fairly strong priors, the Bayes factor converges to the AIC instead of the BIC (Kass and Raftery, 1995).

In practice, computing the marginal likelihood for a particular model can be tricky (Congdon, 2003), involving either complicated multidimensional integrals or some kind of stochastic sampling from the prior distribution. One simple approximation is to calculate the *harmonic mean* of the likelihoods returned from an MCMC run (the harmonic mean is $1/(\sum(1/\mathcal{L})/n)$). Another, the analogue of the quadratic approximations to the likelihood profile described above, is the *Laplace approximation*, which combines the posterior mode (the maximum value of prior × likelihood) with information on the curvature of the posterior probability density near the mode.†

---

* The Bayes factor is based on assuming equal prior probabilities ($p_1 = p_2 = 0.5$) for both models.
† The Laplace approximation is

$$\bar{\mathcal{L}} \approx (2\pi)^{d/2} |\mathbf{V}|^{1/2} \text{Post}_{\text{max}},$$

where d is the number of parameters, $|\mathbf{V}|$ is the determinant of the variance-covariance matrix estimated from the Hessian at the posterior mode, and Post_{max} is the height of the posterior mode.

TABLE 6.5
Log Bayes factor approximations for fir models

	Harmonic Mean	Laplace	BIC
a, b, k all equal	0.0	0.0	0.0
a, b differ	5.2	8.2	10.3
a, b, k differ	24.9	9.5	14.5

Most of these approximations improve as the sample size increases: Kass and Raftery (1995) suggest that the Laplace approximation requires at least 5 times as many samples as parameters, and that the other approximations should be reasonable with 20 times as many samples as parameters. How do these approximations compare for the fir data set, with 242 data points and up to 6 parameters? Table 6.5 shows that the different approximations of the Bayes factor do differ considerably, but the only qualitative difference among them according to the rules of thumb is that the evidence supporting the null model (all parameters equal) over the model with different a and b parameters is "positive" according to the harmonic mean and "strong" according to the Laplace approximation and BIC.

A more recent criterion, conveniently built into WinBUGS, is the DIC, or *deviance information criterion*, which was designed particularly for models containing random effects where even specifying the number of parameters is challenging (see Chapter 10). To compute DIC, start by calculating $\bar{D}$, the average of the deviance over the *posterior* distribution (as contrasted with the marginal likelihood, which is the average over the prior distribution), and $\hat{D}$, which is the deviance calculated at the posterior mean parameters. Then use these two values to estimate an effective number of parameters $p_D = \bar{D} - \hat{D}$; the more spread out the posterior distribution, the bigger the difference between the deviance of the mean parameters and the mean deviance, and the larger the effective number of parameters. Finally, as with AIC and BIC, use this effective number of parameters as a penalty term on the goodness of fit (defined in this case as the deviance at the mean parameters $\hat{D}$): DIC=$\hat{D} + 2p_D$. As with all information criteria, lower values of DIC indicate a better model. The rules of thumb are similar too: differences in DIC from 5 to 10 indicate that one model is clearly better, whereas models with difference in DIC > 10 probably don't need to be considered further (Spiegelhalter et al., 2002).

Two important cautions about the DIC are:

- If the model contains random effects (see Chapter 9), the DIC focuses on the random effects. In the fir tree case, because of a peculiarity of BUGS, we had to parameterize the negative binomial model by assuming that each tree's fecundity is a Poisson variable with a different, Gamma-distributed rate. Since DIC focuses on random effects, it reports the effective number of parameters as > 200 (it takes a lot of information to describe the variation in rates), and the effective number of parameters for the most complex model is actually slightly *smaller* than for the simpler model, because the variation in the rates is slightly lower. This drop in effective model size gives the most complex model the lowest DIC. However, the range of DICs is very small—from 1709.2 to 1710.9—so the DIC is really telling us that the models can't be well distinguished.

- DIC is convenient, and so it is likely to become established as the standard "canned" method of model comparison in Bayesian statistics. It has already begun to appear in ecological journals (Jonsen et al., 2003; McCarthy and Parris, 2004; Morales et al., 2004; Okuyama and Bolker, 2005; Parris, 2006; Vesk, 2006), but statisticians continue to debate its exact meaning and appropriateness (both Spiegelhalter et al. (2002) and Celeux et al. (2006) are accompanied by lively discussions).

The bottom line on Bayesian model selection is that, despite the conceptual simplicity of the Bayes factor (giving the "average" quality of fit to the data, and automatically incorporating a penalty for overfitting), it is difficult to calculate and so is likely to be superseded by the convenient DIC. You should exercise the same care with DIC as you would with any canned model selection procedure.

6.6.4 Model Weighting and Averaging

Bayesians themselves would say that you should not simply select one model. Taking the best model and ignoring the rest is equivalent to assigning a probability of 1.0 to the best and 0.0 to the rest. *Model averaging* methods take the average of the predictions of different models, weighted by the probability of the models or by some other index.

Bayesian model averaging simply takes the probabilities based on the marginal likelihoods or the BIC: the posterior probabilities of a set of models, if they all have equal prior probabilities, are the marginal likelihoods (or BICs) divided by the sum of the marginal likelihoods (or BICs).* If a set of models have BIC values, relative to the best one, of ΔB_i (where $\Delta B_i = BIC_i - \min(BIC)$), then the approximate posterior probabilities of the models, assuming all the prior probabilities are equal, are

$$p_i = \frac{e^{-\Delta B_i/2}}{\sum_{j=1}^{n} e^{-\Delta B_j/2}}. \qquad (6.6.7)$$

To make a weighted prediction, use the posterior probabilities to combine the predictions of the different models (say $C_1, C_2, \ldots, C_n$):

$$\hat{C} = \sum_{i=1}^{n} p_i C_i. \qquad (6.6.8)$$

Of course, you can do the same with marginal likelihoods.

Burnham and Anderson (1998, 2002) have also promoted model averaging, in their case based on *AIC weights*. The AIC weights are analogous to the probabilities calculated from the relative BIC values, but with AIC values substituted for BIC

---

*Equal prior probabilities for all the models usually makes sense, although one does face some of the questions about equal priors raised in Chapter 4; for example, should all of the models incorporating differences between groups in the fir example be treated as subsets of a single model?

values in (6.6.7). AIC weights have no probability interpretation, but they can be used in model averaging.*

Even if you don't do formal model averaging, AIC or BIC weights are a useful way of getting a feel for the relative goodness-of-fit of different models.

6.6.5 Model Criticism and Goodness-of-Fit Tests

If the best model is a poor fit to the data, then *none* of the machinery of model selection and averaging makes sense. You should always check that your model gives a reasonable fit to the data. Goodness-of-fit testing may remind you of the classical Pearson chi-square statistic, adding up ((expected − observed)^2/expected) for all of your data to test whether the variance around the model predictions is greater than expected. However, the chi-square test works only for simple count data where the answers fall in discrete groups. If your data are continuous, or if you are using an overdispersed distribution such as the negative binomial, then your model contains a parameter describing the variance and the chi-square test is no longer useful.[†]

In practice, *model criticism* (a more generic term than goodness-of-fit testing) is simply common sense. Are the predictions reasonable? Are there consistent deviations from the estimates or unexplained outliers? Start with a simple graph of the predictions of the model (Figure 6.15), to see whether the deterministic component of the model works well.

Plots of predicted vs. actual data (Figure 6.18), or of the residuals (actual − predicted) from a model can sometimes be useful. You have already had to figure out how to calculate the predicted values in order to write a likelihood function. Take these values and plot them against the corresponding data points, then use abline(a=0,b=1) to add a predicted = actual line to the plot. However, while the predicted-vs.-actual plot can identify outliers, it really gives a consistency check rather than providing any new information. Ideally, the scatter around the predicted = actual line will be small—in which case the deterministic component of the model explains most of the variation in the data, so that the model is precise as well as accurate—and therefore useful for prediction. Remember, though, that a reasonable amount of unexplained variability does *not* necessarily mean that the model fits badly or is not useful; it just means it can't make precise predictions.[‡] Model criticism is

---

*Akaike weights are widely and incorrectly presented as "the probability that model *i* is the best model for the observed data, given the candidate set of models" (Mazerolle, 2004; Johnson and Omland, 2004). Burnham and Anderson (2004) are more careful: they say that the AIC weights *"are interpreted as* probabilities ..." (emphasis added), but it is clearly a slippery slope. Taking AIC weights as actual probabilities is trying to have one's cake and eat it too; the only rigorous way to compute such probabilities of models is to use Bayesian inference, with its associated complexities (Link and Barker, 2006).

[†] Much of the protocol that Burnham and Anderson (2002) have developed for working with AIC concerns testing and correcting for overdispersion—$\hat{c}$ in their notation. These overdispersion corrections are relevant only when your model uses a simple count distribution such as binomial or Poisson.

[‡] People who are familiar with classical statistical approaches would often like to compute an R^2 statistic (proportion variance explained) for a model. Unfortunately, "despite various analogs for categorical response models, no proposed measure is as widely useful as R and R^2" (Agresti, 2002, p. 390).

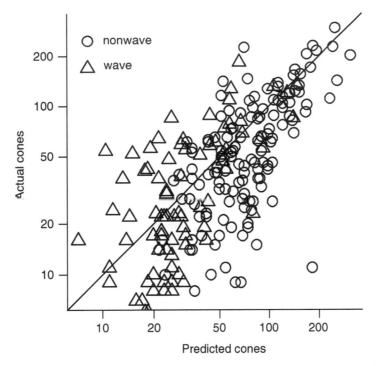

Figure 6.18 Predicted vs. actual cones for the fir data, on a logarithmic scale.

more concerned with systematic deviations that suggest that the form of the model itself is wrong.

Examining the goodness of fit of the stochastic part of a model is harder. If the model contains only discrete groups (factors), you can divide the data into those groups and overlay the observed distribution (described by a histogram or density plot) with the predicted distribution. If it contains continuous covariates, you may have to break the data up into discrete subsets in order to compare the predicted and observed distributions (Figure 6.19).

6.6.6 Model Selection: Comparisons and Conclusions

Deciding what models to use and how to use them is fundamentally difficult. In one form or another, this debate goes all the way back to the early Bayesian/frequentist divide. While statisticians have come a long way in exploring the possible approaches and (to some extent) in providing practical recipes for applying them, we still do not have—and never will have—a single best method.

- *Hypothesis testing based on the Likelihood Ratio test* is well-established, widely used, and simple to implement. At times when we really do want a yes-or-no answer about whether some ecological factor is affecting the system in a way that is distinguishable from randomness, the LRT is appropriate. The LRT becomes unwieldy when there are many possibly interacting factors—one has

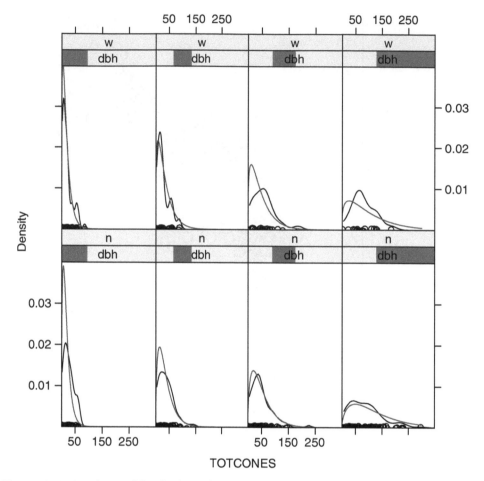

Figure 6.19 Goodness-of-fit checking for the fir model. Panels break data up by wave/ non-wave (rows) and DBH (columns) and plot the density of points for each category along with the predicted negative binomial distribution (gray) for the mean DBH value in the category.

to choose a path through the nested hierarchy of factors (Figure 6.17). Analogous problems in multiple regression analysis led to stepwise model-building approaches, which are widely used by researchers but widely dismissed by statisticians because they encourage data-dredging and because the results depend on the exact thresholds used to include or exclude factors from the model (Whittingham et al., 2006).

If you do find yourself with seemingly inconsistent results from an LRT analysis (e.g., if some parameters are significant only when other parameters are included in the model; Lindsey (1999b) calls these *incompatible* results), examine your data carefully to understand how the fit changes with different sets of parameters. If two parameters explain essentially the same patterns in the data (e.g., if you are using strongly correlated predictors like soil moisture and precipitation), then whichever enters the model first will be selected. On the other hand, the effects of nitrogen availability might be visible only after the effects of soil moisture are accounted for—in this case, nitrogen would be significant only if soil moisture were in the model already. These kinds of interactions are challenging, but handled properly they tell you more about what's going on in your data.

- *Information theoretic (AIC-based) approaches* are also well-established and practical. They neatly avoid the problem of pairwise testing, the need for nested models, and the philosophical issues associated with null hypothesis testing—rather than asking about the probability of a more extreme outcome, they simply try to identify the model with the best predictive ability. They can be used for model averaging, taking the predictions of all reasonable models into account, as well as for model testing. However, AIC-based approaches can also be abused (Guthery et al., 2005). Precisely because of their ease of use, they have led some ecologists down the path of data-dredging and thoughtless model selection (against the warnings of Burnham and Anderson, AIC's main proponents in ecology).

  AIC-based analyses make decisions based on rules of thumb about ΔAIC values or AIC weights, which are in turn based on extensive simulation analysis. The results do not have probabilistic or "statistical significance" interpretations (which may be a good thing). In some theoretical situations (i.e., when sample sizes grow large but the set of candidate models remains fixed), AIC is known to "overfit" data by choosing an inappropriately complex model. Researchers hotly debate the practical relevance of this issue (Spiegelhalter et al., 2002; Burnham and Anderson, 2004; Link and Barker, 2006).

- *Bayesian (marginal likelihood, BIC, DIC) approaches* are philosophically satisfying since they allow us to state results in terms of posterior probabilities of different models. The selection criteria (posterior probabilities) depend on the number of the parameters and on the sample size, which seems sensible. However, Bayesian approaches are also challenging to apply. Marginal likelihood is hard to calculate in a stable way; BIC is an approximation to the marginal likelihood that applies when sample sizes are large *and* the priors are vague (AIC is similarly an approximation to a marginal likelihood with a fairly strongly informative prior). For reasonable sample sizes, BIC will be more conservative than AIC; whether this conservatism is appropriate is still a matter of deep contention. Some researchers cannot accept a method that gives the wrong answer in the limit of large amounts of data, while others are more concerned with the performance of the method in the more realistic, data-limited case.*

  DIC is promising but continues to be controversial among statisticians. According to Spiegelhalter et al. (2002, p. 613), it is "a Bayesian analogue of AIC, with a similar justification but wider applicability." It is similar to AIC in its large-sample behavior. DIC is likely to become increasingly popular among ecologists since it is implemented in WinBUGS. Bayesian approaches are also

---

* Lindsey (1999b) suggests an adjustable penalty term that depends on the sample size and may fall somewhere between the AIC and BIC criteria, but he gives little guidance on choosing such a penalty.

sensitive to the priors used: one may not be able to get away with the common practice of setting a vague prior and forgetting about it.

Should we use formal rules to do model selection (or averaging) at all? Most Bayesians would say that all possible model components really exist in the world, and we ought not throw components away just because they fall below some arbitrary threshold criterion. Gelman et al., (1996) prefer to formulate selection problems as estimating a continuous parameter rather than selecting from discrete choices. Bayesians do recognize the fundamental trade-off between bias and variance, but in general they use less formal methods (such as checking whether the marginal posterior distribution has a peak, indicating that the model component is not just adding noise to the model) to decide what components to include.

A second, more intuitive argument usually comes from biologists, who are unhappy when their favorite bit of biology is dropped from a model even though they *know* that mechanism operates in nature. If you want to evaluate the effects of age structure (or spatial structure, or genetic structure) on population dynamics, you have to include it in the model even if a formal model selection procedure tells you to leave it out (Hilborn and Mangel, 1997, p. 261). What the model selection criterion is warning you, however, is that you may be basing your conclusions on dangerously little information.

A third argument often comes from conservationists who are concerned that adding a biologically relevant but statistically insignificant term to the model changes the predicted dynamics of a species, often for the worse. This is a real problem, but it is also sometimes used dishonestly. Adding complexity to a model often makes its dynamics less stable, and if you're looking to bolster an argument that a species is in trouble and needs to be protected, you'll favor results that show the species is in trouble. How often do we see conservationists arguing for more realistic biological models that suggest that a species is in no real danger and needs no protection? (On the flip side, how often do we see developers arguing that we should sample more thoroughly to make absolutely sure that there are no endangered species on a tract of land before starting construction?)

There are rules of thumb and procedures for model selection, but they don't settle the fundamental questions of model selection. Is parsimony really the most important thing? Is it OK to add more complexity to the model if you're interested in a particular biological mechanism, even if the data don't appear to support it? In the end you have to learn all the rules, but also know when to bend them—and when you do bend them, give a clear justification. The variety of model selection approaches opens a new avenue for data-dredging, by trying many different procedures and choosing the one that gives you the answers you want.

6.7 Conclusion

This chapter has covered an enormous amount of material, starting from the basic ideas of likelihood and maximum likelihood estimation, discussing various ways of estimating confidence intervals, and tackling the contentious issue of hypothesis testing and model selection. The two big ideas to take away are: (1) The geometry of the likelihood surface or posterior probability distribution—where it peaks and how the distribution falls off around the peak—contains essentially all the information you need to estimate parameters and confidence intervals. (2) Deciding which models to use for inference is challenging and cannot be reduced to a simple recipe. Different approaches correspond to different questions about the data.

7 Optimization and All That

This chapter explores the technical methods required to find the quantities discussed in the previous chapter (maximum likelihood estimates, posterior means, and profile confidence limits). The first section covers methods of numerical optimization for finding MLEs and Bayesian posterior modes, the second section introduces Markov chain Monte Carlo, a general algorithm for finding posterior means and credible intervals, and the third section discusses methods for finding confidence intervals for quantities that are not parameters of a given model.

7.1 Introduction

Now we can think about the nitty-gritty details of fitting models to data. Remember that we're trying to find the parameters that give the maximum likelihood for the comparison between the fitted model(s) and the data. (From now on I will discuss the problem in terms of finding the minimum negative log-likelihood, although all the methods apply to finding maxima as well.) The first section focuses on methods for finding minima of curves and surfaces. These methods apply whether we are looking for maximum likelihood estimates, profile confidence limits, or Bayesian posterior modes (which are an important starting point in Bayesian analyses (Gelman et al., 1996)). I will discuss the basic properties of a few common numerical minimization algorithms (most of which are built into R), and their strengths and weaknesses. Many of these methods are discussed in more detail by Press et al. (1994). The second section introduces *Markov chain Monte Carlo* methods, which are the foundation of modern Bayesian analysis. MCMC methods feel a little bit like magic, but they follow simple rules that are not too hard to understand. The last section tackles a more specific but very common problem, that of finding confidence limits on a quantity that is not a parameter of the model being fitted. There are many different ways to tackle this problem, varying in accuracy and difficulty. Having several of these techniques in your toolbox is useful, and learning about them also helps you gain a deeper understanding of the shapes of likelihood and posterior probability surfaces.

7.2 Fitting Methods

7.2.1 Brute Force/Direct Search

The simplest way to find a maximum (minimum) is to evaluate the function for a wide range of parameter values and see which one gives the best answer. In R, you would make up a vector of parameter values to try (perhaps a vector for each of several parameters); use sapply (for a single parameter) or for loops to calculate and save the negative log-likelihood (or posterior log-likelihood) for each value; then use which(x==min(x)) (or which.min(x)) to see which parameter values gave the minimum. (You may be able to use outer to evaluate a matrix of all combinations of two parameters, but you have to be careful to use a vectorized likelihood function.)

The big problem with direct search is speed, or lack of it: the resolution of your answer is limited by the resolution (grid size) and range of your search, and the time needed is the product of the resolution and the range. Suppose you try all values between p_{lower} and p_{upper} with a resolution Δp (e.g., from 0 to 10 by steps of 0.1). Figure 7.1 shows a made-up example—somewhat pathological, but not much worse than some real likelihood surfaces I've tried to fit. Obviously, the point you're looking for must fall in the range you're sampling: sampling grid 2 in the figure misses the real minimum by looking at too small a range.

You can also miss a sharp, narrow minimum, even if you sample the right range, by using too large a Δp—sampling grid 3 in Figure 7.1. There are no simple rules for determining the range and Δp to use. You must know the ecological meaning of your parameters well enough that you can guess at an appropriate order of magnitude to start with. For small numbers of parameters you can draw curves or contours of your results to double-check that nothing looks funny, but for larger models it's difficult to draw the appropriate surfaces.

Furthermore, even if you use an appropriate sampling grid, you will know the answer only to within Δp. If you use a smaller Δp, you multiply the number of values you have to evaluate. A good general strategy for direct search is to start with a fairly coarse grid (although not as coarse as sampling grid 3 in Figure 7.1), find the subregion that contains the minimum, and then "zoom in" on that region by making both the range and Δp smaller, as in sampling grid 4. You can often achieve fairly good results this way, but almost always less efficiently than with one of the more sophisticated approaches covered in the rest of the chapter.

The advantages of direct search are (1) it's simple and (2) it's so dumb that it's hard to fool: provided you use a reasonable range and Δp, it won't be led astray by features like multiple minima or discontinuities that will confuse other, more sophisticated approaches. The real problem with direct search is that it's slow because it takes no advantage of the geometry of the surface. If it takes more than a few seconds to evaluate the likelihood for a particular set of parameters, or if you have many parameters (which leads to many *many* combinations of parameters to evaluate), direct search may not be feasible.

For example, to do direct search on the parameters of the Gamma-distributed myxomatosis data, we need to set the range and grid size for shape and scale. In

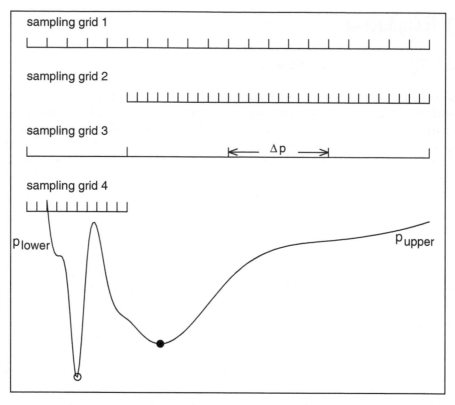

Figure 7.1 Direct search grids for a hypothetical negative log-likelihood function. Grids 1 and 4 will eventually find the correct minimum (open point). Grids 2 and 3 will miss it, finding the false minimum (closed point) instead. Grid 2 misses because its range is too small; grid 3 misses because its resolution is too small.

Chapter 6, we used the method of moments to determine starting values of shape (53.9) and scale (0.13). We'll try shape parameters from 10 to 100 with Δ shape = 1, and scale parameters from 0.01 to 0.3 with Δ scale = 0.01.

```
> shapevec = 10:100
> scalevec = seq(0.01, 0.3, by = 0.01)
```

Using the gammaNLL1 negative log-likelihood function from p. 175:

```
> surf = matrix(nrow = length(shapevec),
+     ncol = length(scalevec))
> for (i in 1:length(shapevec)) {
+     for (j in 1:length(scalevec)) {
+         surf[i, j] = gammaNLL1(shapevec[i], scalevec[j])
+     }
+ }
```

Draw the contour plot:

```
> contour(shapevec, scalevec, log10(surf))
```

Or you can do this more automatically with the `curve3d` function from the `emdbook` package:

```
> curve3d(log10(gammaNLL1(x, y)), from = c(10, 0.01),
+       to = c(100, 0.3), n = c(91, 30), sys3d = "contour")
```

The `gridsearch2d` function (also in `emdbook`) will let you zoom in on a negative log-likelihood surface:

```
> gridsearch2d(gammaNLL1, v1min = 10, v2min = 0.01,
+       v1max = 100, v2max = 0.3, logz = TRUE)
```

7.2.2 Derivative-Based Methods

The opposite extreme from direct search is to make strong assumptions about the geometry of the likelihood surface: typically, that it is smooth (continuous with continuous first and second derivatives) and has only one minimum. At this minimum point the *gradient*, the vector of the derivatives of the surface with respect to each parameter, is a vector of all zeros. Most numerical optimization methods other than direct search use some variant of the criterion that the derivative must be close to zero at the minimum in order to decide when to stop. So-called *derivative-based* methods also use information about the first and second derivatives to move quickly to the minimum.

The simplest derivative-based method is *Newton's method*, also called the *Newton-Raphson* method. Newton's method is a general algorithm for discovering the places where a function crosses zero, called its *roots*. In general, if we have a function $f(x)$ and a starting guess x_0, we calculate the value $f(x_0)$ and the value of the derivative at x_0, $f'(x_0)$. Then we extrapolate linearly to try to find the root: $x_1 = x_0 - f(x_0)/f'(x_0)$ (Figure 7.2). We iterate this process until we reach a point where the absolute value of the function is "small enough"—typically 10^{-6} or smaller.

Although calculating the derivatives of the objective function analytically is the most efficient procedure, approximating the derivatives numerically using finite differences is often convenient and is sometimes necessary:

$$\frac{df(x)}{dx} = \lim_{\Delta x \to 0} \frac{\Delta f(x)}{\Delta x} \approx \frac{f(x + \Delta x) - f(x)}{\Delta x}, \qquad \text{for small } \Delta x. \qquad (7.2.9)$$

R's `optim` function uses finite differences by default, but it sometimes runs into trouble with both speed (calculating finite differences for an n-parameter model requires an additional n function evaluations for each step) and stability. Calculating finite differences requires you to pick a Δx; `optim` uses $\Delta x = 0.001$ by default, but you can change this with `control=list(ndeps=c(...))` within an `optim` or `mle2` call, where the dots stand for a vector of Δx values, one for each parameter. You can also change the effective value of Δx by changing the parameter scale, `control=list(parscale=c(...))`; Δx_i is defined relative to the parameter scale, as `parscale[i]*ndeps[i]`. If Δx is too large, the finite difference approximation will be poor; if it is too small, round-off error will lower its accuracy.

In minimization problems, we actually want to find the root of the *derivative* of the objective function, which means that Newton's method will use the second

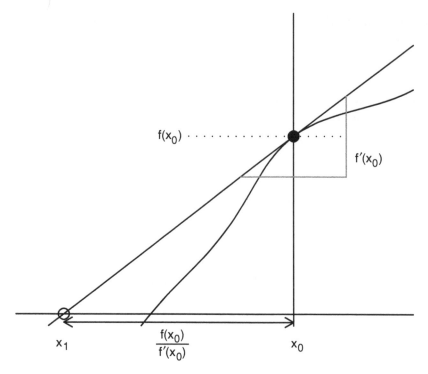

Figure 7.2 Newton's method: schematic.

derivative of the objective function. That is, instead of taking $f(x)$ and calculating $f'(x)$ by differentiation or finite differencing to figure out the slope and project our next guess, Newton's method for minima takes $f'(x)$ and calculates $f''(x)$ (the curvature) to approximate where $f'(x) = 0$.

Using the binomial seed predation data from the last chapter and starting with a guess of $p = 0.6$, Figure 7.3 and Table 7.1 show how Newton's method converges quickly to $p = 0.75$ (for clarity, the figure shows only the first three steps of the process). Newton's method is simple and converges quickly. The precision of the answer rapidly increases with additional iterations. It also generalizes easily to multiple parameters: just calculate the first and second partial derivatives with respect to all the parameters and use linear extrapolation to look for the root. However, if the initial guess is poor or if the likelihood surface is oddly shaped, Newton's method can misbehave—overshooting the right answer or oscillating around it. Various modifications of Newton's method mitigate some of these problems (Press et al., 1994), and similar methods called "quasi-Newton" methods use the general idea of calculating derivatives to iteratively approximate the root of the derivatives. The *Broyden-Fletcher-Goldfarb-Shanno* (BFGS) algorithm built into R's optim code is probably the most widespread quasi-Newton method.

Use BFGS whenever you have a relatively well-behaved (i.e., smooth) likelihood surface, you can find reasonable starting conditions, and efficiency is important. If you can calculate an analytical formula for the derivatives, write an R function to compute it for a particular parameter vector, and supply it to optim via the gr argument (see the examples in ?gr), you will avoid the finite difference calculations and get a faster and more stable solution.

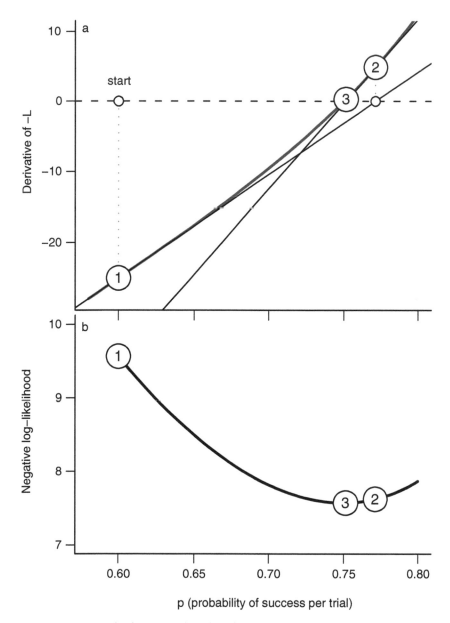

Figure 7.3 Newton's method: (a) Numbered circles represent sequential guesses for the parameter p (starting from guess 1 at 0.6); a dotted gray line joins the current guess with the value of the derivative for that value of the parameter; solid lines "shoot" over to the horizontal axis to find the next guess for p. (b) Likelihood curve.

As with all optimization methods, you *must* be able to estimate reasonable starting parameter values. Sometimes a likelihood surface will become flat for really bad fits—once the parameters are sufficiently far off the correct answer, changing them may make little difference in the goodness of fit. Since the log-likelihood will be nearly constant, its derivative will be nearly zero. Derivative-based methods that start from implausible values (or any optimization procedure that uses a "flatness"

TABLE 7.1
Newton's method

	Guess (x)	$f'(x)$	$f''(x)$
1	0.600000	−25.000	145.833
2	0.771429	4.861	241.818
3	0.751326	0.284	214.856
4	0.750005	0.001	213.339
5	0.750000	0.000	213.333

criterion to decide when to stop, including most of those built into optim) may find this worst-case scenario instead of the minimum you sought.

More often, specifying ridiculous starting values will give infinite or NA values, which R's optimization routines will choke on. Although most of the optimization routines can handle occasional NAs, the negative log-likelihood must be finite for the starting values. You should always test your negative log-likelihood functions at the proposed starting conditions to make sure they give finite answers; also try tweaking the parameters in the direction you think might be toward a better fit, and see if the negative log-likelihood decreases. If you get nonfinite values (Inf, NA, or NaN), check that your parameters are really sensible. If you think they should be OK, check for NAs in your data, or see if you have made any numerical mistakes like dividing by zero, taking logarithms of zero or negative numbers, or exponentiating large numbers (R thinks exp(x) is infinite for any x > 710). Exponentiating negative numbers of large magnitude is not necessarily a problem, but if they "underflow" and become zero (R thinks exp(x) is 0 for any x < −746), you may get errors if you divide by them or calculate a likelihood of a data value that has zero probability. Some log-likelihood functions contain terms like $x \log (x)$, which we can recognize should be zero when $x = 0$ but R treats as NaN. You can use if or ifelse in your likelihood functions to work around special cases, for example, ifelse(x==0,0,x*log(x)). If you have to, break down the sum in your negative log-likelihood function and see which particular data points are causing the problem (e.g., if L is a vector of negative log-likelihoods, try which(!is.finite(L))).

If your surface is *not* smooth—if it has discontinuities or if round-off error or noise makes it "bumpy"—then derivative-based methods will work badly, particularly with finite differencing. When derivative-based methods hit a bump in the likelihood surface, they often project the next guess to be very far away, sometimes so far away that the negative log-likelihood calculation makes no sense (e.g., negative parameter values). In this case, you will need to try an optimization method that avoids derivatives.

7.2.3 Derivative-Free Methods

Between the brute force of direct search and the sometimes delicate derivative-based methods are *derivative-free* methods, which use some information about the surface but do not rely on smoothness.

7.2.3.1 ONE-DIMENSIONAL ALGORITHMS

One-dimensional minimization is easy because once you have bracketed a minimum (i.e., you can find two parameter values, one of which is above and one of which is below the parameter value that gives the minimum negative log-likelihood) you can always find the minimum by interpolation. R's optimize function is a one-dimensional search algorithm that uses *Brent's method*, which is a combination of *golden-section search* and *parabolic interpolation* (Press et al., 1994). Golden-section search attempts to "sandwich" the minimum, based on the heights (negative log-likelihoods) of a few points; parabolic interpolation fits a quadratic function (a parabola) to three points at a time and extrapolates to the minimum of the parabola. If you have a one-dimensional problem (i.e., a one-parameter model), optimize can usually solve it quickly and precisely. The only potential drawback is that optimize, like optim, can't easily calculate confidence intervals. If you need confidence intervals, first fit the model with optimize and then use the answer as a starting value for mle2.*

7.2.3.2 NELDER-MEAD SIMPLEX

The simplest and probably most widely used derivative-free minimization algorithm that works in multiple dimensions (it's optim's default) is the *Nelder-Mead simplex*, devised by Nelder and Mead in 1965.[†]

Rather than starting with a single parameter combination (which you can think of as a point in *n*-dimensional parameter space) Nelder-Mead picks $n + 1$ parameter combinations that form the vertices of an initial *simplex*—the simplest shape possible in *n* dimensions.[‡] In two dimensions, a simplex is three points (each of which represents a pair of parameter values) forming a triangle; in three dimensions, a simplex is four points (each of which is a triplet of parameter values) forming a pyramid or tetrahedron; in higher dimensions, it's $n + 1$ points, which we call an *n*-dimensional simplex. The Nelder-Mead algorithm then evaluates the likelihood at each vertex, which is the "height" of the surface at that point, and moves the worst point in the simplex according to a simple set of rules (Figure 7.4):

- Start by going in what seems to the best direction by reflecting the high (worst) point in the simplex through the face opposite it.
- If the goodness-of-fit at the new point is better than the best (lowest) other point in the simplex, double the length of the jump in that direction.
- If this jump was bad—the height at the new point is worse than the second-worst point in the simplex—then try a point that's only half as far out as the initial try.
- If this second try, closer to the original, is also bad, then contract the simplex around the current best (lowest) point.

---

* mle and mle2 use method="BFGS" by default. Nelder-Mead optimization (see below) is unreliable in one dimension and R will warn you if you try to use it to optimize a single parameter.

[†] The Nelder-Mead simplex is completely unrelated to the simplex method in linear programming, which is a method for solving high-dimensional *linear* optimization problems with constraints.

[‡] However, you need to specify only a single starting point; R automatically creates a simplex around your starting value.

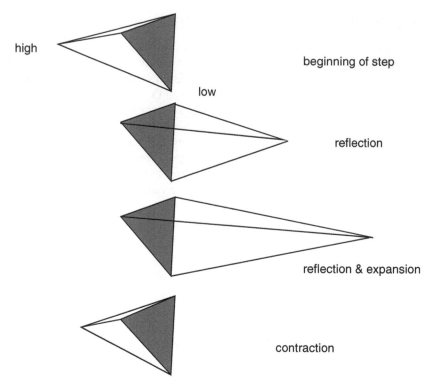

beginning of step

low

reflection

reflection & expansion

contraction

Figure 7.4 Graphical illustration (after Press et al. (1994)) of the Nelder-Mead simplex rules applied to a tetrahedron (a three-dimensional simplex, used for a three-parameter model).

The Nelder-Mead algorithm works well in a wide variety of situations, although it's not foolproof (nothing is) and it's not particularly efficient.

When we use the Nelder-Mead algorithm to fit a Gamma distribution to the myxomatosis data (Figure 7.5), the algorithm starts with a series of steps alternating between simple reflection and expanded reflection, moving rapidly downhill across the contour lines and increasing both shape and scale parameters. Eventually it finds that it has gone too far, alternating reflections and contractions to "turn the corner." Once it has turned, it proceeds very rapidly down the contour line, alternating reflections again; after a total of 50 cycles the surface is flat enough for the algorithm to conclude that it has reached a minimum.

Nelder-Mead can be considerably slower than derivative-based methods, but it is less sensitive to discontinuities or noise in the likelihood surface, since it doesn't try to use fine-scale derivative information to navigate across the likelihood surface.

7.2.4 Stochastic Global Optimization: Simulated Annealing

Stochastic global optimizers are a final class of optimization techniques, even more robust than the Nelder-Mead simplex and even slower. They are *global* because unlike most other optimization methods they may be able to find the right answer even when the likelihood surface has more than one local minimum (Figure 7.1).

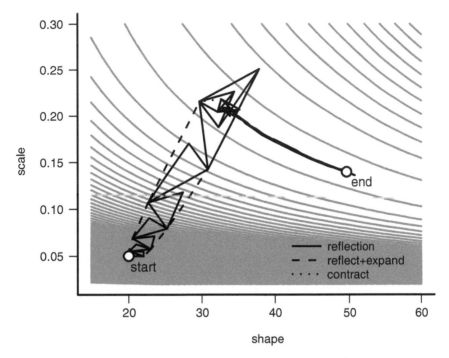

Figure 7.5 Track of Nelder-Mead simplex for the Gamma model of the myxomatosis titer data. Triangles indicating some moves are obscured by subsequent moves.

They are stochastic because they rely on adding random noise to the surface as a way of avoiding being trapped at one particular minimum.

The classic stochastic optimization algorithm is the *Metropolis algorithm*, or *simulated annealing* (Kirkpatrick et al., 1983; Press et al., 1994). The physical analogy behind simulated annealing is that gradually cooling a molten metal or crystal allows it to form a solid with few defects. Starting with a crude ("hot") solution to a problem and gradually refining the solution allows us to find the global minimum of a surface even when it has multiple minima.

The rules of simulated annealing are:

- Pick a starting point (set of parameters) and calculate the negative log-likelihood for those parameters.
- Until your answer is good enough or you run out of time:

  - **A.** pick a new point (set of parameters) at random, somewhere near your old point.
  - Calculate the value of the negative log-likelihood there.
  - If the new value is better than the old negative log-likelihood, accept it and start again at **A.**
  - If it's worse than the old value, calculate the difference in negative log-likelihood $\Delta(-L) = -L_{\mathrm{new}} - (-L_{\mathrm{old}})$. Pick a random number between 0 and 1 and accept the new value if the random number is less than $e^{-\Delta(-L)/k}$, where k is a constant called the *temperature*. Otherwise, go back to the

previous value. The higher the temperature and the smaller $\Delta(-L)$ (i.e., the less bad the new fit), the more likely you are to accept the new value. In mathematical terms, the acceptance rule is

$$\text{Prob(accept)} = \begin{cases} e^{-\frac{\Delta(-L)}{k}} & \text{if } \Delta(-L) > 0 \\ 1 & \text{if } \Delta(-L) < 0. \end{cases} \qquad (7.2.10)$$

– Return to **A** and repeat.

• Periodically (e.g., every 100 steps) lower the value of k to make it harder and harder to accept bad moves.

One variant of simulated annealing is available in R as the SANN method for optim or mle2.

Another variant of the Metropolis algorithm (Metropolis-Szymura-Barton, MSB, metropSB in emdbook; Szymura and Barton, 1986) varies the size of the change in parameters (the scale of the *candidate distribution* or *jump size*) rather than the temperature, and changes the jump size adaptively rather than according to a fixed schedule. Every successful jump increases the jump size, while every unsuccessful jump decreases the jump size. This makes the algorithm good at exploring lots of local minima (every time it gets into a valley, it starts trying to get out) but bad at refining estimates (it has a hard time getting all the way to the bottom of a valley).

To run MSB on the myxomatosis data:

```
> MSBfit = metropSB(fn = gammaNLL2, start = c(20, 0.05),
+       nmax = 2500)
```

Figure 7.6 shows a snapshot of where the MSB algorithm goes on our now-familiar likelihood surface for the myxomatosis Gamma model, with unsuccessful jumps marked in gray and successful jumps marked in black. The MSB algorithm quickly moves "downhill" from its starting point to the central valley, but then drifts aimlessly back and forth along the central valley. It does find a point close to the minimum. After 376 steps, it finds a minimum of 37.66717, equal for all practical purposes to the Nelder-Mead simplex value of 37.66714—but Nelder-Mead took only 70 function evaluations to get there. Since MSB increases its jump size when it is successful, and since it is willing to take small uphill steps, it doesn't stay near the minimum. While it always remembers the best point it has found so far, it will wander indefinitely looking for a better solution. In this case it didn't find anything better by the time I stopped it at 2500 iterations.

Figure 7.7 shows some statistics on MSB algorithm performance as the number of iterations increases. The top two panels show the values of the two parameters (shape and scale), and the best-fit parameters so far. Both of the parameters adjust quickly in the first 500 iterations, but from there they wander around without improving the fit. The third panel shows a scaled version of the jump-size parameter, which increases initially and then varies around 1.0, and the running average of the fraction of jumps accepted, which rapidly converges to a value around 0.5. The fourth and final panel shows the achieved value of the negative log-likelihood: almost all of the gains occur early. The MSB algorithm is inefficient for this problem, but it can be a lifesaver when your likelihood surface is complex and you have the patience to use brute force.

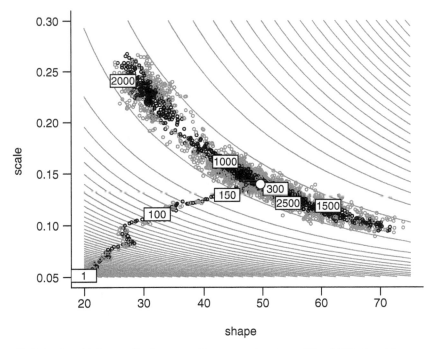

Figure 7.6 Track of Metropolis-Szymura-Barton evaluations. The MSB algorithm starts at (20,0.05) (step 1), and moves quickly up to the central valley, but then wanders aimlessly back and forth along the valley.

There are many other stochastic global optimization algorithms. For example, Press et al. (1994) suggest a hybrid of simulated annealing and the Nelder-Mead simplex where the vertices of the simplex are perturbed randomly but with decreasing amplitudes of noise over time. Other researchers suggest using a stochastic algorithm to find the right peak and finishing with a local algorithm (Nelder-Mead or derivative-based) to get a more precise answer. Various adaptive stochastic algorithms (e.g., Ingber, 1996) attempt to "tune" either the temperature or the jump size and distribution for better results. Methods like genetic algorithms or differential evolution use many points moving around the likelihood surface in parallel, rather than a single point as in simulated annealing. If you need stochastic global optimization, you will probably need a lot of computer time (many function evaluations are required) and you will almost certainly need to tune the parameters of whatever algorithm you choose rather than using the default values.

7.3 Markov Chain Monte Carlo

Bayesians are normally interested in finding the means of the posterior distribution rather than the maximum likelihood value (or analogously the mode of the posterior distribution). Previous chapters suggested that you can use WinBUGS to compute posterior distributions but gave few details. *Markov chain Monte Carlo* (MCMC) is an extremely clever, general approach that uses stochastic jumps in parameter

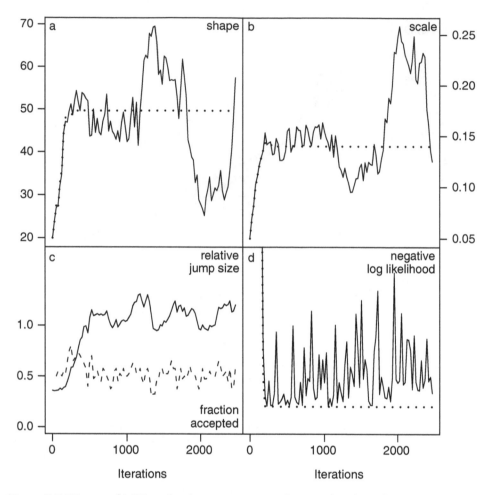

Figure 7.7 History of MSB evaluations: parameters (shape and scale), relative jump size and fraction of jumps accepted, and current and minimum negative log-likelihood. The minimum negative log-likelihood is achieved after 376 steps; thereafter the algorithm remembers its best previous achievement (horizontal dotted line) but fails to improve on it.

space to find the distribution. MCMC is similar to simulated annealing in the way it picks new parameter values sequentially but randomly. The main difference is that MCMC's goal is not to find the best parameter combination (posterior mode or MLE) but to sample from the posterior distribution.

Like simulated annealing, many variants of MCMC use different rules for picking new parameter values (i.e., different candidate distributions) and for deciding whether to accept the new choice. However, all variants of MCMC must satisfy the fundamental rule that the ratio of successful jump probabilities ($P_{jump} \times P_{accept}$) is proportional to the ratio of the posterior probabilities:

$$\frac{\text{Post}(A)}{\text{Post}(B)} = \frac{P(\text{jump } B \to A)P(\text{accept } A|B)}{P(\text{jump } A \to B)P(\text{accept } B|A)}. \tag{7.3.1}$$

If we follow this rule (and if other technical criteria are satisfied),* the long run the chain will spend a lot of time occupying areas with high probability and will visit (but not spend much time in) areas with low probability, so that the long-term distribution of the sampled points will match the posterior probability distribution.

7.3.1 Metropolis-Hastings

The *Metropolis-Hastings* MCMC updating rule is very similar to the simulated annealing rules discussed above, except that the temperature does not decrease over time to make the algorithm increasingly picky about accepting uphill moves. The Metropolis updating rule defined above for simulated annealing (p. 230) can use any *symmetric* candidate distribution (i.e., $P(\text{jump } B \to A) = P(\text{jump } A \to B)$). For example, the MSB algorithm (p. 232) picks values in a uniform distribution around the current set of parameters, while most MCMC algorithms use normal distributions. The critical part of the Metropolis algorithm is the acceptance rule, which is the simulated annealing rule (7.2.10) with the temperature parameter k set to 1 and the posterior probability substituted for the likelihood.[†] The Metropolis-Hastings rule generalizes the Metropolis algorithm by multiplying the acceptance probability by the ratio of the jump probabilities in each direction, $P(\text{jump } B \to A)/P(\text{jump } A \to B)$:

$$P(\text{accept } B|A) = \min\left(1, \frac{\text{Post}(B)}{\text{Post}(A)} \cdot \frac{P(\text{jump } B \to A)}{P(\text{jump } A \to B)}\right) \qquad (7.3.2)$$

This equation reduces to the Metropolis rule for symmetric distributions but allows for asymmetric candidate distributions, which is particularly useful when you need to adjust candidate distributions so that a parameter does not become negative.

As in simulated annealing, if a new set of parameters has a higher posterior probability than the previous parameters (weighted by the asymmetry in the probability of moving between the parameter sets), then the ratio in (7.3.2) is greater than 1 and we accept the new parameters with probability 1. If the new set has a lower posterior probability (weighted by jump probabilities), we accept them with a probability equal to the weighted ratio. If you work this out for $P(\text{accept } A|B)$ in a similar way, you'll see that the rule fits the basic MCMC criterion (7.3.1). In fact, in the MSB example above the acceptance probability was set equal to the ratio of the likelihoods of the new and old parameter values (the `scale` parameter in `metropMSB` was left at its default value of 1), so that analysis also satisfied the Metropolis-Hasting rule (7.3.2). Since it used negative log-likelihoods rather than multiplying by an explicit prior probability to compute posterior probabilities, it assumed a completely flat prior (which can be dangerous, leading to unstable estimates or slow convergence, but seems to have been OK in this case).

The `MCMCpack` package provides another way to run a Metropolis-Hastings chain in R. Given a function that computes the log posterior density (if the prior

---

* The chain must be *irreducible* (it must be possible eventually to move from any point in parameter space to any other) and *aperiodic* (it should be impossible for it to get stuck in a loop).

[†] In the simulated annealing rule we exponentiated $-k$ times the log-likelihood difference, which gave us the likelihood ratio raised to the power $-k$; if we set $k = 1$, then we have $\mathcal{L}_{\text{old}}/\mathcal{L}_{\text{new}}$, which corresponds to $\text{Post}(A)/\text{Post}(B)$.

is completely flat, this is just the (*positive*) log-likelihood function), the MCMC-metrop1R function first uses optim to find the posterior mode, then uses the approximate variance-covariance matrix at the mode to scale a multivariate normal candidate distribution, then runs a Metropolis-Hastings chain based on this candidate distribution.

For example:

```
> gammaNLL2B = function(p) {
+       sum(dgamma(myxdat$titer, shape = p[1], scale = p[2],
+           log = TRUE))
+ }
> m3 <- MCMCmetrop1R(gammaNLL2B, theta.init = c(shape = 20,
+       scale = 0.05), thin = 30, mcmc = 30000,
+       optim.lower = rep(0.004, 2),
+       optim.method = "L-BFGS-B", tune = 3)
```

When I initially ran this analysis with the default value of tune=1 and used plot(m3) to view the results, I saw that the chain took long excursions to extreme values. Inspecting the contour plot of the surface, and slices (using calcslice from the emdbook package), didn't suggest that there was another minimum that the chain was visiting during these excursions. The authors of the package suggested that MCMC-metrop1R was having trouble because of the banana shape of the posterior density (Figure 7.6), and that increasing the tune parameter, which increases the scale of the candidate distribution, would help.* Setting tune=3 seems to be enough to make the chains behave better. (Increasing tune still more would make the Metropolis sampling less efficient.) Another option, which might take more thinking, would be to transform the parameters to make the likelihood surface closer to quadratic, which would make a multivariate normal candidate distribution a better fit. Since the likelihood contours approximately follow lines of constant mean (shape · scale; Figure 7.5), changing the parameterization from {shape, scale} to {mean, variance} makes the surface approximately quadratic and should make MCMCmetrop1R behave better.

Using colnames(m3) = c("shape","scale") to set the parameter names is helpful when looking at summary(m3) or plot(m3) since MCMCmetrop1R doesn't set the names itself.

7.3.2 Burn-In and Convergence

Metropolis-Hastings updating, and any other MCMC rule that satisfies (7.3.1), is guaranteed to reach the posterior distribution eventually, but we usually have to

---

* They specifically suggested:

1. Set the tuning parameter much larger than normal so that the acceptance rate is actually below the usual 20–25% rule of thumb. This will fatten and lengthen the proposal distribution so that one can jump from one tail to the other.
2. Forgo the proposal distribution based on the large sample var-cov matrix. Set the V parameter in MCMCmetrop1R to something that will work reasonably well over the entire parameter space.
3. Use an MCMC algorithm other than the random walk Metropolis algorithm. You'll need to use something other than MCMCmetrop1R to do this, but this option will be the most computationally efficient.

discard the iterations from a *burn-in* period before the distribution converges to the posterior distribution. For example, during the first 300 steps in the MSB optimization above (Figures 7.6 and 7.7), the algorithm approaches the minimum from its starting points and bounces around the minimum thereafter. Treating this analysis as an MCMC, we would drop the first 300 steps (or 500 to be safe) and focus on the rest of the data set.

Assessing convergence is simple for such a simple model but can be difficult in general. Bayesian analysts have developed many *convergence diagnostics*, but you need to know about only a few.

The *Raftery-Lewis* (RL) diagnostic (raftery.diag in the coda package) takes a pilot run of an MCMC and estimates, based on the variability in the parameters, how long the burn-in period should be and how many samples you need to estimate the parameters to within a certain accuracy. The parameters for the Raftery-Lewis diagnostic are the quantile that you want to estimate (2.5% by default, i.e., the standard two-sided tails of the posterior distribution), the accuracy with which you want to estimate the quantile (±0.005 by default), and the desired probability that the quantile is in the desired range (default 0.95). For the MSB/myxomatosis example above, running the Raftery-Lewis diagnostic with the default accuracy of $r = 0.005$ said the pilot run of 2500 was not even long enough to estimate how long the chain should be, so I relaxed the accuracy to $r = 0.01$:

```
Quantile (q) = 0.025
Accuracy (r) = +/- 0.01
Probability (s) = 0.95
```

	Burn-in (M)	Total (N)	Lower bound (Nmin)	Dependence factor (I)
p1	44	10100	937	10.8
p2	211	29839	937	31.8

The first column gives the estimated burn-in time for each parameter—take the maximum of these values as your burn-in time. The next two columns give the required total sample size and the sample size that would be required if the chain were uncorrelated. The final column gives the dependence factor, which essentially says how many steps the chain takes until it has "forgotten" about its previous value. In this case, RL says that we would need to run the chain for about 30,000 samples to get a sufficiently good estimate of the quantiles for the scale parameter, but that (because the dependency factor is close to 30) we could take every 30th step in the chain and not lose any important information.

Another way of assessing convergence is to run multiple chains that start from widely separated (*overdispersed*) points and see whether they have run long enough to overlap (which is a good indication that they have converged). The starting points should be far enough apart to give a good sample of the surface, but they should be sufficiently reasonable to give finite posterior probabilities. The *Gelman-Rubin* (G-R, gelman.diag in the coda package; Gelman et al., 1996) diagnostic takes this approach. G-R provides a *potential scale reduction factor* (PRSF), estimating how much the between-chain variance could be reduced if the chains were run longer. The closer to 1 the PRSFs are, the better. The rule of thumb is that they should be less than 1.2.

Running a second chain (m2) for the myxomatosis data starting from (shape = 70, scale = 0.2) instead of (shape = 20, scale = 0.05) and running G-R diagnostics on the two chains gives

```
> gelman.diag(mcmc.list(m1, m2))
```

```
Potential scale reduction factors:
```

```
Point  est.   97.5% quantile
p1     1.15         1.48
p2     1.31         2.36
```

```
Multivariate psrf
```

```
1.28
```

The upper confidence limits for the PRSF for parameter 1 (shape) and the estimated value for parameter 2 (scale) are both greater than 1.2. Apparently we need to run the chains longer.

7.3.3 Gibbs Sampling

The major alternative to Metropolis-Hastings sampling is *Gibbs sampling* (or the *Gibbs sampler*), which works for models where we can figure out the posterior probability distribution of one parameter (and pick a random sample from it), *conditional* on the values of all the other parameters in the model. For example, to estimate the mean and variance of normally distributed data we can cycle back and forth between picking a random value from the posterior distribution for the mean, assuming a particular value of the variance, and picking a random value from the posterior distribution for the variance, assuming a particular value of the mean. The Gibbs sampler obeys the MCMC criterion (7.3.1) because the candidate distribution *is* the posterior distribution, so the jump probability ($P(\text{jump } B \to A)$) is equal to the posterior distribution of A. Therefore, the Gibbs sampler can always accept the jump ($p_{\text{accept}}) = 1$ and still satisfy

$$\frac{\text{Post}(A)}{\text{Post}(B)} = \frac{P(\text{jump } B \to A)}{P(\text{jump } A \to B)}. \tag{7.3.3}$$

Gibbs sampling works particularly well for hierarchical models (Chapter 10). Whether we can do Gibbs sampling or not, we can do *block sampling* by breaking the posterior probability up into a series of conditional probabilities. A complicated posterior distribution $\text{Post}(p_1, p_2, \ldots, p_n | y) = L(y | p_1, p_2, \ldots, p_n)\text{Prior}(p_1, p_2, \ldots, p_n)$, which is hard to compute in general, can be broken down in terms of the marginal posterior distribution of a single parameter (p_1 in this case), assuming all the other parameters are known:

$$\text{Post}(p_1 | p_2, \ldots, p_n, y)$$
$$= L(y | p_1, p_2, \ldots, p_n) \cdot P(p_1 | p_2, \ldots, p_n) \cdot \text{Prior}(p_1, p_2, \ldots, p_n). \tag{7.3.4}$$

This decomposition allows us to sample parameters one at a time, either by Gibbs sampling or by Metropolis-Hastings. The advantage is that the posterior distribution of a single parameter, conditional on the rest, may be simple enough so that we can sample directly from the posterior.

BUGS (**B**ayesian inference **U**sing **G**ibbs **S**ampling) is an amazing piece of software that takes a description of a statistical model and automatically generates a Gibbs sampling algorithm.* WinBUGS is the Windows version, and R2WinBUGS is the R interface for WinBUGS.

Some BUGS models have already appeared in Chapter 6. BUGS's syntax closely resembles R's, with the following important differences:

- BUGS is not vectorized. Definitions of vectors of probabilities must be specified using a `for` loop.
- R uses the = symbol to assign values. BUGS uses <- (a stylized left-arrow, e.g., a <- b+1 instead of a=b+1).
- BUGS uses a tilde (~) to mean "is distributed as." For example, to say that x comes from a standard normal distribution (with mean 0 and variance 1: $x \sim N(0, 1)$) tell BUGS x~dnorm(0,1).
- While many statistical distributions have the same names as in R (e.g., normal = dnorm, Gamma = dgamma), watch out! BUGS often uses a different parameterization. For example, where R uses dnorm(x,mean,sd), BUGS uses x~dnorm(mean,prec) where prec is the *precision*—the reciprocal of the variance. Also note that x is included in the dnorm command in R, whereas in BUGS it is on the left side of the ~ operator. Read the BUGS documentation (included in WinBUGS) to make sure you understand BUGS's definitions.

The model definition for BUGS should include the priors as well as the likelihoods. Here's a very simple input file, which defines a model for the posterior of the myxomatosis titer data:

```
model {
  for (i in 1:n) {
    titer[i] ~ dgamma(shape,rate)
  }
  shape ~ dunif(0,150)
  rate ~ dunif(0,20)
}
```

After making sure that this file is saved as a text file called myxogamma.bug in your working directory (use a text editor such as Wordpad or Tinn-R to edit BUGS files), you can run this model in BUGS by way of R2WinBUGS as follows:

```
> library(R2WinBUGS)

> titer = myxdat$titer
> n = length(titer)
> inits = list(list(shape = 100, rate = 3),
+   list(shape = 20, rate = 10))
```

*I will focus on a text file description, and on the R interface to WinBUGS implemented in the R2WinBUGS package, but many different variants of automatic Gibbs samplers are springing up. These vary in interface, degree of polish, and supported platforms. (1) WinBUGS runs on Windows and under WINE on Linux and Intel Macs; models can be defined either graphically or as text files; R2WinBUGS is the R interface. (2) OpenBUGS (http://mathstat.helsinki.fi/openbugs/) is a new version of WinBUGS that runs on the same platforms as WinBUGS and has an R interface, BRugs. (3) JAGS is an alternative automatic sampler that runs on Linux, Windows, and MacOS and has an R interface, rjags.

```
> testmyxo.bugs = bugs(data = list("titer", "n"),
+     inits, parameters.to.save = c("shape", "rate"),
+     model.file = "myxogamma.bug", n.chains =
+     length(inits), n.iter = 5000)
```

Printing out the value of testmyxo.bugs gives a summary including the mean, standard deviation, quantiles, and the Gelman-Rubin statistic (Rhat) for each variable. It also gives a DIC estimate for the model. By default this summary uses a precision of only 0.1, but you can use the digits argument to get more precision, e.g., print(testmyxo.bugs,digits=2).

```
> testmyxo.bugs
```

```
Inference for Bugs model at "myxogamma.bug", fit using winbugs,
2 chains, each with 5000 iterations (first 2500 discarded),
n.thin = 5 n.sims = 1000 iterations saved
```

	mean	sd	2.5%	25%	50%	75%	97.5%	Rhat	n.eff
shape	54.6	16.9	28.5	43.0	51.7	63.9	92.8	1.1	41
rate	7.9	2.5	4.1	6.2	7.5	9.3	13.5	1.1	42
deviance	77.7	2.3	75.4	76.0	76.9	78.7	83.8	1.1	18

```
pD = 2.3 and DIC = 80.0 (using the rule, pD = Dbar-Dhat) DIC is an
estimate of expected predictive error (lower deviance is better).
```

The standard diagnostic plot for a WinBUGS run (plot.bugs(testmyxo.bugs)) shows the mean and credible intervals for each variable in each chain, as well as the Gelman-Rubin statistics for each variable.

You can get slightly different information by turning the result into a coda object:

```
> testmyxo.coda = as.mcmc(testmyxo.bugs)
```

summary(testmyxo.coda) gives similar information as printing testmyxo.bugs. HPDinterval gives the credible interval for each variable computed from MCMC output.

Plotting testmyxo.coda gives *trace plots* (similar to Figure 7.7) and *density plots* of the posterior density (Figure 7.8). Other diagnostic plots are available; see especially densityplot.mcmc.

This information should be enough to get you started using WinBUGS. A growing number of papers—some in ecology, but largely focused in conservation and management (especially in fisheries) provide example models for specific systems (Millar and Meyer, 2000; Jonsen et al., 2003; Morales et al., 2004; McCarthy and Parris, 2004; Clarke et al., 2006).*

In summary, the basic procedure for fitting a model via MCMC (using MCMC-pack, WinBUGS, or rolling your own) is: (1) design and code your model; (2) enter the data; (3) pick priors for parameters; (4) initialize the parameter values for several chains (overdispersed, or by a random draw from priors); (5) run the chains for "a long time" (R2WinBUGS's default is 2000 steps); (6) check convergence; (7) run longer if necessary; (8) discard burn-in and thin the chains; (8)

---

* In a few years this list of citations will probably be too long to include!

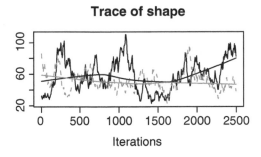

Trace of shape

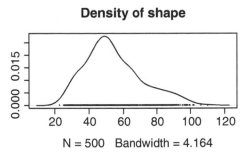

Density of shape

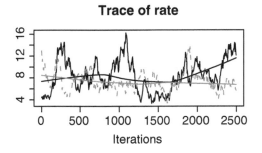

Trace of rate

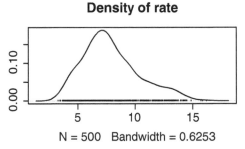

Density of rate

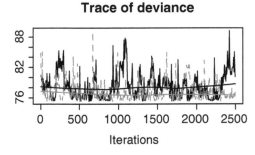

Trace of deviance

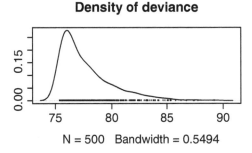

Density of deviance

Figure 7.8 WinBUGS output plot: default coda plot, showing trace plots (left) and density plots (right).

compute means, 95% intervals, correlations among parameters, and other values of interest.

7.4 Fitting Challenges

Now that we've reviewed the basic techniques for maximum likelihood and Bayesian estimation, I'll go over some of the special characteristics of problems that make fitting harder.

7.4.1 High Dimensional/Many-Parameter Models

Finding the MLE for a one-parameter model means finding the minimum of the likelihood curve; finding the MLE for a two-parameter model means finding the minimum of a 2D surface; finding the MLE for models with more parameters means finding the minimum on a multidimensional "surface." Models with more than a few parameters suffer from the *curse of dimensionality*: the number of parameter combinations, or derivatives, or directions you have to consider increases as a power law of the sampling resolution. For example, if you want find the MLE for a five-parameter model (a pretty simple model) by direct search and you want to subdivide the range of each parameter into 10 intervals (which is quite coarse), you already need 10^5 parameter combinations. Combine this with function evaluations that take more than a fraction of a second and you're into the better part of a day to do a single optimization run. Direct search is usually just not practical for models with more than two or three parameters.

If you need to visualize a high-dimensional likelihood surface (e.g., examining the region around a putative MLE to see if the algorithm has found a reasonable answer), you'll probably need to look at 2D slices (varying two parameters at a time over reasonable ranges, calculating the objective function for each combination of values while holding all the other parameters constant) or profiles (varying two parameters at a time over reasonable ranges and optimizing over all the other parameters for each combination of values). You are more likely to have to fall back on the information matrix–based approach described in the previous chapter for finding approximate variances and covariances (or correlations) of the parameter estimates; this approach is more approximate and gives you less information than fitting profiles, but extends very simply to any number of parameters.

MCMC fitting adapts well to large models. You can easily get univariate (using HPDinterval from coda for credible intervals or summary for quantiles) and bivariate confidence intervals (using HPDregionplot from emdbook).

7.4.2 Slow Function Evaluations

Since they require many function evaluations, high-dimensional problems also increase the importance of speed in the likelihood calculations. Many of the models you'll deal with take only microseconds to calculate a likelihood, so running tens of thousands of function evaluations can still be relatively quick. However, fitting a high-dimensional model using simulated annealing or other stochastic optimization approaches, or finding confidence limits for such models, can sometimes require *millions* of evaluations and hours or days to fit. In other cases, you might have to run a complicated population dynamics model for each set of parameters and so each likelihood function evaluation could take minutes or longer (Moorcroft et al., 2006).

Some possible solutions to this problem:

- Use more efficient optimization algorithms, such as derivative-based algorithms instead of Nelder-Mead, if you can.
- Derive an analytical expression for the derivatives and write a function to compute it. optim and mle2 can use this function (via the gr argument) instead of computing finite differences.

- Rewrite the code that computes the objective function more efficiently in R. Vectorized operations are almost always faster than for loops. For example, filling a 1000×2000 matrix with normally distributed values one at a time takes 30 seconds, whereas picking 2 million values and then reformatting them into a matrix takes only 0.75 second. Calculating the column sums of the matrix by looping over rows and columns takes 20.5 seconds; using apply(m,1,sum) takes 0.13 second; and using colSums(m) takes 0.005 second.

- If you can program in C or FORTRAN, or have a friend who can, write your objective function in one of these faster, lower-level languages and link it to R (see the R Extensions Manual for details).

- For really big problems, you may need to use tools beyond R. One such tool is AD Model Builder, which uses *automatic differentiation*—a very sophisticated algorithm for computing derivatives efficiently—which can speed up computation a lot (R has a very simple form of automatic differentiation built into its deriv function).

- Compromise by allowing a lower precision for your fits, increasing the reltol parameter in optim. Do you really need to know the parameters within a factor of 10^{-8}, or would 10^{-3} do, especially if you know your confidence limits are likely to be much larger? (Be careful: increasing the tolerance in this way may also allow algorithms to stop prematurely at a flat spot on the way to the true minimum.)

- Find a faster computer, or break the problem up and run it on several computers at once, or wait longer for the answers.

7.4.3 Discontinuities and Thresholds

Models with sudden changes in the log-likelihood (discontinuities) or derivatives of the log-likelihood, or perfectly flat regions, can cause real trouble for general-purpose optimization algorithms.* Discontinuities in the log-likelihood or its derivative can make derivative-based extrapolations wildly wrong. Flat or almost-flat regions can make most methods (including Nelder-Mead) falsely conclude that they've reached a minimum.

Flat regions are often the result of threshold models, which in turn can be motivated on simple phenomenological grounds or as the result (e.g.) of some optimal-foraging theories (Chapter 3). Figure 7.9 shows simulated "data" and a likelihood profile/slice for a very simple threshold model. The likelihood profile for the threshold model has discontinuities at the x value of each data point. These breaks occur because the likelihood changes only when the threshold parameter is changed from just below an observed value of x to just above it; adjusting the threshold parameter anywhere in the range between two observed x values has no effect on the likelihood.

The logistic profile, in addition to being smooth rather than choppy, is lower (representing a better fit to the data) for extreme values because the logistic function can become essentially linear for intermediate values, while the threshold function is flat. For optimum values of the threshold parameter, the logistic and threshold models

---

* Specialized algorithms, such as those included in the segmented package on CRAN, can handle certain classes of piecewise models (Muggeo, 2003).

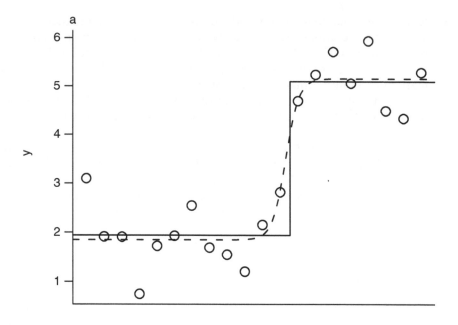

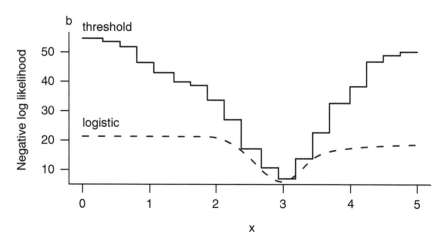

Figure 7.9 Threshold and logistic models. (a) Data, showing the data (generated from a threshold model) and the best threshold and logistic fits to the data. (b) Likelihood profiles.

give essentially the same answer. Since the logistic is slightly more flexible (having an additional parameter governing steepness), it gives marginally better fits—but these would not be significantly better according to the Likelihood Ratio test or any other model selection criterion. Both profiles become flat for extreme values (the fit doesn't get any worse for ridiculous values of the threshold parameter), which could cause trouble with an optimization method that is looking for flat regions of the profile.

Some ways to deal with thresholds:

- If you know a priori where the threshold is, you can fit different models on either side of the threshold.
- If the threshold occurs for a single parameter, you can compute a log-likelihood profile for that parameter. For example, in Figure 7.9 only the parameter for the location of the threshold causes a problem, while the parameters for the values before and after the threshold are well-behaved. This procedure reduces to direct search for the threshold parameter while still searching automatically for all the other parameters (Barrowman and Myers, 2000). This kind of profiling is also useful when a parameter needs to be restricted to integer values or is otherwise difficult to fit by a continuous optimization routine.
- You can adjust the model, replacing the sharp threshold by some smoother behavior. Figure 7.9 shows the likelihood profile of a logistic model fitted to the same data. Many fitting procedures for threshold models replace the sharp threshold with a smooth transition that preserves most of the behavior of the model but alleviates fitting difficulties (Bacon and Watts, 1974; Barrowman and Myers, 2000).

7.4.4 Multiple Minima

Even if a function is smooth, it may have multiple minima (e.g., Figure 7.1): alternative sets of parameters that each represent better fits to the data than any nearby parameters. Multiple minima may occur in either smooth or jagged likelihood surfaces.

Multiple minima are a challenging problem, and they are particularly scary because they're not always obvious—especially in high-dimensional problems. Figure 7.10 shows a slice through parameter space connecting two minima that occur in the negative log-likelihood surface of the modified logistic function that Vonesh and Bolker (2005) used to fit data on tadpole predation as a function of size (the function calcslice in the emdbook package will compute such a slice). Such a pattern strongly suggests, although it does not guarantee, that the two points really are local minima. When we wrote the paper, we were aware only of the left-hand minimum, which seemed to fit the data reasonably well. In preparing this chapter, I reanalyzed the data using BFGS instead of Nelder-Mead optimization and discovered the right-hand fit, which is actually slightly better ($-L = 11.77$ compared to 12.15 for the original fit). Since they use different rules, the Nelder-Mead and BFGS algorithms found their way to different minima despite starting at the same point. This is alarming. While the log-likelihood difference (0.38) is not large enough to reject the first set of parameters, and while the fit corresponding to those parameters still seems more biologically plausible (a gradual increase in predation risk followed by a slightly slower decrease, rather than a very sharp increase and gradual decrease), we had no idea that the second minimum existed. Etienne et al. (2006b) pointed out a similar issue affecting a paper by Latimer et al. (2005) about diversification patterns in the South African fynbos: some estimates of extremely high speciation rates turned out to be spurious minima in the model's likelihood surface (although the basic conclusions of the original paper still held).

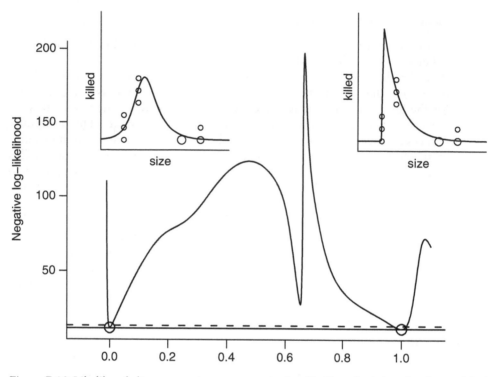

Figure 7.10 Likelihood slice connecting two negative log-likelihood minima for the modified logistic model of Vonesh and Bolker (2005). The x axis is on an arbitrary scale where $x = 0$ and $x = 1$ represent the locations of the two minima. Subplots show the fits of the curves to the frog predation data for the parameters at each minimum; the right-hand minimum is a slightly better fit ($-L = 11.77$ (right) vs. 12.15 (left)). The horizontal solid and dashed lines show the minimum negative log-likelihood and the 95% confidence cutoff ($-L + \chi_1^2(0.95)/2$). The 95% confidence region includes small regions around both $x = 0$ and $x = 1$.

No algorithm can promise to deal with the pathological case of a very narrow, isolated minimum as in Figure 7.1. To guard against multiple-minimum problems, try to fit your model with several different reasonable starting points, and check to make sure that your answers are reasonable.

If your results suggest that you have multiple minima—that is, you get different answers from different starting points or from different optimization algorithms— check the following:

- Did both fits really converge properly? The fits returned by mle2 from the bbmle package will warn you if the optimization did not converge; for optim results you need to check the $convergence term of results (it will be zero if there were no problems). Try restarting the optimizations from both of the points where the optimizations ended up, possibly resetting parscale to the absolute value of the fitted parameters. (If O1 is your first optim fit, run the second fit with control=list(parscale=abs(O1$par)). If O1 is an mle2 fit, use control=list(parscale=abs(coef(O1))).) Try different optimization methods (BFGS if you used Nelder-Mead, and vice versa). Calculate slices or profiles around the optima to make sure they really look like local minima.

- Use `calcslice` to compute a likelihood slice between the two putative fits to make sure that the surface is really higher between them.

If your surface contains several minima, the simplest solution may be to use a simple, fast method (like BFGS) but to start it from many different places. This will work if the surface is smooth, but with two (or many) valleys of approximately the same depth.*You will need to decide how to assign starting values (randomly or on a grid? along some transect?), and how many starting values you can afford to try. You may need to tune the optimization parameters so that each individual optimization runs as fast and smoothly as possible. Researchers have also developed hybrid approaches based on multiple starts (Tucci, 2002).

When multiple minima occur it is possible, although unusual, for the 95% confidence limits to be discontinuous—that is, for there to be separate regions around each minimum that are supported by the data. This does happen in Figure 7.10, although on the scale of that figure the confidence intervals in the regions around $x = 0$ and $x = 1$ would be almost too small to see. More frequently, either one minimum will be a lot deeper than the other so that only the region around one minimum is included in the confidence region, or the minima will be about the same height but the two valleys will join at the height of the 95% cutoff so that the 95% confidence interval is continuous.

If the surface is jagged instead of smooth, or if you have a sort of fractal surface— valleys within valleys, of many different depths—a stochastic global method such as simulated annealing is your best bet. Markov chain Monte Carlo can in principle deal with multiple modes, but convergence can be slow—you need to start chains at different modes and allow enough time for each chain to wander to all of the different modes (see Mossel and Vigoda, 2006; Ronquist et al., 2006, for a related example in phylogenetics).

7.4.5 Constraints

The last technical detail covered here is the problem of constraining parameter values within a particular range. Constraints occur for many reasons, but the most common constraints in ecological models are that some parameters make sense only when they have positive values (e.g., predation or growth rates) or values between 0 and 1 (e.g., probabilities). The three important characteristics of constraints concern:

1. *Equality vs. inequality*: Must a parameter or set of parameters be exactly equal to some value, or just within specified boundaries? Constraints on individual parameters are always inequality constraints (e.g., $0 < p < 1$). The most common equality constraint is that probabilities must sum to 1 ($\sum_{i=1}^{N} p_i = 1$).
2. *Individual parameters vs. combinations*: Are parameters subject to independent constraints, or do they interact? Inequality constraints on individual parameters ($a_1 < p_1 < b_1$, $a_2 < p_2 < b_2$) are called *box constraints*. Constraints on linear combinations of parameters ($a_1 p_1 + a_2 p_2 < c$) are called *linear constraints*.

---

*The many-valley case, or rather its inverse the many-peaks case (if we are maximizing rather than minimizing), is sometimes known as a "fakir's bed" problem after the practice of sitting on a board full of nails (Swartz, 2003).

3. *Solving constraint equations*: Can the constraint equations be solved analytically in terms of one of the parameters? For example, you can restate the constraint $p_1 p_2 = C$ as $p_1 = C/p_2$.

In Chapter 8 of the *Ecological Detective*, Hilborn and Mangel constrain the equilibrium of a complex wildebeest population model to have a particular value. This is the most difficult kind of constraint; it's an equality constraint, a nonlinear function of the parameters, and there's no way to solve the constraint equation analytically.

The simplest approach to inequality constraints is to ignore them completely and hope that your optimizing routine will find a minimum that satisfies the constraint without running into trouble. You can often get away with this if your minimum is far from the boundary, although you may get warning messages that look something like `NaNs produced in: dnbinom(x, size, prob, log)`. If your answers make sense, you can often ignore the warnings, but you should definitely test the results by restarting the optimizer from near its ending point to verify that it still finds the same solution. You may also want to try some of the other constrained approaches listed below to double-check.

The next simplest approach to optimization constraints is to find a canned optimization algorithm that can incorporate constraints in its problem definition. The `optim` function (and its `mle2` wrapper) can accommodate box constraints if you use the L–BFGS-B method. So can `nlminb`, which was introduced to R more recently and uses a different algorithm. R also provides a `constrOptim` function that can handle linear constraints. Algorithms that can fit models with general nonlinear equality and inequality constraints do exist, but they have not been implemented in R: they are typically large FORTRAN programs that cost hundreds or thousands of dollars to license (see below for the cheapskate ecologist's approach to nonlinear constraints).

Constrained optimization is finicky, so it's useful to have additional options when one method fails. In my experience, constrained algorithms are less robust than their unconstrained counterparts. For example, L-BFGS-B, the constrained version of BFGS, is (1) more likely to crash than BFGS; (2) worse at handling NAs or infinite values than BFGS; and (3) will sometimes try parameter values that violate the constraints by a little bit when it's calculating finite differences. You can work around the last problem by reducing `ndeps` and setting boundaries that are a little bit tighter than the theoretical limits, for example, a lower bound of 0.002 instead of 0.

The third approach to constraint problems is to add a penalty to the negative log-likelihood that increases as parameter values stray farther outside the allowed region. Instead of minimizing the negative log-likelihood $-L$, try minimizing $-L + P \times |C - C(p)|^n$ where P is a penalty multiplier, n is a penalty exponent, C is the desired value of the constraint, and $C(p)$ is the value of the constraint at the current parameter values (Hilborn and Mangel, 1997). For example, if you were using $P = 1000$ and $n = 2$ (a quadratic penalty, the most common type) and the sum of probabilities for a set of parameters was 1.2 instead of the desired value of 1.0, you would add a penalty term of $1000(1 - 1.2)^2 = 40$ to the negative log-likelihood. The penalty term will tend to push minimizers back into the allowed region. However, you need to implement such penalties carefully. For example, if your likelihood calculation is nonsensical outside the allowed region (e.g., if some parameters lead to negative probabilities), you may need to use the value of the negative log-likelihood at the closest boundary rather than trying to compute $-L$ for parameters outside the boundary. If your penalties make the surface nonsmooth at the boundary, derivative-based minimizers are likely to fail. You will often need to tune the penalty multiplier and exponent, especially for equality constraints.

The fourth, often most robust, approach is to transform your parameters to avoid the constraints entirely. For example, if you have a rate or density parameter λ that must be positive, rewrite your function and minimize with respect to $x = \log \lambda$ instead. Every value of x between $-\infty$ and ∞ translates to a positive value of λ; negative values of x correspond to values of $\lambda < 1$. As x approaches $-\infty$, λ approaches zero; as x approaches ∞, λ also approaches ∞.

Similarly, if you have a parameter p that must be between 0 and 1 (such as a parameter representing a probability), the *logit transformation* of p, $q = \log (p/(1-p))$, will be unconstrained (its value can be anywhere between $-\infty$ and ∞). You can use qlogis in R to calculate the logit. The inverse transformation is the logistic transformation, $\exp (q)/(1 + \exp (q))$ (plogis).

The log and logit transformations are by far the most common transformations. Many classical statistical methods use them to ensure that parameters are well defined; for example, logistic regression fits probabilities on a logit scale. Another less common but still useful transformation is the *additive log ratio* transformation (Aitchison, 1986; Billheimer et al., 1998; Okuyama and Bolker, 2005). When you're modeling proportions, you often have a set of parameters $p_1, \ldots, p_n$ representing the probabilities or proportions of a variety of outcomes (e.g., predation by different predator types). Each p_i must be between 0 and 1, and $\sum p_i = 1$. The sum-to-one constraint means that the constraints are not box constraints (which would apply to each parameter separately), and even though it is linear, it is an equality constraint rather than a inequality constraint—so constrOptim can't handle it. The additive log ratio transformation takes care of the problem: the vector $y = (\log (p_1/p_n), \log (p_2/p_n), \ldots, \log (p_{n-1}/p_n))$ is a set of $n-1$ unconstrained values from which the original values of p_i can be computed. There is one fewer additive log-ratio-transformed parameter because if we know $n-1$ of the values, then the nth is determined by the summation constraint. The inverse transformation (the additive logistic) is $p_i = \exp (y_i)/(1 + \sum_{j=1}^{n-1} \exp (y_j))$ for $i < n$, $p_n = 1 - \sum_{i=1}^{n-1} p_i$.

The major problem with transforming constraints this way is that sometimes the best estimates of the parameters, or the null values you want to test against, actually lie on the boundary—in mixture or composition problems, for example, the best fit may set the contribution from some components equal to zero. For example, the best estimate of the contribution of some turtle nesting beaches (rookeries) to a mixed foraging-ground population may be exactly zero (Okuyama and Bolker, 2005). If you logit-transform the proportional contributions from different nesting beaches, you will move the lower boundary from 0 to $-\infty$. Any optimizer that tries to reach the boundary will have a hard time, resulting in warnings about convergence and/or large negative estimates that differ depending on starting conditions. One option is simply to set the relevant parameters to zero (i.e., construct a reduced model that eliminates all nesting beaches that seem to have minimal contributions), estimate the minimum negative log-likelihood, and compare it to the best fit that the optimizer could achieve. If the negative log-likelihood is smaller with the contributions set to zero (e.g., the negative log-likelihood for contribution = 0 is 12.5, compared to a best-achieved value of 12.7 when the log-transformed contribution is -20), then you can conclude that zero is really the best fit. You can also compute a profile (negative log-)likelihood on one particular contribution with values ranging upward from zero and see that the minimum really is at zero. However, going to all this trouble every time you have a parameter or set of parameters that appear to have their best fit on the boundary is quite tedious.

One final issue with parameters on the boundary is that the standard model selection machinery discussed in Chapter 6 (Likelihood Ratio Test, AIC, etc.) always assumes that the null (or nested) values of parameters do not lie on the boundary of their feasible range. This issue is well-known but still problematic in a wide range of statistical applications, for example, in deciding whether to set a variance parameter to zero. For the specific case of linear mixed-effect models (i.e., models with linear responses and normally distributed random variables), the problem is relatively well studied. Pinheiro and Bates (2000) suggest the following approaches (listed in order of increasing sophistication):

- Simply ignore the problem, and treat the parameter as though it were not on the boundary—i.e., use a likelihood ratio test with 1 degree of freedom. Analyses of linear mixed-effect models (Self and Liang, 1987; Stram and Lee, 1994) suggest that this procedure is conservative; it will reject the null hypothesis less often (sometimes much less often) than the nominal type I error rate α.[*]
- Some analyses of mixed-effect models suggest that the distribution of the log-likelihood ratio under the null hypothesis when n parameters are on the boundary is a mixture of χ_n^2 and χ_{n-1}^2 distributions rather than a χ_n^2 distribution. If you are testing a single parameter, as is most often the case, then $n = 1$ and χ_{n-1}^2 is χ_0^2—defined as a spike at zero with area 1. For most models, the distribution is a 50/50 mixture of χ_n^2 and χ_{n-1}^2, which Goldman and Whelan (2000) call the $\bar{\chi}_n^2$ distribution. For $n = 1$, $\bar{\chi}_1^2(1 - \alpha) = \chi_1^2(1 - 2\alpha)$. In this case the 95% critical value for the likelihood ratio test would thus be $\bar{\chi}_1^2(0.95)/2 =$ qchisq(0.9,1)/2=1.35 instead of the usual value of 1.92. The qchibarsq function in the emdbook package will compute critical values for $\bar{\chi}_n^2$.
- The distribution of deviances may *not* be an equal mixture of χ_n^2 and χ_{n-1}^2 (Pinheiro and Bates, 2000). The "gold standard" is to simulate the null hypothesis and determine the distribution of the log-likelihood ratio under the null hypothesis; see Section 7.6.1 for a worked example.

7.5 Estimating Confidence Limits of Functions of Parameters

Quite often, you estimate a set of parameters from data, but you actually want to say something about a value that is not a parameter (e.g., about the predicted population size some time in the future). It's easy to get the point estimate—you just feed the parameter estimates into the population model and see what comes out. But how do you estimate the confidence limits on that prediction?

There are many possibilities, ranging in accuracy, sophistication, and difficulty. The data for an extended example come from J. Wilson's observations of "death"

---

[*] Whether this is a good idea or not, it is the standard approach—as far as I can tell it is *always* what is done in ecological analyses, although some evolutionary analyses are more sophisticated.

(actually disappearance, which may also represent emigration) times of juvenile reef gobies in a variety of experimental treatments. The gobies' times of death are assumed to follow a *Weibull* distribution:

$$f(t) = \frac{a}{b} \left(\frac{t}{b}\right)^{a-1} e^{-(t/b)^a}. \tag{7.5.1}$$

The Weibull distribution (dweibull in R), common in survival analysis, allows for a per capita mortality rate that either increases or decreases with time. It is shaped like the Gamma distribution, ranging from L-shaped to humped (normal). It has two parameters, shape (*a* above) and scale (*b* above): when shape = 1 it reduces to an exponential. It's straightforward to calculate the univariate or bivariate confidence limits of these parameters, but what if we want to calculate the confidence interval of the mean survival time, which is likely to be more meaningful to the average ecologist or manager?

First, pull in the data and take a useful subset:

```
> library(emdbookx)*
> data(GobySurvival)
> dat = subset(GobySurvival, exper == 1 &
+ density == 9 & qual > median(qual))
```

Define the death time as the midpoint between the last time the fish was observed (d1) and the first time it was *not* observed (d2):[†]

```
> time = (dat$d1 + dat$d2)/2
```

Set up a simple likelihood function:

```
> weiblikfun = function(shape, scale) {
+       -sum(dweibull(time, shape = shape, scale = scale,
+            log = TRUE))
+ }
```

Fit the model starting from an exponential distribution (if shape = a = 1, the distribution is an exponential with rate $1/b$ and mean b):

```
> w1 <- mle2(weiblikfun, start = list(shape = 1,
+     scale = mean(time)))
```

---

* The emdbookx package contains data that have not yet been generally released; please contact me for access.

† Survival analyses usually assume that the time of death is known exactly. With these data, as is common in ecological studies, we have a range of days during which the fish disappeared. To handle this so-called *interval censoring* properly in the likelihood function, we would have to find the probability of dying after day d1 but before day d2, which is (probability of dying before d2 − probability of dying before d1). In R the negative log-likelihood function would be

```
> weiblikfun <- function(shape, scale) {
+ -sum(log(pweibull(dat$d2, shape, scale) - pweibull(dat$d1,
+ shape, scale)))
+ }
```

For this example, I've used the cruder, simpler approach of averaging d1 and d2.

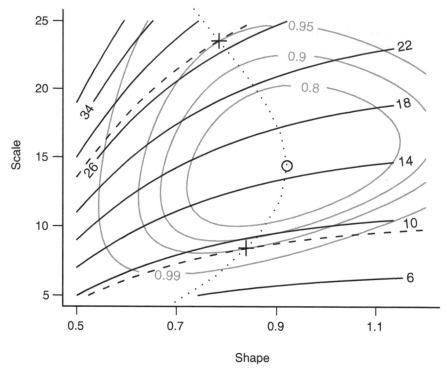

Figure 7.11 Geometry of confidence intervals on mean survival time. Gray contours: univariate (80%, 90%, 95%, 99%) confidence intervals for shape and scale. Black contours: mean survival time. Dotted line: likelihood profile for mean survival time.

The parameter estimates (coef(w1)) are shape = 0.921 and scale = 14.378; the estimate of the mean survival time (using the information on p. 253) is 14.945 days.

7.5.1 Profile Likelihood

Now we'd like confidence intervals for the mean that take variability in both shape and scale into account. The most rigorous way to estimate confidence limits on a nonparameter is to calculate the profile likelihood for the value and find the 95% confidence limits, using almost the same procedure as if you were finding the univariate confidence limits of one of the parameters.

Figure 7.11 illustrates the basic geometry of this problem: the underlying contours of the height of the surface (contours at 80%, 90%, 95%, and 99% *univariate* confidence levels) are shown in gray. The black contours show the lines on the plot that correspond to different constant values of the mean survival time. The dotted line is the likelihood profile for the mean, which passes through the minimum negative log-likelihood point on each mean contour, the point where the mean contour is tangent to a likelihood contour line. We want to find the intersections of the Likelihood Ratio Test contour lines with the likelihood profile for the mean; looking at the 95% line, we can see that the confidence intervals of the mean are approximately 9 to 27.

7.5.1.1 THE VALUE CAN BE EXPRESSED IN TERMS OF OTHER PARAMETERS

When the value for which you want to estimate confidence limits has a formula that you can solve in terms of one of the parameters, calculating its confidence limits is easy.

For the Weibull distribution the mean μ is given by

$$\mu = \text{scale} \cdot \Gamma(1 + 1/\text{shape}), \tag{7.5.2}$$

or, translating to R,

```
> meanfun - function(shape, scale) {
+       scale * gamma(1 + 1/shape)
+ }
```

How do we actually calculate the profile for the mean? We can solve (7.5.2) for one of the parameters:

$$\text{scale} = \mu/\Gamma(1 + 1/\text{shape}). \tag{7.5.3}$$

Therefore, we can find the likelihood profile for the mean in almost the same way we would for one of the parameters. Fix the value of μ; then, for each value of the shape that R tries on its way to estimating the parameter, it will calculate the value of the scale that must apply if the mean is to be fixed at μ. The constraint means that, even though the model has two parameters (shape and scale), we are really doing a one-dimensional search, which just happens to be a search along a specified constant-mean contour.

To calculate the confidence interval on the mean, we have to rewrite the likelihood function in terms of the mean:

```
> weiblikfun2 = function(shape, mu) {
+       scale = mu/gamma(1 + 1/shape)
+       -sum(dweibull(time, shape = shape, scale = scale,
+           log = TRUE))
+ }
```

Find the maximum again, and calculate the confidence intervals—this time for the shape and the mean.

```
> w2 = mle2(weiblikfun2, start = list(shape = 1,
+       mu = mean(time)))
> confint(w2)
```

	2.5 %	97.5 %
shape	0.6248955	1.281101
mu	9.1826049	27.038785

We could also draw the univariate likelihood profile, the minimum negative log-likelihood achievable for each value of the mean, and find the 95% confidence limits in the same way as before by creating a likelihood profile for μ. We would use 1 degree of freedom to establish the critical value for the LRT because we are varying only one value, even though it represents a combination of two parameters.

7.5.1.2 CONSTRAINED/PENALIZED LIKELIHOOD

What if we can't solve for one of the parameters (e.g., scale) in terms of the value we are interested in (e.g., mean), but still want to calculate a likelihood profile and profile confidence limits for the mean? We can use a penalized likelihood function to constrain the mean to a particular value, as described above in the section on constraints.

While this approach is conceptually the same as the one we took in the previous section—we are calculating the profile by sliding along each mean contour to find the minimum negative log-likelihood on that contour, then finding the values of the mean for which the minimum negative log-likelihood equals the LRT cutoff—the problem is much fussier numerically. (The complicated code is presented on p. 259.) To use penalties effectively we usually have to play around with the strength of the penalty. Too strong, and our optimizations will get stuck somewhere far away from the real minimum. Too weak, and our optimizations will wander off the line we are trying to constrain them to. I tried a variety of penalty coefficients P in this case (penalty $= P \times$ (deviation of mean survival from target value)^2) from 0.1 to 10^6. The results were essentially identical for penalties ranging from 1 to 10^4 but varied for weaker or stronger penalties. One might be able to tweak the optimization settings some more to make the answers better, but there's no simple recipe—you just have to keep returning to the pictures to see if your answers make sense.

7.5.2 The Delta Method

The delta method provides an easy approximation for the confidence limits on values that are not parameters of the model. To use it you must have a formula for $\mu = f(a, b)$ that you can differentiate with respect to a and b. Unlike the first likelihood profile method, you don't have to be able to solve the equation for one of the parameters.

The formula for the delta method comes from a Taylor expansion of the formula for μ, combined with the definitions of the variance ($V(a) = E[(a - \bar{a})^2]$) and covariance ($C(a, b) = E[(a - \bar{a})(b - \bar{b})]$):

$$V(f(a, b)) \approx V(a) \left(\frac{\partial f}{\partial a} \right)^2 + V(b) \left(\frac{\partial f}{\partial b} \right)^2 + 2C(a, b) \frac{\partial f}{\partial a} \frac{\partial f}{\partial b}. \qquad (7.5.4)$$

See the appendix or Lyons (1991) for details.

We can obtain approximate variances and covariances of the parameters by taking the inverse of the information matrix: vcov does this automatically for mle2 fits.

We also need the derivatives of the function with respect to the parameters. In this example these are the derivatives of $\mu = b\Gamma(1 + 1/a)$ with respect to shape $= a$ and scale $= b$. The derivative with respect to b is easy—$\partial\mu/\partial b = \Gamma(1 + 1/a)$—but $\partial\mu/\partial a$ is harder. By the chain rule

$$\frac{\partial(\Gamma(1 + 1/a))}{\partial a} = \frac{\partial(\Gamma(1 + 1/a))}{\partial(1 + 1/a)} \cdot \frac{\partial(1 + 1/a)}{\partial a} = \frac{\partial(\Gamma(1 + 1/a))}{\partial(1 + 1/a)} \cdot -\frac{1}{a^2}, \qquad (7.5.5)$$

but in order to finish this calculation you need to know that $d\Gamma(x)/dx = \Gamma(x) \cdot$ digamma(x), where digamma is a special function (defined as the derivative of the log-gamma function). The good news is that R knows how to compute this function, so

```
> shape.deriv <- -1/shape^2 * gamma(1 + 1/shape) *
+       digamma (1 + 1/shape)
```

will give you the right numeric answer. The emdbook package has a built-in deltavar function that uses the delta method to compute the variance of a function:

```
> dvar <- deltavar(fun = scale * gamma(1 + 1/shape),
+       meanval = coef(w1), Sigma = vcov(w1))
```

Once you find the variance of the mean survival time, you can take the square root to get the standard deviation σ and calculate the approximate confidence limits $\mu \pm 2\sigma$ (use meanfun, defined on p. 253, to compute the value of μ).

```
> sdapprox <- sqrt(dvar)
> mlmean <- meanfun(coef(w1)["shape"], coef(w1)["scale"])
> ci.delta <- mlmean + c(-2, 2) * sdapprox
```

If you can't compute the derivatives analytically, R's numericDeriv function will compute them numerically (p. 261).

7.5.3 Population Prediction Intervals (PPIs)

Another simple procedure for calculating confidence limits is to draw random samples from the estimated sampling distribution (approximated by the information matrix) of the parameters. In the approximate limit where the information matrix approach is valid, the distribution of the parameters will be multivariate normal with a variance-covariance matrix given by the inverse of the information matrix. The MASS package in R has a function, mvrnorm,*for selecting multivariate normal random deviates. With the mle2 fit w1 from above,

```
> vmat = mvrnorm(1000, mu = coef(w1), Sigma = vcov(w1))
```

will select 1000 sets of parameters drawn from the appropriate distribution (if there are n parameters, the answer is a $1000 \times n$ matrix). (If you have used optim instead of mle2—suppose opt1 is your result—then use opt1$par for the mean and solve(opt1$hessian) for the variance.) You can then use this matrix to calculate the estimated value of the mean for each of the sets of parameters, treat this distribution as a distribution of means, and find its lower and upper 95% quantiles (Figure 7.12). In the context of population viability analysis, Lande et al. (2003) refer to confidence intervals computed this way as "population prediction intervals."

This procedure is easy to implement in R, as follows:

```
> dist = numeric(1000)
```

---

*mvrnorm should really be called rmvnorm for consistency with R's other distribution functions, but S-PLUS already has a built-in function called rmvnorm, so the MASS package had to use a different name.

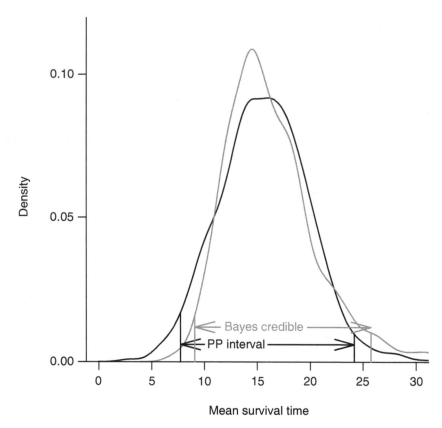

Figure 7.12 Population prediction distribution and Bayesian posterior distribution of mean survival time, with confidence and credible intervals.

```
> for (i in 1:1000) {
+     dist[i] = meanfun(vmat[i, 1], vmat[i, 2])
+ }
> quantile(dist, c(0.025, 0.975))
```

2.5%	97.5%
6.830397	24.526036

Calculating population prediction intervals in this way has two disadvantages:

- It blurs the line between frequentist and Bayesian approaches. Many papers (including some of mine, e.g., Vonesh and Bolker (2005)) have used this approach, but I have yet to see a solidly grounded justification for propagating the sampling distributions of the parameters in this way.
- Since it uses the asymptotic estimate of the parameter variance-covariance matrix, it inherits whatever inaccuracies that approximation introduces. It makes one fewer assumption than the delta method (it doesn't assume the variance is so small that the functions are close to linear), but it may not be much more accurate.

7.5.4 Bayesian analysis

Finally, you can use a real Bayesian method: construct either an exact Bayesian model or, more likely, a Markov chain Monte Carlo analysis for the parameters. Then you can calculate the posterior distribution of any function of the parameters (such as the mean survival time) from the posterior samples of the parameters, and get the 95% credible interval.

The hardest part of this analysis is converting between R and WinBUGS versions of the Weibull distribution: where R uses $f(t) = (a/b)(t/b)^{a-1} \exp(-(t/b)^a)$, Win-BUGS uses $f(t) = \nu \lambda t^{\nu-1} \exp(-\lambda t^\nu)$. Matching up terms and doing some algebra shows that $\nu = a$ and $\lambda = b^{-a}$ or $b = \lambda^{-1/a}$.

The BUGS model is

```
model {
  for (i in 1:n) {
    time[i] ~ dweib(shape,lambda)
  }
  scale <- pow(lambda,-1/shape)
  mean <- scale*exp(loggam(1+1/shape))
  ## priors
  shape ~ dunif(0,5)
  lambda ~ dunif(0,1)
}
```

Other differences between R and WinBUGS are that BUGS uses pow(x,y) instead of x^y and has only a log-gamma function loggam instead of R's gamma and lgamma functions. The model includes code to convert from WinBUGS to R parameters (i.e., calculating scale as a function of lambda) and to calculate the mean survival time, but you could also calculate these values in R.

Set up three chains that start from different, overdispersed values of shape and λ:

```
> lval <- coef(w1)["scale"]^(-coef(w1)["shape"])
> n <- length(time)
> inits <- list(list(shape = 0.8, lambda = lval),
+     list(shape = 0.4, lambda = lval * 2),
+     list(shape = 1.2, lambda = lval/2))
```

Run the chains assuming that the model is stored in a file called reefgobysurv.bug:

```
> reefgoby.bugs <- bugs(data = list("time", "n"), inits,
+     parameters.to.save = c("shape", "scale", "lambda",
+         "mean"), model.file = "reefgobysurv.bug",
+     n.chains = length(inits), n.iter = 5000)
```

Finally, use HPDinterval or summary to extract credible intervals or quantiles from the MCMC output. Figure 7.12 compares the marginal posterior density of the mean and the credible intervals computed from it with the distribution of the mean derived from the sampling distribution of the parameters and the population prediction intervals (Section 7.5.3).

TABLE 7.2
Confidence intervals on mean survival for goby data

Method	Lower	Upper
Exact profile	9.183	27.039
Profile:penalty	9.180	27.025
Delta method	7.446	22.445
PPI	6.830	24.526
Bayes credible	9.086	25.750

7.5.5 Confidence Interval Comparison

Table 7.2 presents a head-to-head comparison of all the methods we've applied so far. All methods give approximately the same answers. Despite answering a different question, the Bayes credible interval is in the same range as the other confidence intervals. The point to take away from this comparison is that *all* methods for estimating confidence limits use approximations, some cruder than others. Use the most accurate feasible approach, but don't expect estimates of confidence limits to be very precise. To paraphrase a comment of Press et al. (1994), if the difference between confidence-interval approximations ever matters to you, "then you are probably up to no good anyway—e.g., trying to substantiate a questionable hypothesis with marginal data."*

7.6 R Supplement

7.6.1 Testing Hypotheses on Boundaries by Simulating the Null Hypothesis

Suppose you want to test the hypothesis that the data set

```
> x = c(0, 0, 0, 0, 0, 0, 0, 1, 1, 1, 1, 1, 2, 2, 2,
+     2, 2, 3, 4, 5)
```

comes from a negative binomial distribution against the null hypothesis that it is Poisson distributed with $\lambda = \bar{x} = 1.35$.

A negative binomial fit (`fit.nb=fitdistr(x,"negative binomial")`) gives a negative log-likelihood (`-logLik(fit.nb)`) of 31.38, while a Poisson fit (`fit.pois=fitdistr(x,"Poisson")`) gives a negative log-likelihood of 32.12. The Likelihood Ratio test

```
> devdiff = 2 * (logLik(fit.nb) - logLik(fit.pois))
> pchisq(devdiff, df = 1, lower.tail = FALSE)
```

---

* Their original statement referred to whether to divide by n or $n-1$ when estimating a variance.

says that the p-value is 0.22, but the corrected ($\bar{\chi}^2_1$) test (pchibarsq(devdiff,df=1, lower.tail=FALSE)) says that p is only 0.12—still not significant at $p = 0.05$ but stronger evidence.

To evaluate the hypothesis more thoroughly by simulation, we will set up a function with no arguments that (1) simulates Poisson-distributed values with the appropriate mean; (2) fits a negative binomial and Poisson distributions (returning NA if the negative binomial fit should happen to crash); and (3) returns the deviance (twice the log-likelihood ratio):

```
> simulated.dev = function() {
+     simx = rpois(length(x), lambda = mean(x))
+     simfitnb = try(fitdistr(simx, "negative binomial"))
+     if (inherits(simfitnb, "try-error"))
+         return(NA)
+     simfitpois = fitdistr(simx, "Poisson")
+     dev = c(2 * (logLik(simfitnb) - logLik(simfitpois)))
+ }
```

Now simulate 3000 such values, throw out the NAs, and count the number of replicates remaining:

```
> set.seed(1001)
> devdist = replicate(3000, simulated.dev())
> devdist = na.omit(devdist)
> nreps = length(devdist)
```

Calculate the proportion of simulated values that exceed the observed deviance: this is the best estimate of the "true" p-value we can get.

```
> obs.dev = 2 * (logLik(fit.nb) - logLik(fit.pois))
> sum(devdist >= obs.dev)/nreps
```

```
[1] 0.06247912
```

So, in this case where we have two reasons—small sample size and a boundary condition—to doubt the assumptions of Likelihood Ratio Test, the classical LRT turns out to be nearly four times too conservative, while the boundary-corrected version ($\bar{\chi}^2$) is only twice as conservative as it should be.

7.6.2 Nonlinear Constraints by Penalization

Using penalties to implement an equality constraint or a nonlinear constraint (neither of which can be done with built-in functions in R) is reasonably straightforward: just add a penalty term to the negative log-likelihood. For best results, the penalty should start small and increase with increasing violation of the constraint, to avoid a discontinuity in the negative log-likelihood surface.

For example, to find the best shape and scale parameters for the fish survival data while constraining the mean to equal a particular value target.mu (use the fixed argument in mle2 to specify the target value):

```
> weiblikfun3 = function(shape, scale, target.mu,
+     penalty = 1000) {
+     mu = meanfun(shape, scale)
+     NLL = -sum(dweibull(time, shape = shape, scale = scale,
+         log = TRUE))
+     pen = penalty * (mu - target.mu)^2
+     NLL + pen
+ }
> w3 = mle2(weiblikfun3, start = list(shape = 0.9,
+     scale = 13), fixed = list(target.mu = 13))
```

If you have a problem where the function behaves badly (generates infinite or NaN values) when the constraint is violated, then you don't want to calculate the likelihood for values outside the constraints. For example, if we had to restrict shape to be greater than zero, we could use the following code snippet:

```
> if (shape > 0) {
+     NLL = -sum(dweibull(time, shape = shape, scale = scale,
+         log = TRUE))
+     pen = 0
+ } else {
+     NLL = -sum(dweibull(time, shape = 1e-04, scale = scale,
+         log = TRUE))
+     pen = penalty * shape^2
+ }
> NLL + pen
```

In other words, if the shape parameter is beyond the constraints, then use the likelihood value at or near the boundary of the feasible region and then add the penalty.

To use this constrained likelihood function to calculate confidence limits on the mean, first calculate the critical value of the negative log-likelihood:

```
> critval = -logLik(w1) + qchisq(0.95, 1)/2
```

Second, define a function that finds the best fit for a specified value of the mean and returns the distance above the critical value (use the data argument in mle2 so that you can try out different values of the penalty):

```
> pcritfun = function(target.mu, penalty = 1000) {
+     mfit = mle2(weiblikfun3, start = list(shape = 0.85,
+         scale = 12.4), fixed = list(target.mu = target.mu),
+         data = list(penalty = penalty))
+     lval = -logLik(mfit)
+     lval - critval
+ }
```

Third, define the range of mean values in which you think the lower confidence limit lies and use uniroot to search within this range for the point where the negative log-likelihood is exactly equal to the critical value:

```
> lowx = c(5, 13)
> penlower = uniroot(pcritfun, lowx)$root
```

Do the same for the upper confidence limit:

```
> upx = c(14, 30)
> penupper = uniroot(pcritfun, upx)$root
```

Try with a different value of the penalty:

```
> uniroot(pcritfun, lowx, penalty = 1e+06)$root
```

7.6.3 Numeric Derivatives

Analytical derivatives are always faster and more numerically stable, but R can compute numeric derivatives for you. For example, to compute the derivatives of the mean survival time at the maximum likelihood estimate:

```
> shape = coef(w1)["shape"]
> scale = coef(w1)["scale"]
> n1 = numericDeriv(quote(scale * gamma(1 + 1/shape)),
+      c("scale", "shape"))
> attr(n1,"gradient")
```

(The `quote` inside the `numericDeriv` command prevents R from evaluating the expression prematurely.) Of course, you can always do the same thing yourself by hand:

```
> dshape = 1e-04
> x2 = scale * gamma(1 + 1/(shape + dshape))
> x1 = scale * gamma(1 + 1/shape)
> (x2 - x1)/dshape
```

7.6.4 Extracting Information from BUGS and CODA Output

R2WinBUGS returns its results as a bugs object, which can be plotted or printed. The `as.mcmc` function in the emdbook package will turn this object into an mcmc.list object for a multichain run or an mcmc object for a single-chain run. `read.bugs` in the R2WinBUGS package does the same thing, but it requires an extra step. The mcmc and mcmc.list objects are more flexible—they can be plotted and summarized in a variety of ways (summary, HPDinterval, densityplot ...; see the help for the coda package). Once you ensure that the chains in a multichain R2WinBUGS run have converged, you can use lump.mcmc.list in the emdbook package to collapse the mcmc.list object so you can draw inferences from the combined chains.

Using the reefgoby.bugs object derived from the WinBUGS run on p. 257, calculate the Bayesian credible interval:

```
> reefgoby.coda = as.mcmc(reefgoby.bugs)
> reefgoby.coda = lump.mcmc.list(reefgoby.coda)
> ci.bayes = HPDinterval(reefgoby.coda)["mean", ]
```

Appendix: Trouble-Shooting Optimization

- Make sure you understand the model you're fitting.
- Check starting conditions.
- Check convergence conditions.
- Adjust `parscale`/restart from previous best fit.
- Switch from constraints to transformed parameters.
- Adjust finite-difference tolerances (`ndeps`).
- Switch to more robust methods (Nelder-Mead, SANN), or even just alternate methods.
- Optimization stops with NAs: `debug` objective function, constrain parameters, put `if` clauses in objective function.
- Results depend on starting conditions: check slices between and around answers: multiple minima or just convergence problems?
- Convergence problems: try restarting from previous stopping point, resetting `parscale`.
- Examine profile likelihoods.

This chapter combines all the methods we've considered so far to carry out more complete analyses of some of the example data sets, specifically the data of Vonesh and Bolker (2005) on tadpole predation, Wilson (2004) on goby survival, and Duncan and Duncan (2000) on seed removal.

8.1 Tadpole Predation

8.1.1 Introduction

The goal of Vonesh and Bolker's (2005) tadpole predation study was to quantify the effects of prey size and density on predation rate, and to use the results along with data on growth rates to understand the trade-offs between growth and survival. The response variable in all of the data we will consider here is the number of tadpoles killed by a given number or density of predators in a specified amount of time; the covariates are changing (initial) number of tadpoles (which gives rise to a *functional response* curve) and the size of tadpoles (estimating the presence of a "size refuge").

The binomial distribution is an obvious choice as a stochastic model for predation data, because the data are a discrete sample with a fixed upper limit. The challenge for the frog predation data is to decide on deterministic models that adequately describe the changes in predation probability with tadpole size and density.

8.1.2 Fitting the Size-Predation Curve

Vonesh and Bolker (2005) used the function

$$\gamma(S) = \frac{e^{\epsilon(\phi - S)}}{1 + e^{\beta\epsilon(\phi - S)}} \qquad (8.1.1)$$

to represent the dependence of predation probability $\gamma(S)$ on prey size S (Figure 8.1).

The location parameter ϕ represents a baseline prey size at which 50% of tadpoles are eaten; ϵ is the rate of change of mortality with size, controlling the steepness of the curve; and β determines the asymmetry of the curve—the extent to which prey

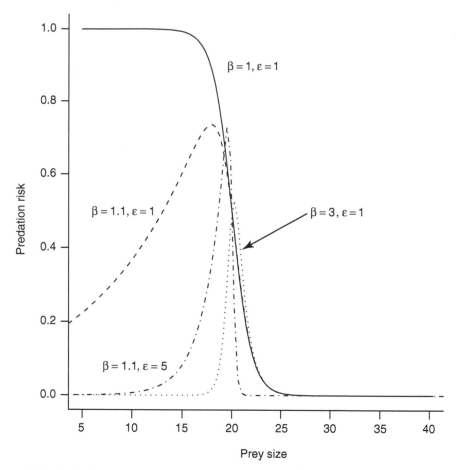

Figure 8.1 Modified logistic function from Vonesh and Bolker (2005) (eq. 8.1.1). Location parameter $\phi = 20$ for all curves.

escape predation at both small and large sizes. If $\beta = 1$, then (8.1.1) describes a logistic predation function that decreases (if $\epsilon > 0$) or increases (if $\epsilon < 0$) with size.

Some slightly tedious calculus establishes that the most vulnerable size is $\hat{S} = \phi + \log{(\beta - 1)}/(\epsilon\beta)$, which gives a predation probability

$$(\beta - 1)^{(-1/\beta)}/(1 + 1/(\beta - 1)).$$

The peak predation probability depends only on β. If $\beta < 1$, then the function is monotonically decreasing, with no peak. (To find $\hat{S}$, solve $d\gamma/dS = 0$ for S, using the quotient and chain rules to calculate the derivative, and remembering that in this case you only need to find where the numerator is zero. Then plug $\hat{S}$ back into $\gamma(S)$ to find the predation probability.)

A more traditional function to describe a humped (unimodal) dependence of predation on size is the *generalized Ricker function* (Persson et al., 1998),

$$y = b \left(\frac{S}{a} \exp\left(1 - \frac{S}{a}\right) \right)^{\alpha}. \tag{8.1.2}$$

This function is basically a reparameterization of the Ricker function ($y = axe^{-bx}$) with an added power parameter α that can broaden or narrow the peak. When $\alpha = 1$, the generalized Ricker reduces to the standard Ricker function.

A third possibility is another modification of the Ricker, which I will call the truncated Ricker: this function shifts the Ricker's origin away from zero by a distance t, and sets the function to zero below t so that it doesn't become negative:

$$y = \begin{cases} 0 & \text{if } S < t \\ b\left(\frac{S-t}{a}e^{1-\left(\frac{S-t}{a}\right)}\right) & \text{if } S \geq t. \end{cases} \tag{8.1.3}$$

All of these functions are phenomenological rather than mechanistic: while ecologists have ideas about the mechanisms leading to low predation at small size (poor detectability and being of little value to the predator) and large size (escape speed and predator gape limitation), they don't know enough about these mechanisms to guess at an appropriate functional form.

Load the data and `attach` it (don't forget to detach it later):

```
> data(ReedfrogSizepred)
> attach(ReedfrogSizepred)
```

Define the functions (`modlogist` for the modified logistic, `powricker` and `tricker` for the generalized (power) and truncated Ricker):

```
> modlogist = function(x, eps, beta, phi) {
+     exp(eps * (phi - x))/(1 + exp(beta * eps * (phi -
+         x)))
+ }
> powricker = function(x, a, b, alpha) {
+     b * (x/a * exp(1 - x/a))^alpha
+ }
> tricker = function(x, a, b, t, min = 1e-04) {
+     ifelse(x < t, min, b * ((x - t)/a * exp(1 - (x -
+         t)/a)))
+ }
```

Set up negative log-likelihood functions for each model, including one for the modified logistic that uses a beta-binomial distribution (p. 126) of numbers killed (`NLL.modlogist.bb`, with overdispersion parameter θ) instead of a binomial in order to account for possible overdispersion.*

```
> NLL.modlogist = function(eps, beta, phi) {
+     p.pred = modlogist(TBL, eps, beta, phi)
```

---

* A quick and dirty way to check for overdispersion is to compute the *residual deviance*, which is $-2\times$ the log-likelihood for the most complex model you fit. For sufficiently large data sets the scaled residual deviance should be χ^2 distributed with degrees of freedom equal to the residual degrees of freedom. However, Venables and Ripley (2002, p. 208) warn that this estimate can be misleading for moderate-size data sets (e.g., expected Poisson means less than 5 or expected number of successes in a binomial trial (Np) less than 10). For this data set, the quick and dirty approach suggests that there is overdispersion, but the likelihood fit below shows more accurately that there isn't.

```
+       -sum(dbinom(Kill, size = 10, prob = p.pred,
+           log = TRUE))
+ }
> NLL.modlogist.bb = function(eps, beta, phi, theta) {
+       p.pred = modlogist(TBL, eps, beta, phi)
+       -sum(dbetabinom(Kill, size = 10, prob = p.pred,
+           theta = theta, log = TRUE))
+ }
> NLL.powricker = function(a, b, alpha) {
+       p.pred = powricker(TBL, a, b, alpha)
+       -sum(dbinom(Kill, size = 10, prob = p.pred,
+           log = TRUE))
+ }
> NLL.tricker = function(a, b, t) {
+       p.pred = tricker(TBL, a, b, t)
+       -sum(dbinom(Kill, size = 10, prob = p.pred,
+           log = TRUE))
+ }
```

Eyeballing the data (Figure 8.2) gives approximate starting parameters for the modified logistic of $\{\phi = 15, \beta = 1.1, \epsilon = 5\}$ (compare Figure 8.1, and use ϕ to shift the peak to approximately $S = 15$). I'll start the beta-binomial version at the best-fit parameters for the binomial model and add $\theta = 1000$ (representing very little overdispersion—the beta-binomial becomes binomial as $\theta \to \infty$), setting the parscale control option to let R know that we expect this parameter to be larger than the others. (In an initial exploration with worse starting parameter guesses, I also played around with options like method="Nelder-Mead" and setting the maxit control parameter larger in order to get the optimization to work.)

```
> FSP.modlogist = mle2(NLL.modlogist, start = list(eps = 5,
+       beta = 1.1, phi = 15))
> FSP.modlogist.bb = mle2(NLL.modlogist.bb, start = as.list
+       (c(coef(FSP.modlogist), list(theta = 1000))),
+       control = list(parscale = c(1, 1, 1, 1000)))
```

The beta-binomial fit estimates $\theta = 6865$, evidence that the beta-binomial model is not really necessary; the decrease in negative log-likelihood is only 0.003.

We hardly need to run the likelihood ratio test (anova(FSP.modlogist,FSP.modlogist.bb)) or the AIC calculation (AICtab(FSP.modlogist,FSP.modlogist.bb)). Even dividing the p-value for the Likelihood Ratio test by 2 to account for the fact that the null hypothesis is on the boundary (i.e., the beta-binomial model reduces to the binomial model when $\theta \to \infty$) makes no difference to the conclusions.

If we try to get confidence limits on θ, however, we run into trouble:

```
> confint(FSP.modlogist.bb, which = "theta")
```

```
Profiling has found a better solution, so original
fit had not converged:
New minimum= 12.13806
```

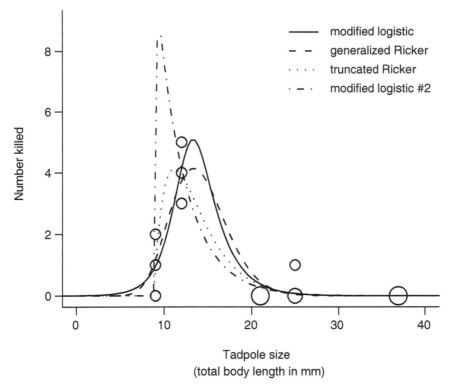

Figure 8.2 Size-predation relationship for *Hyperolius spinigularis* tadpoles: modified logistic, generalized and truncated Ricker fits.

```
Parameter values:
        eps       beta.beta        phi          theta
   0.3577257      8.9873023     9.7457033    3405.1429647
Error in onestep(step) : try restarting fit from values above
```

Refitting the parameters from this new starting point (using `modlogist` instead of `modlogist.bb`, and extending the maximum number of iterations):

```
> FSP.modlogist2 = mle2(NLL.modlogist, start = list
+      (eps = 0.357, beta = 8.99, phi = 9.75),
+      control = list(maxit = 1000))
```

The parameters of this fit are quite different:

```
> rbind(coef(FSP.modlogist), coef(FSP.modlogist2))

              eps            beta          phi
   [1,] 0.4042309       2.470003    12.908932
   [2,] 0.3045399      67.080841     9.109064
```

and the negative log-likelihood is slightly lower (11.77 vs. 12.15). You can use `plot(calcslice(FSP.modlogist,FSP.modlogist2))` to calculate and plot the

TABLE 8.1
Comparison of fits for tadpole size-dependent predation models

	AIC	df	Weight
Modified logistic (fit 2)	29.5	3	0.350
Truncated Ricker	29.9	3	0.297
Modified logistic (fit 1)	30.3	3	0.238
Generalized Ricker	31.8	3	0.115

negative log-likelihoods along a "slice" through parameter space, showing that the two different fits probably do represent distinct local minima (Figure 7.10).

However, despite fitting the data a little better the fit seems unrealistic, spiking up abruptly to a high predation rate and then dropping exponentially (Figure 8.2).

Fitting the generalized and truncated Ricker models:

```
> FSP.powricker = mle2(NLL.powricker, start = list(a = 0.4,
+     b = 0.3, alpha = 1))
> FSP.tricker = mle2(NLL.tricker, start = list(a = 0.4,
+     b = 0.3, t = 8))
```

The confidence limits on α for the generalized Ricker (`confint(FSP.powricker, parm="alpha")`) are $\{7.18, 31.69\}$—the standard Ricker ($\alpha = 1$) is clearly not competitive.

Calculating AIC values with `AICtab`, we get the results presented in Table 8.1. None of the models is nested (indeed, all have the same number of parameters), and all the fits are (almost) within 2 log-likelihood units of each other. Burnham and Anderson would recommend using the weighted predictions of all the models in subsequent analyses, but in this case (where we are just trying to gain qualitative insights into life-history trade-offs) this extra complication feels unnecessary. In this case, I would be willing to override the narrow definition of "best fit" and discard the first two models because I don't believe that predation risk is going to increase sharply as tadpoles grow bigger than 9 mm, as suggested by the truncated Ricker or by the second fit to the modified logistic. I might even choose the generalized Ricker, the worst-fitting model, over the first fit of the modified logistic, because the generalized Ricker is better established in the literature. The lesson here is that the sparser the data, the more you have to use your judgment in selecting a model—whether or not you are explicitly Bayesian.

8.1.3 Fitting the Functional Response Curve

The other data set we will examine from Vonesh and Bolker (2005) is the functional response experiment, which varied the density of tadpoles (with total body length ≈ 12.8 mm). As many as 67% (10/15) of the tadpoles in an experiment were eaten, suggesting that we should allow for the effect of depletion over the course of the experiment. The standard model for saturating functional responses is the Holling type II response, $N = aPTN_0/(1 + abN_0)$, where N is the number eaten, N_0 is the starting number/density, a and h are baseline attack rate and handling time, P is the predator number or density, and T is the total exposure time.* The Rogers random-predator equation, which allows for depletion, is

$$N = N_0(1 - e^{a(Nh-PT)}).$$ (8.1.4)

The Rogers random-predator equation (8.1.4) contains N on both the left- and right-hand sides of the equation; traditionally, one has had to use iterative numerical methods to compute the function (Vonesh and Bolker, 2005). However, the *Lambert W* function (Corless et al., 1996), which gives the solution to the equation $W(x)e^{W(x)} - x$, can be used to compute the Rogers equation efficiently: in terms of the Lambert W the Rogers equation is

$$N = N_0 - \frac{W(ahN_0e^{-a(PT-hN_0)})}{ah}.$$ (8.1.5)

Implement this equation (using the `lambertW` function in the `emdbook` package) in R, as well as the Holling type II function for comparison:

```
> rogers.pred = function(N0, a, h, P, T) {
+     N0 - lambertW(a * h * N0 * exp(-a * (P * T -
+         h * N0)))/(a * h)
+ }
> holling2.pred = function(N0, a, h, P, T) {
+     a * N0 * P * T/(1 + a * h * N0)
+ }
```

Load and attach the data (detaching is up to you):

```
> data(ReedfrogFuncresp)
> attach(ReedfrogFuncresp)
```

Write the likelihood functions:

```
> NLL.rogers = function(a, h, T, P) {
+     if (a < 0 || h < 0)
+         return(NA)
+     prop.exp = rogers.pred(Initial, a, h, P, T)/Initial
+     -sum(dbinom(Killed, prob = prop.exp, size = Initial,
+         log = TRUE))
+ }
> NLL.holling2 = function(a, h, P = 1, T = 1) {
+     -sum(dbinom(Killed, prob = a * T * P/(1 + a *
+         h * Initial), size = Initial, log = TRUE))
+ }
```

---

*P and T are usually ignored in the Holling equation, giving the function units of "number eaten per predator per unit time," but we include them here for consistency with the Rogers equation.

In the Rogers likelihood function I constrained the range of the function by simply returning NA if $a < 0$ or $h < 0$, rather than using constrained optimization; if you are not using L-BFGS-B, this shortcut sometimes works.

What about initial values? Eyeballing the data (Figure 8.3), we see the initial slope of the functional response curve is about 0.5 (50% of tadpoles are killed at low densities) and the asymptote is about 50. These values correspond to $aPT = 0.5$ or $a = 0.5/(PT) \approx 0.012$ and $PT/h = 50$ or $h \approx 0.84$. These values will be overestimates, but still usable, as starting points for the Rogers estimation as well:

```
> FFR.rogers = mle2(NLL.rogers, start = list(a = 0.012,
+     h = 0.84), data = list(T = 14, P = 3))
> FFR.holling2 = mle2(NLL.holling2, start = list(a = 0.012,
+     h = 0.84), data = list(T = 14, P = 3))
```

Running AICtab(FFR.rogers,FFR.holling2,weights=TRUE) shows that the Holling type II is a marginally better fit (0.3 log-likelihood unit difference):

	AIC	df	Weight
Holling type II	97.4	2	0.536
Rogers	97.7	2	0.464

The best-fit Holling and Rogers curves are practically indistinguishable in the plot (Figure 8.3) as well: However, we strongly prefer the Rogers curve on biological grounds, because we know that predators are depleting tadpole prey significantly over the course of the experiment. The "Rogers (no depletion)" curve in Figure 8.3 shows that depletion decreases the effect of predation by about two tadpoles across the board—as much as a 40% effect at low numbers. It will be important to take depletion into account when we compare experiments with different exposure times and predator densities below.

	a	h
Rogers	0.0171	0.814
Holling type II	0.0126	0.704

Taking depletion into account leads to a 36% increase in the estimated attack rate and a 16% increase in the estimated handling time.

8.1.4 Combined Effects of Size and Density

Vonesh and Bolker (2005) combined the effects of size and density by algebraically combining the parameters of the separate size and density fits. Here, we will instead combine all the data in a single likelihood function, estimating the functional response parameter (h) and the size-dependent attack rate parameters (α, β, and ϵ) at the same time.* The only thing we need to sort out is that the experiments were run in different volumes, as well as with different numbers of predators and for different lengths of time. The functional response experiments were run in 300-L tanks ($1.2 \times 0.8 \times 0.4$ m

---

*It would be realistic to make the handling time vary as a function of size as well (Persson et al., 1998), but unfortunately we don't have enough data.

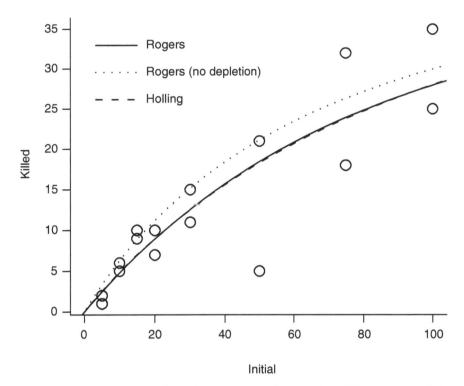

Figure 8.3 Functional response fit to frog predation data. Both Holling type II and Rogers random-predator fits are shown, but are barely distinguishable. "Rogers (no depletion)" curve plots the expected functional response from the estimated Rogers parameters in the absence of depletion.

high) filled to 220 L; the size experiments were run in 35-L plastic tubs (0.32 m in diameter) filled to 25 L. Based on the way that predators foraged, Vonesh and Bolker (2005) assumed that predation success depended on the area of the foraging arena $(1.2 \cdot 0.8 = 0.96 \text{ m}^2$ vs. $\pi((0.32)/2)^2 = 0.080 \text{ m}^2)$ rather than its volume. To make the predation probabilities match, we have to divide the predator numbers by area.* It is convenient to collect the auxiliary parameters for each experiment (number of predators, area, exposure time, etc.) in a couple of lists:

```
> xpars.Funcresp = list(T = 14, P = 3, vol = 220,
+     area = 1.2 * 0.8, size = 12.8)
> xpars.Sizepred = list(T = 3, P = 2, vol = 25, area = pi *
+     0.16^2, initprey = 10)
```

Put together a combined data set representing the initial numbers, size, number killed, predator density, and exposure time for both experiments, using rep to repeat values where necessary:

```
> n.Funcresp = nrow(ReedfrogFuncresp)
> n.Sizepred = nrow(ReedfrogSizepred)
```

---

* But *not* the prey numbers—figuring this out reminded me of an old riddle, "If a hen and a half lays an egg and a half in a day a half, how many eggs can one hen lay in a day?"

```
> combInit = c(ReedfrogFuncresp$Initial,
+     rep(xpars.Sizepred$initprey, n.Sizepred))
> combSize = c(rep(xpars.Funcresp$size, n.Funcresp),
+     ReedfrogSizepred$TBL)
> combKilled = c(ReedfrogFuncresp$Killed,
+     ReedfrogSizepred$Kill)
> combP = rep(c(xpars.Funcresp$P/xpars.Funcresp$area,
+     xpars.Sizepred$P/xpars.Sizepred$area), c(n.Funcresp,
+     n.Sizepred))
> combT = rep(c(xpars.Funcresp$T, xpars.Sizepred$T),
+     c(n.Funcresp, n.Sizepred))
```

Write a combined function for the expected proportion eaten, computing the attack rate a from the parameters ϵ, β, and ϕ and combining it with the handling time h:

```
> prop.eaten = function(N0, S, h, P, T, eps, beta,
+     phi, minprop = .Machine$double.eps) {
+     a = modlogist(S, eps = eps, beta = beta, phi = phi)
+     N.eaten = rogers.pred(N0, a = a, h = h, P = P,
+         T = T)
+     prop = N.eaten/N0
+     prop[prop <= 0] = minprop
+     prop[prop >= 1] = 1 - minprop
+     prop
+ }
```

The value `.Machine$double.eps` is a built-in constant corresponding to the smallest difference between numeric values your computer can keep track of without rounding (it is 2.22×10^{-16} on the machine I am using). Using `minprop` to adjust values that are ≤ 0 or ≥ 1 takes care of the cases where the `rogers.pred` function returns an expected proportion eaten slightly less than zero, or exactly equal to 1 (which causes an infinite negative log-likelihood if no tadpoles are eaten); these minor errors happen because of round-off error.

A negative log-likelihood function incorporating the proportion eaten:

```
> NLL.rogerscomb = function(a, h, eps, beta, phi, T = combT,
+     P = combP) {
+     if (h < 0)
+         return(NA)
+     prob = prop.eaten(combInit, combSize, h, P, T,
+         eps, beta, phi)
+     dprob = dbinom(combKilled, prob = prob, size =
+         combInit, log = TRUE)
+     -sum(dprob)
+ }
```

Set the starting values by combining h from the Rogers fit (which has to be put inside its own `list`) with the attack rates from the size-dependence fit (which will be a slight underestimate since they don't incorporate the effects of handling time):

```
> startvals = c(list(h = coef(FFR.rogers)["h"]),
+       as.list(coef(FSP.modlogist)))
```

Finding the optimum, avoiding the alternate fit (fit 2 above) when profiling, and avoiding overflow errors is quite finicky in this case. The easiest way to avoid the alternate fit is to restrict β, but using the L-BFGS-B optimizer leads to lots of headaches with NAs being produced in the Lambert W function. I used a two-stage method—first, optimizing with method="Nelder-Mead" and using confint (FPcomb,method="quad") to get approximate confidence limits:

```
> FPcomb = mle2(NLL.rogerscomb, start = startvals,
+       method = "Nelder-Mead")
> confint(FPcomb, method = "quad")
```

Then I used slightly larger values for the upper and lower bounds to refit the model and get more precise confidence limits (confint must use the same optimization rules that were used in the original fit). Getting this to work took some frustrating trial and error, including incorporating debugging statements like

```
> cat(h, eps, beta, phi, "\n")
```

or

```
> if (any(!is.finite(prob))) cat("NAs:", h, eps, beta,
+       phi, "\n")
```

or

```
> if (any(!is.finite(dprob))) {
+       browser()
+ }
```

into the NLL.rogerscomb function to track down where the problems were occurring in order to set bounds that would prevent NAs. cat prints a list of variables to the screen in the middle of a function evaluation ("\n" specifies a new line), while browser stops the function and lets you examine the values of different variables. In the course of this exploration I also went back and incorporated the minimum and maximum bounds in prop.eaten, which I had initially left out.

```
> FPcomb = mle2(NLL.rogerscomb, start = startvals,
+       method = "L-BFGS-B", lower = c(0.7, 0.5, 1, 14),
+       upper = c(1.8, 2.25, 2, 20), control = list(parscale =
+           c(1, 1, 1, 10)))
> FPcomb.ci = confint(FPcomb)
```

What is the combined estimate of the proportion eaten under the conditions of the size-predation experiment (12.8 mm body length, 2 predators in an area of 0.08 m^2 for 3 days)? How well does it match the estimate based only on the size-predation experiment? (That is, does combining the data change the baseline estimate from the size-predation experiment?)

Figure 8.4 is mildly alarming at first sight, showing that the estimate of the size refuge changes markedly when we incorporate the data from the functional response experiment. This suggests a major difference between the two experiments. A closer

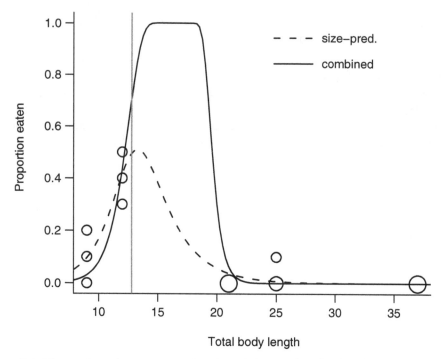

Figure 8.4 Observed number eaten as a function of size; predicted values from size-predation experiment only and from all data combined.

look, however, shows that the major difference between the results falls in a region where we have no data, between 12.8 and 21 mm body length. The slightly higher predation rate in the functional response experiment (even corrected for predator exposure) pulls the curve up.

How would we go about quantifying the uncertainty in the two curves and convincing ourselves that they're not (statistically) significantly different?

Calculating the estimates of the proportion eaten at size 12.8 mm from the size-predation fit alone:

```
> c1 = coef(FSP.modlogist)
> FSP.expprop.mean = modlogist(12.8, c1["eps"], c1["beta"],
+      c1["phi"])
```

and from the combined fit:

```
> c2 = coef(FPcomb)
> FP.expprop.mean = prop.eaten(N0 = 10, S = 12.8, c2["h"],
+      P = 2/0.08, T = 3, eps = c2["eps"], beta = c2["beta"],
+      phi = c2["phi"])
```

The estimated predation proportions are 0.49 for the size-predation experiment alone and 0.7 for the combined data—a difference that certainly might be biologically significant, if it were statistically significant.

As discussed in Chapter 7, population prediction intervals are a simple way to calculate the confidence intervals of a quantity of interest that is not a parameter in the model. Using mvrnorm to generate 5000 values from the sampling distribution of the parameters and calculating the 95% population prediction intervals of the size-predation data:

```
> set.seed(1001)
> FSP.expprop.pars = mvrnorm(5000, mu = c1,
+       Sigma = vcov(FSP.modlogist))
> FSP.expprop.val = numeric(5000)
> for (i in 1:5000) {
+       FSP.expprop.val[i] = modlogist(12.8, FSP.expprop.pars
+           [i, 1], FSP.expprop.pars[i, 2], FSP.expprop.pars
+           [i, 3])
+ }
> FSP.expprop.ppi = quantile(FSP.expprop.val, c(0.025,
+       0.975))
```

Doing the same thing for the combined fit:

```
> FP.expprop.pars = mvrnorm(5000, mu = c2,
+       Sigma = vcov(FPcomb))
> FP.expprop.val = numeric(5000)
> for (i in 1:5000) {
+       FP.expprop.val[i] = prop.eaten(N0 = 10, S = 12.8,
+           P = 2/0.08, T = 3, h = FP.expprop.pars[i,
+               "h"], eps = FP.expprop.pars[i, "eps"],
+           beta = FP.expprop.pars[i, "beta"],
+               phi = FP.expprop.pars[i, "phi"])
+ }
> FP.expprop.ppi = quantile(FP.expprop.val, c(0.025,
+       0.975))
```

	Mean	Low	High
Size-predation	0.494	0.397	0.852
Combined	0.702	0.641	0.992

The results show that the uncertainty in the estimates is large enough that at least the confidence limits of the size-predation estimates (0.4, 0.85) overlap with the estimate from the combined data (0.7), if not vice versa.

Vonesh and Bolker (2005) took results like these (although they did not try fitting the combined data as we have done here) and used them together with size-dependent growth rate estimates from a growth experiment to simulate the survival of tadpoles hatching at different sizes. They found that because smaller-starting tadpoles grew faster through the window of vulnerability between 10 and 20 mm, their overall survival was comparable to tadpoles that hatched at a larger size.

This analysis suggests several more questions:

- Because it must compromise between two sets of data with slightly different survival rates, the fit of the combined curve to the size-predation data is slightly worse than the fit of the size-predation curve itself (Figure 8.4). We initially rejected the need for a beta-binomial model to account for overdispersion, but the larger deviations suggest that it might be worth testing again.

- Following Vonesh and Bolker (2005), we assumed that predator efficiency scaled with area, not volume; this approach may have understated the predator threat in the functional response experiment, leading to an inflation of the expected proportion eaten per unit of exposure. The total predator exposure ($P \times T$) in the functional response experiment was $14 \times 3 = 42$ predator-days, in contrast to $3 \times 2 = 6$ predator-days for the size-predation experiment. If we calculate PT/area for each experiment and take the ratio, we get a relative risk of $43.8/74.6 = 0.6$; overall predator pressure per unit area was lower in the functional response experiment. On the other hand, repeating the same calculation but scaling by volume instead gives a risk ratio of $0.19/0.24 = 0.8$—less difference, leading to less inflation of the per-predator risk in the functional response. We could adjust the model by adding a scaling factor to account for the differences between the experiments, and tentatively interpret it in terms of the geometry of the foraging arena (Petersen et al., 1999). While we clearly don't have enough data to make a decision just from these two experiments, the discrepancy between the results of the two experiments does open up some interesting questions.

8.2 Goby Survival

Next, we will take a look at the effects of density and "quality" (spatial variation in habitat quality correlated with natural rates of immigration) on the survival of the small marine gobies *Elacatinus evelynae* and *E. prochilos* in field experiments (Wilson, 2004).

The questions here are straightforward: How fast do fish die (or disappear) at different levels of density and quality? Do quality, density, or their interaction (i.e., an effect of quality on the density-dependent mortality rate) have significant effects on mortality?

As a reminder, the data contain information on the survival of marine gobies in experiments where ambient density was manipulated on coral heads with different background settlement rates. Settlement rates were suspected to be correlated with some unknown aspect of environmental quality, such as flow patterns or availability of refuges (Wilson and Osenberg, 2002), which revealed itself through lower mortality rates (Figure 8.6).

8.2.1 Preliminaries

Load and attach the data, remembering to detach them later:

```
> library(emdbookx)
> data(GobySurvival)
> attach(GobySurvival)
```

In the data, time starts from day 1 (the day the fish were put on the reef) and runs until day 12; any fish that survived past the end of the experiment (i.e., that were still present on day 12) were given a "last day seen" (d2) value of 70 in the original data set. For the following analysis, time should start from zero and run to ∞ (the cumulative distribution functions we will be using can handle infinite values), so we will subtract 1 from d1 and d2 and set the ending value of d2 to Inf:

```
> day1 = d1 - 1
> day2 = ifelse(d2 == 70, Inf, d2 - 1)
```

As discussed in Chapter 7, we will use the Weibull distribution to fit the data, allowing for the observed decrease in mortality rate over time. We are interested in whether mortality is density-dependent, and whether quality affects either the density-independent or the density-dependent mortality rate. We may need to allow for the possibility that different experiments show different results (this data set combines the results from five experiments run over the course of three years).

The most complete model of the survival time of an individual fish in experiment x with density (number of neighboring fish) d and quality (background settlement rate) q would be

$$T \sim \text{Weibull}(a_x(d, q), s_x(d, q))$$

$$a_x(d, q) = \exp\left(\alpha_{a,x} + \beta_{a,x} \cdot q + (\gamma_{a,x} + \delta_{a,x} \cdot q)d\right) \tag{8.2.1}$$

$$s_x(d, q) = \exp\left(\alpha_{s,x} + \beta_{s,x} \cdot q + (\gamma_{s,x} + \delta_{s,x} \cdot q)d\right).$$

In other words, we are fitting the shape and scale parameters on the log scale. For both the (log) shape and scale parameter we are allowing for a baseline or intercept value (α), a linear effect of quality (β), a linear effect of density (γ), and an interaction between density and quality (δ)—i.e., a linear effect of quality on the density-dependent mortality coefficient. As indicated by the x in the subscripts, we are also allowing each parameter to be different in each experiment, for a total of 40 (!) parameters. Given that we have only 369 observations, unevenly divided among experiments (with as few as 11 observations in an experiment), and that each observation tells us fairly imprecisely when a fish disappeared, this model is certainly more complex than we can hope to fit.

We might try anyway, fitting all possible submodels and using model-selection rules to decide which pieces really belong in the model,[*] but even so there would be far too many submodels to consider. There are two possibilities for the intercept parameter α (the same for all experiments or different among experiments), and three for each of the other parameters β, γ, and δ (zero for all experiments, meaning no effect of density or quality or their interaction; nonzero but the same for all experiments; or different for different experiments). There are 34 possible models for shape and 34 for scale,[†] or $34^2 = 1156$ models in total, even for this moderate-sized problem!

We must make some a priori decisions about which parameters to drop—decisions made harder by the difficulty of graphically representing the dependence of survival on continuous covariates. Figure 8.5 shows the effects of the shape

---

[*] The statistical equivalent of the advice of a crusading abbot who when asked how to tell the innocents and the heretics apart said, "Kill them all, God will recognize his own."

[†] You might expect $2 \times 3 \times 3 \times 3 = 54$ models for each parameter of the Weibull, but there are a few combinations that don't make sense—specifically, fitting the δ (density-quality interaction parameter) if either the density or quality effect is set to zero.

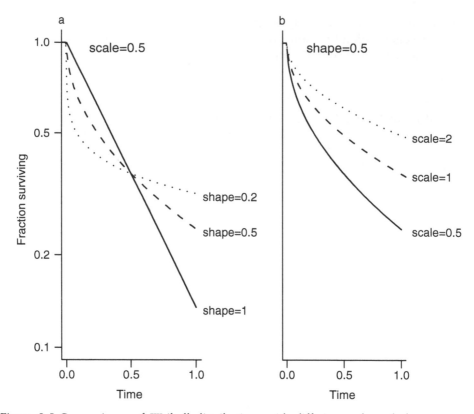

Figure 8.5 Comparisons of Weibull distributions with differing scale and shape parameters. The R commands to plot the curves are variations on `curve(pweibull(x,shape=...,scale=...,lower.tail=FALSE))`.

and scale parameter on the Weibull distribution. Comparing these differences to the survival curves in Figure 8.6 suggests that the scale, but not the shape, of the Weibull distribution varies between density and quality categories. Figure 8.6 also suggests an interaction between quality and density categories, because survival in the low-quality/high-density category is considerably below that in any other category. Figure 8.6 does not separate the results of different experiments. Drawing this figure to check might be worth while, but for now we will assume that the only possible difference among experiments is in the baseline scale parameter, not in the effects of density and quality. Wilson (2004) used a standard survival analysis to demonstrate nonsignificant interactions between experiment and density/quality, supporting this assumption.

These simplifications reduce our most complex model to

$$T \sim \text{Weibull}(a, s_x(d, q))$$
$$s_x(d, q) = \exp\left(\alpha_{s,x} + \beta_s \cdot q + (\gamma_s + \delta_s \cdot q)d\right), \tag{8.2.2}$$

with nine parameters (five for differences in scale among experiments, three for the effects of density and quality and their interaction on scale, and one for the shape parameter). Our suite of models reduces to 10. If we denote the simplest model (a single shape and scale parameter) by 0; the presence of experiment effects ($\alpha_i \neq \alpha_j$ for

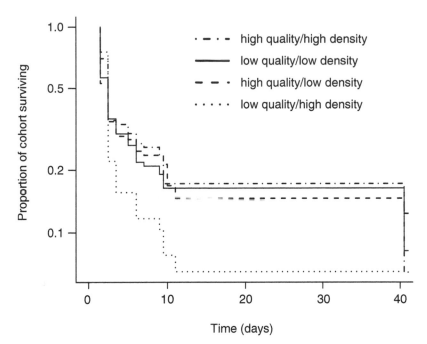

Figure 8.6 Goby survival curves by quality and density categories (above/below median values), based on mean survival time $(d1+d2)/2$.

at least one pair of experiments) by x; a quality effect ($\beta_s \neq 0$) by q; a density effect ($\gamma_s > 0$) by d; and a quality-density interaction ($\delta_s > 0$) by i, then our remaining models with their numbers of parameters are

$$
\begin{array}{lllll}
0\ (2) & x\ (6) & xq\ (7) & xqd\ (8) & xqdi\ (9) \\
& q\ (3) & xd\ (7) & qdi\ (5) & \\
& d\ (3) & qd\ (4) & &
\end{array}
$$

(These are all combinations of x, q, d, and i, with the restriction that i cannot be included without both q and d.) If we wanted to allow the shape parameter to vary with quality and density but not experiment, we would have a most-complex model with 12 parameters and a total of 40 (4×10) model possibilities.

Here is the most complex model, which fits scale and shape parameters that differ with quality, density, and their interaction:

```
> NLL.GS.xqdi = function(lscale0, lscale.q, lscale.d,
+     lscale.i, lscale.x2, lscale.x3, lscale.x4, lscale.x5,
+     lshape) {
+     lscalediff = c(0, lscale.x2, lscale.x3, lscale.x4,
+         lscale.x5)
+     scale = exp(lscale0 + lscalediff[exper] + lscale.q *
+         qual + (lscale.d + lscale.i * qual) * density)
+     shape = exp(lshape)
+     -sum(log(pweibull(day2, shape, scale) - pweibull(day1,
+         shape, scale)))
+ }
```

The only unusual thing here is that we've parameterized the difference among experiments so that the baseline parameter (lscale0) represents the log of the scale parameter (at density $= 0$ and quality $= 0$) in experiment 1, while the experiment parameters (lscale.x2, etc.) represent the differences between experiment 1 and the other experiments. This parameterization, which is consistent with the way that other functions in R define parameters, makes it possible to test the hypothesis that all experiments are the same by setting lscale.x2 and the other experiment parameters to zero. The differences among parameters are indexed by exper and added to the baseline value along with the effects of density and quality.

Since we don't know exactly when (between day1 and day2) a given fish disappeared, we calculate the probability that it disappeared somewhere between day1 and day2 taking the difference between the probability that it disappeared before day2 (pweibull(day2,...)) and the probability that it disappeared before day1 (pweibull(day1,...)); we take the log only *after* calculating the difference.

What about starting values for this model? The mean of the Weibull distribution with shape a and scale s is $s\Gamma(1 + 1/a)$, which for an exponential ($a = 1$) is equal to s. We'll start $\log(s)$ from the log of the overall mean survival time (calculated from d1 and d2 rather than day1 and day2 because day2 contains infinite values that will mess up the mean calculation), and $\log(a)$ from 0, which represents an exponential distribution. Since the rest of the parameters represent differences from the baseline case, we'll try starting them all from zero.

```
> totmeansurv = mean((d1 + d2)/2)
> startvals.GS = list(lscale0 = log(totmeansurv),
+     lscale.x2 = 0, lscale.x3 = 0, lscale.x4 = 0,
+     lscale.x5 = 0,
+     lscale.q = 0, lscale.d = 0, lscale.i = 0, lshape = 0)
> GS.xqdi = mle2(NLL.GS.xqdi, startvals.GS)
```

Looking at the estimates of the parameters and their approximate *p*-values:

```
> summary(GS.xqdi)
```

Maximum likelihood estimation

Call:
mle2(minuslogl = NLL.GS.xqdi, start = startvals.GS)

Coefficients:

	Estimate	Std. Error	z value	Pr(z)	
lscale0	1.9506010	0.7450665	2.6180	0.008844	**
lscale.q	-0.0137277	0.0993038	-0.1382	0.890051	
lscale.d	-0.2198680	0.0973726	-2.2580	0.023945	*
lscale.i	0.0126382	0.0130451	0.9688	0.332644	
lscale.x2	-1.0707399	0.5000217	-2.1414	0.032243	*
lscale.x3	-0.7677602	0.3830876	-2.0041	0.045055	*
lscale.x4	-0.1315136	1.0460335	-0.1257	0.899949	
lscale.x5	0.0048526	0.9516556	0.0051	0.995932	
lshape	-1.0016188	0.0944042	-10.6099	< 2.2e-16	***

Signif. codes: 0 '***' 0.001 '**' 0.01 '*' 0.05 '.' 0.1 ' ' 1

-2 log L: 886.122

From this summary, it appears that there may be an effect of experiment (experiments 2 and 3 both show significantly shorter survival times than experiment 1), an effect of density, and a shape parameter that is significantly less than 1 ($\log(a) < 0$)—that is, per capita mortality declines significantly with time.

In a stepwise analysis, we would continue by dropping the interaction term from the model (dropping the parameters for experiments 4 and 5 doesn't really make sense, since they are part of the overall difference among experiments). One shortcut for dropping terms from an mle2 fit, rather than writing another likelihood function that is missing one term, is to use the fixed argument to set a subset of the parameters to zero. For example, to drop the interaction term from the model:

```
> GS.xqd = mle2(NLL.GS.xqdi, startvals.GS,
+      fixed = list(lscale.i = 0))
```

We can use the Likelihood Ratio Test on particular series of nested hypotheses to test specific conclusions. For example, we might be most interested in testing whether quality and density have an effect. We attempt to drop the interaction term first, then quality, then density. Because the differences among experiments are potentially important, and an unavoidable part of the experimental design, we leave them in the model. Therefore we want to test the sequence of models xqdi → xqd → xd → x.

Fitting the remaining two models in the sequence:

```
> GS.xd = mle2(NLL.GS.xqdi, startvals.GS,
+    fixed = list (lscale.i = 0, lscale.q = 0))
> GS.x = mle2(NLL.GS.xqdi, startvals.GS,
+    fixed = list (lscale.i = 0, lscale.q = 0, lscale.d = 0))
```

Applying anova to run the Likelihood Ratio test:

```
> anova(GS.xqdi, GS.xqd, GS.xd, GS.x)
```

	Tot Df	Deviance	Chisq	Df	Pr(>Chisq)	
1	9	886.12				
2	8	887.04	0.9139	1	0.33907	
3	7	890.77	3.7384	1	0.05318	.
4	6	895.30	4.5210	1	0.03348	*

Signif. codes: 0 '***' 0.001 '**' 0.01 '*' 0.05 '.' 0.1 ' ' 1

This analysis confirms the results of summary on the most complex model to some extent. It finds that the effect of the interaction (model 1 vs. 2) is insignificant and the effect of density is significant at $p = 0.03$ (model 3 vs. 4). The effect of quality (when added to a model that already accounts for density) is weakly significant. The parameter values (coef(GS.xqd)) show the positive effect of quality (0.076) to be about half the negative effect of density (-0.149) on the log scale; adding one competitor to a reef decreases the scale parameter (and hence survival) by a factor of $e^{-0.149} = 0.86$, while an additional background settler indicates some element

TABLE 8.2
Comparison of goby survival models

Model	Parameters	ΔAIC	AIC Weights	ΔAIC$_c$	ΔBIC
qd	4	0.00	0.23	0.00	2.54
xqd	8	0.25	0.20	0.54	18.44
qdi	5	0.92	0.15	0.98	7.38
xqdi	9	1.34	0.12	1.73	23.44
d	3	1.37	0.12	1.32	0.00
xd	7	1.99	0.09	2.19	16.27
xq	7	4.02	0.03	4.22	18.30
x	6	4.51	0.02	4.63	14.88
q	3	4.85	0.02	4.80	3.48
0	2	5.50	0.02	5.42	0.22

of quality that increases survival on average by a factor of $e^{0.076} = 1.08$. (You can interpret small coefficients on the log scale as approximate percentage differences: $0.076 \approx 8\%$ increase and $-0.149 \approx 15\%$ decrease.)

Alternatively, we can simply fit the remaining six models (qdi, qd, xq, d, q, 0—not shown) and use information criteria (AICtab, AICctab, or BICtab) to compare the results. Table 8.2 shows perhaps too much information—because of the different weighting used by the different information criteria, they give qualitatively different answers. AIC and AIC$_c$ prefer the model that incorporates the effects of quality and density, with all the models considered plausible (ΔAIC, ΔAIC$_c$ < 6 for all candidate models) but with the simplest models weighted very little: In contrast, BIC prefers the simplest models (0, d), ruling out the most complex ones (ΔBIC > 10 for xqd, xqdi, xd, xq, x).

What should one conclude in this situation, with too many possible answers? There isn't really a good answer, except that one should decide *in advance* which model selection approach (if any) comes closest to answering the kind of question you have, rather than trying several and then having to choose among the answers. Here there is fairly strong evidence that density decreases survival, and that the effect of quality is about half as strong (per fish present) as that of density. In terms of the range of values used in the experiment, density and quality have approximately equivalent effects (density has a range of 9, from 2 to 11, while quality ranges from 1 to 18).

This particular analysis leaves few loose ends, but there are a number of possible directions for further exploration:

- We have followed standard survival analysis in making the mortality rate an exponential function of covariates such as density. Fisheries biologists commonly model mortality as a linear (additive) function of density instead (i.e., Prob(survival to t) proportional to $e^{-a+b\cdot d}$ rather than Prob(survival to t) proportional to $e^{-e^{a+b\cdot d}}$). The exponential analysis is more convenient because it guarantees that the mortality rate will always be positive regardless of the parameters, thus avoiding the need for constrained optimization. For small mortality rates the analysis will give approximately the same answers, since by Taylor expansion the exponential is approximately linear near zero. It would be interesting to redo the analysis with a linear model and see how similar the answers were. More challengingly, one could explore the dependence of survival on density and quality in greater detail—perhaps graphically—and see if a more flexible function could give a better answer, although with this small a data set greater flexibility might not be warranted.

- We ignored differences in shape parameter. Returning to explore the possibilities of differences in shape (representing the differences in change in mortality over time) some more, would be interesting as would exploring with a wider variety of data; does the shape parameter vary with the mode of mortality?

8.3 Seed Removal

For the Duncan seed predation/seed removal data, some of the ecological questions are: How does the probability of seed removal vary as a function of distance from the forest edge (10 or 25 m)? With species, possibly as a function of seed mass? By time?

Since most of the predictor variables are categorical in this case (species; distance from forest), the deterministic models are relatively simple—simply different probabilities for different levels of the factors. On the other hand, the distribution of the number of seeds taken is unusual, so most of the initial modeling effort will go into finding an appropriate stochastic model.

8.3.1 Preliminaries

Load the data:

```
> data(SeedPred)
```

Drop NAs and records where there are zero seeds available; attach the results, remembering to detach them later.

```
> SeedPred = na.omit(subset(SeedPred, available > 0))
> attach(SeedPred)
```

About 90% of the data consist of "zero taken" entries. We don't want to ignore these data, but sometimes we can see more if we look only at the nonzero cases; we'll use nz for that case.

```
> nz = subset(SeedPred, taken > 0)
```

8.3.2 Stochastic Model: Which Distribution?

I used barchart from the lattice package to look at the data in a variety of different ways—rearranging the order of the factors in the table to get different arrangements of panels and bars, plotting data with zero-taken data included and excluded, and dropping factors from the `table` command to see coarser views of the data:

```
> barchart(table(nz$taken, nz$available, nz$dist,
+     nz$species), stack = FALSE)
> barchart(table(nz$taken, nz$species, nz$dist,
+     nz$available), stack = FALSE)
> barchart(table(nz$species, nz$available, nz$dist,
+     nz$taken), stack = FALSE)
> barchart(table(nz$available, nz$dist, nz$taken),
+     stack = FALSE)
> barchart(table(nz$available, nz$species, nz$taken),
+     stack = FALSE)
```

I could also have included the argument `subset=taken>0`, instead of defining `nz` beforehand, to restrict the plots to nonzero data.

Plot all data (not just cases where some seeds are taken):

```
> barchart(table(available, dist, taken), stack = FALSE)
```

Plot by date:

```
> tcumfac = cut(nz$tcum, breaks = c(0, 20, 40, 60,
+     180))
> barchart(table(nz$available, tcumfac, nz$taken),
+     stack = FALSE)
> barchart(table(available, tcumfac, taken), stack = FALSE)
```

Two additional useful arguments are `auto.key=TRUE`, to draw a legend for the bar colors, and `scales=list(relation="free")`, to allow different scales in each panel.

As with the reed frog predation experiment, the data are discrete and the results have an upper limit (i.e., the number of seeds available for removal at the beginning of the interval). The zero-inflated binomial introduced in Chapter 4 might make sense, if there were more zeros in the data set than expected from the binomial sampling process (e.g., if the probability distribution had modes both at zero and away from zero). This distribution would be appropriate if predators sometimes missed the site entirely. However, Figure 8.7 shows that the seed removal data set doesn't look like a zero-inflated binomial either, because the distribution is lowest in the middle and increases gradually for higher or lower values. Compare that with Figure 4.1 (p. 107), which shows that the probability distribution function of the zero-inflated binomial distribution usually drops toward zero, then has a spike at zero ($p(0) > p(1)$, $p(1) < p(2)$).

Next I tried the beta-binomial distribution, which allows for variability in the underlying probabilities per trial and can be bimodal at 0 and N for extreme values of the overdispersion parameter, and a zero-inflated beta-binomial distribution.

One should really test the fits of distributions on a small piece of the data set or allowing for different parameters for each combination of factors; variation among groups can mask the shape of the underlying distribution. However, trying to fit parameters for an unknown distribution for all combinations of factors simultaneously can be tedious, and the exploratory graphical analysis described

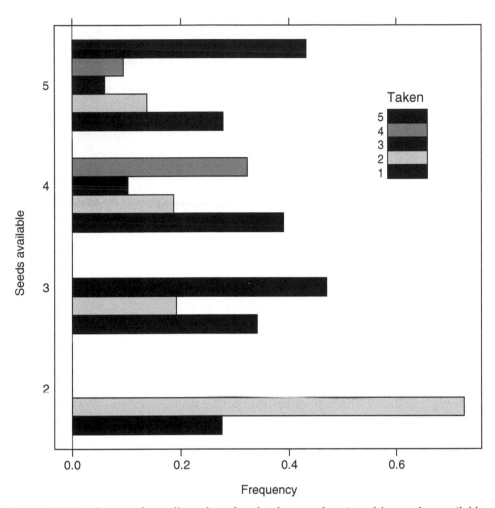

Figure 8.7 Distribution of overall number of seeds taken as a function of the number available, when number available > 1 and number taken > 0.

above convinced me that the pattern shown in Figure 8.7 holds up even when the data are disaggregated by species, distance, or date.

Using the dzinbinom function in the emdbook package as a model, I constructed probability density functions for the zero-inflated binomial (dzibinom) and zero-inflated beta-binomial (dzibb):

```
> dzibinom = function(x, prob, size, zprob, log = FALSE) {
+       logv = log(1 - zprob) + dbinom(x, prob = prob,
+           size = size, log = TRUE)
+       logv = ifelse(x == 0, log(zprob + exp(logv)),
+           logv)
+       if (log)
+           logv
+       else exp(logv)
+ }
```

```
> dzibb = function(x, size, prob, theta, zprob, log = FALSE)
+     { logv = ifelse(x > size, NA, log(1 - zprob) +
+         dbetabinom(x, prob = prob, size = size,
+             theta = theta, log = TRUE))
+     logv = ifelse(x == 0, log(zprob + exp(logv)),
+         logv)
+     if (log)
+         logv
+     else exp(logv)
+ }
```

Next I took a shortcut and used the formula interface to mle2 rather than writing an explicit negative log-likelihood function. Since the zero-inflation probability must be between 0 and 1, I fitted it on a logit scale, using plogis to transform it on the fly:

```
> SP.zibb = mle2(taken ~ dzibb(size = available, prob,
+     theta, plogis(logitzprob)), start = list(prob = 0.5,
+     theta = 1, logitzprob = 0))
```

There were warnings about NaNs in lbeta, but the final answers look reasonable. I was surprised to see that the zero-inflation probability was so small: plogis(-2.33) = 0.089. I suspected that the zero-inflation parameter and the overdispersion parameter (θ) might both be affecting the number of zeros, so I checked the correlations among the parameters:

```
> cov2cor(vcov(SP.zibb))
```

	prob	theta	logitzprob
prob	1.0000000	0.2885011	0.9867901
theta	0.2885011	1.0000000	0.3436282
logitzprob	0.9867901	0.3436282	1.0000000

Indeed, logitzprob and prob are 99% correlated—suggesting that we could drop the zero-inflation parameter from the model.

```
> SP.bb = mle2(taken ~ dbetabinom(size = available,
+     prob, theta), start = list(prob = 0.5, theta = 1))

> logLik(SP.bb) - logLik(SP.zibb)

'log Lik.' 0.07956568 (df=2)
```

The log-likelihood difference is only about 0.08. Even allowing for the fact that the null value of the zero-inflation parameter is on the boundary, so that the appropriate $\bar{\chi}^2_1$ p-value is half the usual χ^2_1 p-value, this difference is certainly not significant.

Just for completeness, I fitted the zero-inflated binomial too (although I didn't think it would fit well):

```
> SP.zib = mle2(taken ~ dzibinom(size = available,
+     prob = p, zprob = plogis(logitzprob)), start = list
+     (p = 0.2, logitzprob = 0))
```

Using AIC to compare all three distributions confirmed my suspicions:

```
> AICtab(SP.zib, SP.zibb, SP.bb, sort = TRUE, weights =
+     TRUE)
```

	AIC	df	weight
SP.bb	3626.1	2	0.746
SP.zibb	3628.3	3	0.254
SP.zib	4045.6	2	<0.001

Figure 8.8 compares the predictions of the different distributions, with stacked barplots showing the breakdown of different numbers of seeds taken for each number of seeds available.

The R code to calculate this distribution for the data first computes the table of number-taken-by-number-available, then uses sweep to divide each column (margin 2) by its sum:

```
> comb = table(taken, available)
> pcomb = sweep(comb, 2, colSums(comb), "/")
```

The equivalent computation for the zero-inflated beta-binomial sets up an empty matrix with six rows (for 0 to 5 seeds taken) and five columns (for 1 to 5 seeds available). For each number available N, it then sets the first $N + 1$ rows in column N of the matrix to the predicted probability of taking 0 to N seeds.

```
> mtab = matrix(0, nrow = 6, ncol = 5)
> for (N in 1:5) {
+     cvals = coef(SP.zibb)
+     mtab[1:(N + 1), N] = dzibb(0:N, size = N,
+     prob = cvals["prob"],
+         theta = cvals["theta"], zprob = plogis(cvals
+         ["logitzprob"]))
+ }
```

Similar calculations work for the other two distributions.

As we would expect from the statistical results so far, the zero-inflated beta-binomial and beta-binomial predictions look nearly identical, and much closer than the zero-inflated binomial results. However, there are still visible discrepancies for the cases of 4 and 5 seeds available—the predicted distributions are more regular, and have more even distributions, than the observed. None of the three models capture the increased probability that one seed would be taken (dark blocks) when four or five seeds were available.

We can calculate standard Pearson χ^2 p-values for the probability of the observed numbers taken for each number of seeds available:

```
> pval = numeric(5)
> for (N in 1:5) {
```

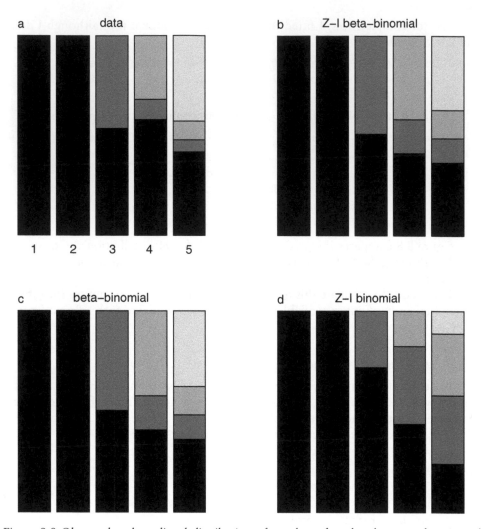

Figure 8.8 Observed and predicted distribution of number of seeds taken as a function of number available. (Zero-taken results are omitted, and columns are rescaled to add to 1.)

```
+        obs = comb[1:(N + 1), N]
+        prob = mtab[1:(N + 1), N]
+        pval[N] = chisq.test(obs, p = prob)$p.value
+ }
```

The p-values are:

1	2	3	4	5
0.53	0.29	0.81	0.01	<0.001

There are still statistically significant discrepancies between the expected and observed distributions when 4 or 5 seeds are available. We could try to find a way to make the stochastic model more complex and accurate, but we have reached the limit of what we can do with simple models, and we may also have reached the limit of what we can do with the data. The mechanism for the pattern remains obscure. While I can imagine mechanisms that would lead to all seeds or none being taken, it's hard to see why it's least likely that 3 out of 5 available seeds would be taken. I suspect that there is some disaggregation of the data by species, date, etc., that would divide stations into those where few or many seeds were taken, with an extreme pattern in each case that combines to create the observed bimodal pattern, but I haven't been able to find it.

8.3.3 Deterministic Model: Differences among Species, Distance, Space, and Time

Now we can check for differences among distances from the forest, species, and possibly differences in space and time: How does the distribution of number of seeds removed vary? Does p, the overall probability that a seed will be removed, vary? Does θ (the overdispersion parameter, which in this case is more related to the probability that any seeds will be removed) vary? Do they both vary?

8.3.3.1 DIFFERENCES AMONG TRANSECTS (DISTANCE FROM EDGE)

mle2's formula interface allows us to specify that some parameters vary among groups, by giving a parameters argument which is a list of the formulas for each group (p. 206). Here I wanted to parameterize the model so that mle2 would estimate the probability and overdispersion parameter for each distance, rather than estimating the parameters for the first transect and the difference between the first and second transect, so I used the formulas prob~dist-1 and theta~dist-1 to fit the model without an intercept.

```
> SP.bb.dist = mle2(taken ~ dbetabinom(prob, theta,
+      size = available),
+ parameters = list(prob ~ dist - 1, theta ~
+      dist - 1), start = as.list(coef(SP.bb)))
```

A Likelihood Ratio test on the two models suggests a significant difference between transects:

```
> anova(SP.bb, SP.bb.dist)

Likelihood Ratio Tests
Model 1: SP.bb, taken~dbetabinom(prob,theta,size=available)
Model 2: SP.bb.dist,
        taken~dbetabinom(prob,theta,size=available):
        prob~dist-1, theta~dist-1
Tot   Df    Deviance    Chisq    Df    Pr(>Chisq)
1     2     3622.1
2     4     3615.6      6.4823   2     0.03912    *
---
Signif. codes:  0 '***' 0.001 '**' 0.01 '*' 0.05 '.' 0.1 ' ' 1
```

Reparameterizing the model in terms of differences between the 10-m and 25-m transect rather than the p and θ values for each transect (i.e., dropping the -1 in the parameter formulas) allows us to calculate confidence limits on the differences between transects. At the same time, I decided to switch to fitting p on a logit scale and θ on a log scale. With the formula interface, I can do the inverse transformations on the fly with plogis and exp.

Set up starting values, using qlogis (the logit transform) and log to transform the estimated values of the p and θ parameters from above.

```
> startvals = list(lprob =
+     qlogis(coef(SP.bb.dist)["prob.dist10"]),
+     ltheta = log(coef(SP.bb.dist)["theta.dist10"]))

> SP.bb.dist2 = mle2(taken ~ dbetabinom(plogis(lprob),
+     exp(ltheta),size = available), parameters =
+     list(lprob ~ dist, ltheta ~ dist), start = startvals)
```

The summary of the model now gives us approximate p-values on the parameters, showing that the difference between transects is caused by a change in p and not a change in θ.

```
> summary(SP.bb.dist2)

Maximum likelihood estimation

Call:
mle2(minuslogl = taken ~ dbetabinom(plogis(lprob),
    exp(ltheta), size = available), start = startvals,
    parameters = list(lprob ~ dist, ltheta ~ dist))

Coefficients:
                     Estimate    Std. Error   z value    Pr(z)
lprob.(Intercept)   -2.7968262   0.0813997   -34.3592   < 2e-16   ***
lprob.dist25         0.2663037   0.1110270     2.3985   0.01646   *
ltheta.(Intercept)  -1.1255457   0.1261399    -8.9230   < 2e-16   ***
ltheta.dist25       -0.0035835   0.1719498    -0.0208   0.98337
---
Signif. codes:  0 '***' 0.001 '**' 0.01 '*' 0.05 '.' 0.1 ' ' 1

-2 log L: 3615.627
```

(The highly significant p-values for lprob.10 and ltheta.10 are not biologically significant: they merely show that $\mathrm{logit}(p_{10}) \neq 0$ (i.e., $p_{10} \neq 0.5$) and $\log\theta_{10} \neq 0$ ($\theta_{10} \neq 1$), neither of which is ecologically interesting.)

Now reduce the model, allowing only p to vary between transects:

```
> SP.bb.probdist = mle2(taken ~ dbetabinom(plogis(lprob),
+     exp(ltheta), size = available), parameters =
+     list(lprob ~ dist), start = startvals)
```

Both the LRT and the AIC approaches suggest that the best model is one in which p varies between transects but θ does not (although the AIC table suggests that the more complex model with differing θ should be kept in consideration):

```
> anova(SP.bb, SP.bb.probdist, SP.bb.dist)

Likelihood Ratio Tests
Model 1: SP.bb,taken~dbetabinom(prob,theta,size=available)
Model 2: SP.bb.probdist,
         taken~dbetabinom(plogis(lprob),exp(ltheta),size=available):
         lprob~dist
Model 3: SP.bb.dist,
         taken~dbetabinom(prob,theta,size=available):
         prob~dist-1, theta~dist-1
Tot    Df    Deviance    Chisq    Df    Pr(>Chisq)
1      2     3622.1
2      3     3615.6      6.4819   1     0.01090    *
3      4     3615.6      0.0004   1     0.98341
---
Signif. codes:  0 '***' 0.001 '**' 0.01 '*' 0.05 '.' 0.1 ' ' 1

> AICtab(SP.bb, SP.bb.probdist, SP.bb.dist, sort = TRUE,
+      weights = TRUE)

                   AIC      df    weight
SP.bb.probdist    3621.6    3     0.678
SP.bb.dist        3623.6    4     0.250
SP.bb             3626.1    2     0.072
```

How big is the difference between transects?

```
> c1 = coef(SP.bb.probdist)
> plogis(c(c1[1], c1[1] + c1[2]))

lprob.(Intercept)    lprob.(Intercept)
   0.05751881           0.07372130
```

The difference is small—6% vs. 7% probability of removal per observation. This difference is unlikely to be ecologically significant, and it reminds us that when we have a big data set (4406 observations) even small differences can be statistically significant. On the other hand, Duncan and Duncan (2000) failed to find a significant difference between the transects—so the likelihood framework is more powerful, and has given us answers in terms (average percent difference in probability of removal) that we can understand.

8.3.3.2 DIFFERENCES AMONG SPECIES

Now I proceeded to test differences among species. First I tried a model with both θ and p varying. (Both parameters are again fitted on transformed scales, logit and log respectively.)

```
> SP.bb.sp = mle2(taken ~ dbetabinom(plogis(lprob),
+       exp(ltheta), size = available), parameters =
+       list(lprob ~ species, ltheta ~species),
+       start = startvals)
```

The parameter estimates (shown in full by summary(SP.bb.sp); here I dropped one column of the table) suggest that, as in the case of differences among transects, differences in p and not in θ are driving the differences among species:

	Estimate	Std. Error	Pr(z)	
lprob.(Intercept)	−1.925509	0.1428	< 2.2e-16	***
lprob.speciescd	0.329247	0.2186	0.1321056	
lprob.speciescor	−1.332956	0.2144	5.090e-10	***
lprob.speciesdio	−0.991505	0.2111	2.645e-06	***
lprob.speciesmmu	−0.432409	0.2130	0.0423696	*
lprob.speciespol	0.413143	0.2098	0.0489483	*
lprob.speciespsd	−1.274415	0.2207	7.704e-09	***
lprob.speciesuva	−1.302890	0.2146	1.266e-09	***
ltheta.(Intercept)	−0.824310	0.2240	0.0002327	***
ltheta.speciescd	−0.560802	0.3473	0.1063536	
ltheta.speciescor	0.016070	0.3292	0.9610611	
ltheta.speciesdio	−0.377969	0.3276	0.2485773	
ltheta.speciesmmu	−0.618604	0.3354	0.0651542	.
ltheta.speciespol	0.152877	0.3331	0.6462837	
ltheta.speciespsd	−0.173435	0.3405	0.6105292	
ltheta.speciesuva	−0.058962	0.3341	0.8599198	

```
---
Signif. codes:  0 '***' 0.001 '**' 0.01 '*' 0.05 '.' 0.1 ' ' 1
```

So I fitted a model with only probability p, and not overdispersion θ, varying by species:

```
> SP.bb.probsp = mle2(taken ~ dbetabinom(plogis(lprob),
+       exp(ltheta), size = available), parameters = list
+       (lprob ~ species), start = startvals)
```

Once again, both LRT and AIC suggest that only the p parameters differ among species:

```
> anova(SP.bb.sp, SP.bb.probsp, SP.bb)
```

```
Likelihood Ratio Tests
Model 1: SP.bb.sp,
        taken~dbetabinom(plogis(lprob), exp(ltheta),
        size=available): lprob~species, ltheta~species
Model 2: SP.bb.probsp,
        taken~dbetabinom(plogis(lprob), exp(ltheta),
        size=available): lprob~species
Model 3: SP.bb, taken~dbetabinom(prob,theta, size=available)
```

```
Tot   Df   Deviance      Chisq    Df   Pr(>Chisq)
1     16   3460.4
2      9   3469.8        9.3894    7   0.2259
3      2   3622.1      152.2873    7   <2e-16   ***
---
Signif. codes:  0 '***' 0.001 '**' 0.01 '*' 0.05 '.' 0.1 ' ' 1
```

```
> AICtab(SP.bb.sp, SP.bb.probsp, SP.bb, sort = TRUE,
+        weights = TRUE)
```

```
                AIC     df    weight
SP.bb.probsp    3487.8   9    0.909
SP.bb.sp        3492.4  16    0.091
SP.bb           3626.1   2    <0.001
```

Now I want to know whether seed mass and p are related. If they were, I could fit a likelihood model where p was treated as a function of seed mass, reducing the number of parameters to estimate and perhaps allowing me to predict removal probabilities for other species on the basis of their seed masses.

```
> SP.bb.probsp0 = mle2(taken ~ dbetabinom(plogis(lprob),
+       exp(ltheta), size = available), parameters = list
+       (lprob ~ species - 1), start = startvals,
+       method = "L-BFGS-B",
+       lower = rep(-10, 9), upper = rep(10, 9))
```

Fitting this model was numerically problematic. In my first attempt, using default methods and parameters, mle2 found a ridiculous answer (all the logit probabilities were strongly negative, giving removal probabilities near zero) and crashed while evaluating the Hessian. I used skip.hessian=TRUE to temporarily stop mle2 from crashing and trace=TRUE to see what was happening. Switching to method="Nelder-Mead" helped stabilize the calculation, but it failed to converge until I increased the number of iterations to 3000 (control=list(maxit=3000)), and even then it got stuck on a solution that was worse than the previous model. (In this case, since all I am doing is reparameterizing the previous model, mle2 ought to be able to achieve an equally good fit.) I then went back to BFGS and tried changing the size of the finite difference interval both down (control=list(ndeps=rep(1e-4,9))) and up (control=list(ndeps=rep(1e-2,9))), neither of which helped. I finally got the model to fit as well as the previous parameterization by switching to L-BFGS-B and setting the parameter boundaries to disallow ridiculous fits.

```
> predprob = plogis(coef(SP.bb.probsp0))[1:8]
> SP.bb.ci = plogis(confint(SP.bb.probsp0,
+       method = "quad"))[1:8, ]
```

Figure 8.9 shows the results: rather than the possible trend toward higher seed removal for larger seeds that I expected, the figure shows elevated removal rates for the three smallest-seeded species (explained by Duncan and Duncan as a possible artifact of small seeds being washed out of the trays by rainfall), and a somewhat elevated rate for species mmu; in this case, I would want to go back and see if there

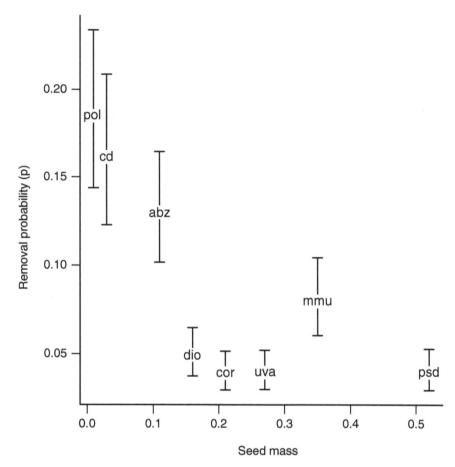

Figure 8.9 Removal probability parameter (*p*) as a function of seed mass; error bars show quadratic confidence intervals.

was something special about this species' characteristics or the way it was handled in the experiment.

8.3.3.3 IS THERE A SPECIES-DISTANCE INTERACTION?

The initial scan of the data suggested that some species might be more sensitive to the distance from the edge: This possibility is certainly biologically sensible (some species might be taken by specialized seed predators that have more restricted movement), and it is the kind of information that could easily be masked by looking at aggregated data.

Using the formula interface, we can simply say `lprob~species*dist` to allow for such an interaction: if you need to code such a model by hand, `interaction(f1,f2)` will create a factor that represents the interaction of factors `f1` and `f2`.

```
> SP.bb.probspdist = mle2(taken ~ dbetabinom(plogis(lprob),
+       exp(ltheta), size = available), parameters = list
```

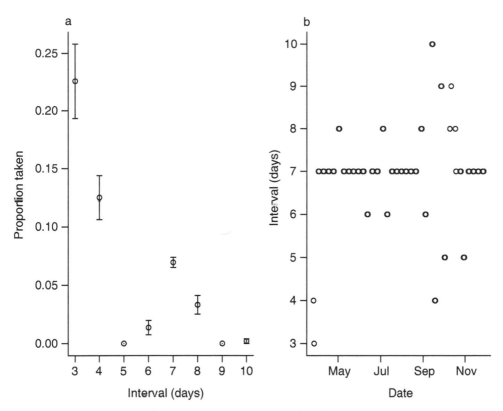

Figure 8.10 Relationships between proportion removed and time interval (Δt), and between Δt and date.

```
+       (lprob ~ species * dist), start = startvals, method =
+       "L-BFGS-B", lower = rep(-10, 9), upper = rep(5, 9))
```

I had to restrict the upper bounds still further, to 5, to make L-BFGS-B happy, since values of 10 gave NaN results for some parameter combinations.

A Likelihood Ratio test (anova(SP.bb.probsp,SP.bb.probspdist)) gives a p-value of 0.054; AIC says that the model without distance × species interaction is best, but only by a little bit:

```
> AICtab(SP.bb, SP.bb.probsp, SP.bb.probspdist, SP.bb.sp,
+       SP.bb.probdist, SP.bb.dist, weights = TRUE,
+       sort = TRUE)
```

	AIC	df	weight
SP.bb.probsp	3487.8	9	0.559
SP.bb.probspdist	3488.6	17	0.386
SP.bb.sp	3492.4	16	0.056
SP.bb.probdist	3621.6	3	<0.001
SP.bb.dist	3623.6	4	<0.001
SP.bb	3626.1	2	<0.001

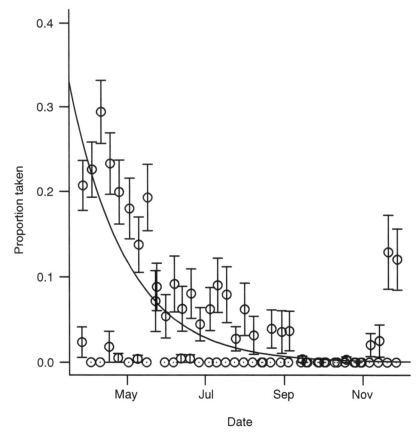

Figure 8.11 Proportion taken as a function of date. The line shows fitted exponential dependence ($p = 0.26 \times e^{-0.023t}$), based on a fitted model that lumps all the species together.

8.3.3.4 OTHER ISSUES: TIME

One issue that I have neglected so far is that the intervals between observations varied between 3 and 14 days. To account for these differences in exposure time, I could use a model like $p = 1 - e^{-r(\Delta t)}$, which assumes that seeds are taken at a constant rate r. Do the predictions improve, or the conclusions change, if I account for the time interval allowed for removal?

Before going to the trouble of building a model, let's look at the data again. Calculate the mean and standard error of the proportion taken, using `tapply` to calculate means and standard deviations of proportions divided up by the time interval (`tint`); then use `table` to calculate the number of observations for each time interval and divide by $\sqrt{n}$ to convert standard deviations to standard errors.

```
> mean.prop.taken = tapply(taken/available, tint, mean,
+     na.rm = TRUE)
> sd.prop.taken = tapply(taken/available, tint, sd,
+     na.rm = TRUE)
```

```
> n.tint = table(tint)
> se.prop.taken = sd.prop.taken/sqrt(n.tint)
```

Figure 8.10a is a surprise: the model $p = 1 - e^{-r(\Delta t)}$ suggests the proportion taken should increase rather than decrease with Δt. What's going on? Figure 8.10b, which plots the time interval between observations against date, gives the answer: the short-interval (3–4 day) observations were mostly made before May, when the removal rate was high, while the longest intervals between observations (10 days) are in September.

This brings us to the issue of temporal variation: we already know from Figure 2.1 in Chapter 2 that the removal rate decreases over time. Figure 8.11 shows the relationship between proportion removed and date, calculated in the same way as the removal vs. Δt relationship. Removal appears to decrease exponentially with time. Replotting the data with a logarithmic y scale suggests that the removal rate might level off above zero, but it's hard to tell. Similarly, it's hard to know what causes the anomalously low proportions for some sampling dates throughout the study and the anomalously high proportions at the very end of the study. Nevertheless, we can add a parameter to the model allowing for exponential decrease in removal rate over time:

```
> SP.bb.probspdate = mle2(taken ~ dbetabinom(plogis(lprob)*
+       exp(-tcum * date), exp(ltheta), size = available),
+       parameters = list(lprob ~ species), start =
+         c(startvals, date = 0), method = "L-BFGS-B",
+         lower = c(rep(-10, 9), 0), upper = c(rep(5, 9),
+         2))
```

The model incorporating date is 237.6 log-likelihood units better—the model should definitely include the effect of date.

We have gotten a lot of mileage from these data, but as always there are more questions we could ask: Do the removal rates of different species drop off at different rates? Can we figure out what causes the anomalous samples in Figure 8.11? Once we have split the data according to these criteria, can we simplify the underlying distribution?

9 Standard Statistics Revisited

This chapter rapidly reviews much of classical statistics, discussing the underlying likelihood models for procedures such as ANOVA, linear regression, and generalized linear models. It also gives brief pointers to the built-in procedures in R that implement these standard techniques. This summary connects maximum likelihood approaches with more familiar classical techniques. If you're already familiar with classical techniques, it may help you understand maximum likelihood better. It also provides a starting point for using efficient, "canned" approaches when they are appropriate for your data. It does not, and cannot, provide full coverage of all these topics. For more details, see Dalgaard (2003), Crawley (2005, 2007), or Venables and Ripley (2002).

9.1 Introduction

So far this book has covered maximum likelihood and Bayesian estimation in some detail. In the course of the discussion I have sometimes mentioned that maximum likelihood analyses give answers equivalent to those provided by familiar, "old-fashioned" statistical procedures. For example, the statistical model $Y \sim$ Normal$(a + bx, \sigma^2)$—specifying that Y is a normally distributed random variable whose mean depends linearly on x—underlies ordinary least-squares linear regression. This chapter will briefly review special cases where our general recipe for finding MLEs for statistical models reduces to standard procedures that are built into R and other statistics packages.

In the best case, your data will match a classical technique like linear regression exactly, and the answers provided by classical statistical models will agree with the results from your likelihood model. Other models you build may be formally equivalent to a classical model that is parameterized in a different way. Most often, the customized model you build will not be exactly equivalent to any existing classical model, but a similar classical model may be close enough that you wouldn't mind changing your model slightly in order to gain the convenience of using a standard procedure.

For example, in Chapter 6 we used the model

$$Y \sim \text{NegBin}(\mu = a \cdot \text{DBH}^b, k) \qquad (9.1.1)$$

to represent cone production by fir trees as a function of diameter at breast height. If we approximated the discrete distribution of cones by a continuous log-normal distribution instead,

$$Y \sim \text{LogNormal}(\mu = a \cdot \text{DBH}^b, \sigma^2), \tag{9.1.2}$$

we could log-transform both sides and fit the linear regression model

$$\log Y \sim \text{Normal}(\log a + b \cdot \log(\text{DBH}), \sigma^2). \tag{9.1.3}$$

Figure 9.1a shows all three models for the DBH–fecundity relationship—power-law with a negative binomial distribution (power/NB), power-law with a log-normal distribution (power/LN), and linear with a normal distribution—fitted to the fir data; all are plausible. Figure 9.1b shows various models for the distribution of cone production, fitted to the individuals with DBH between 6 and 8 cm: a nonparametric density estimate, the negative binomial, lognormal, and normal. The negative binomial is closest to the nonparametric density estimate of the distribution, while the lognormal is more peaked and the normal distribution has an unrealistic negative tail.

Although the power-law/negative binomial is the most realistic model and has a plausible mechanistic interpretation (the data are discrete, positive, and overdispersed; we can imagine individual trees producing cones at an approximately constant rate with variation in fecundity among trees), the difference between the fit of negative binomial and lognormal distributions is small enough that the convenience of linear regression may be worthwhile. When the results of different models are similar on both biological and statistical grounds, you choose among them by balancing convenience, mechanistic arguments, and convention.

Why might you want to use standard, special-case procedures rather than the general MLE approach?

- *Computational speed and stability*: The special-case procedures use special-case optimization algorithms that are faster (sometimes much faster) and less likely to encounter numerical problems. Many of these procedures relieve you of the responsibility of choosing starting parameters.
- *Stable definitions*: The definitions of standard models have often been chosen to simplify parameter estimation. For example, to model a relatively sudden change between two states you could choose between a logistic equation and a threshold model. Both might be equally sensible in terms of the biology, but the logistic equation is easier to fit because it involves smoother changes as parameters change. Similarly, generalized linear models such as logistic or Poisson regression fit parameters on scales (logit- or log-transformed, respectively) that allow unconstrained optimization.
- *Convention*: If you use a standard method, you can just say (e.g.) "we used linear regression" in your Methods section and no one will think twice. If you use a nonstandard method, you need to explain the method carefully and overcome readers' distrust of "fancy" statistics—even if your model is actually simpler and more appropriate than any standard model. Similarly, it may minimize confusion to use the same models, and the same parameterizations, as previous studies of your system.
- *Varying models and comparing hypotheses*: The machinery built into R and other packages makes it easy to compare a variety of models. For example,

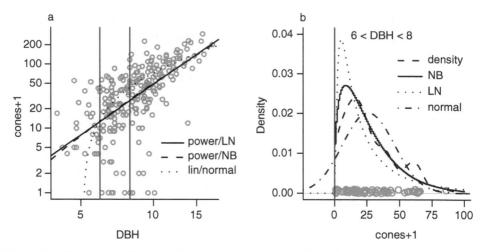

Figure 9.1 Comparing different functional forms for fir fecundity data: power-law with a lognormal (LN) distribution, power-law with a negative binomial (NB) distribution, and linear with a normal distribution. (The linear model appears as a curved line because the data are plotted on a log-log scale.)

when analyzing a factorial growth experiment that manipulates nitrogen (N) and phosphorus (P), you can easily switch between models incorporating the effects of nitrogen only (growth~N), phosphorus only (growth~P), additive effects of N and P (growth~N+P), and the main effects plus interactions between nitrogen and phosphorus (growth~N*P). You can carry out all of these comparisons by hand with your own models, and mle2's formula interface is helpful, but R's built-in functions make the process easy for classical models.

This chapter discusses how a variety of different kinds of models fit together, and how they all represent special cases of a general likelihood framework. Figure 9.2 shows how many of these areas are connected. The chapter also gives *brief* descriptions of how to use them in R; if you want more details on any of these approaches, you'll need to check an introductory (Dalgaard, 2003; Crawley, 2005; Verzani, 2005), intermediate (Crawley, 2002), or advanced (Chambers and Hastie 1992; Venables and Ripley, 2002) reference.

9.2 General Linear Models

General linear models include linear regression, one-way and multiway analysis of variance (ANOVA), and analysis of covariance (ANCOVA); R uses the function lm for all of these procedures. SAS implements this with PROC GLM.* While regression, ANOVA, and ANCOVA are often handled differently, and they are usually taught differently in introductory statistics classes, they are all variants of the same basic model. The assumptions of the general linear model are that all observed values are independent and normally distributed with a constant variance (*homoscedastic*),

---

* This terminology is unfortunate since the rest of the world uses "GLM" to mean general*ized* linear models, which correspond to SAS's PROC GENMOD.

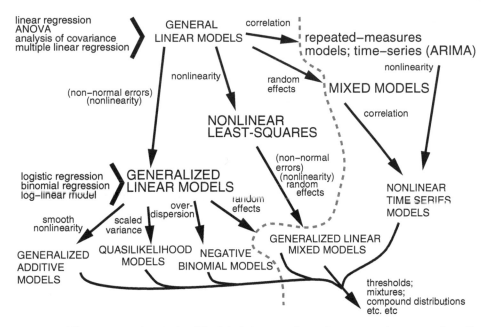

Figure 9.2 All (or most) of statistics. The labels in parentheses (non-normal errors and nonlinearity) imply restricted cases: (non-normal errors) means exponential family (e.g., binomial or Poisson) distributions, while (nonlinearity) means nonlinearities with an invertible linearizing transformation. Models to the right of the gray dashed line involve multiple levels or types of variability; see Chapter 10.

and that any continuous predictor variables (covariates) are measured without error. (Remember that the assumption of normality applies to the variation around the expected value—the residuals—not to the whole data set.)

The "linear" part of "general linear model" means that the models are linear functions *of the parameters*, not necessarily of the independent variables. For example, quadratic regression

$$Y \sim \text{Normal}(a + bx + cx^2, \sigma^2) \tag{9.2.1}$$

is still linear in the parameters (a, b, c), and thus is a form of multiple linear regression. Another way to think about this is to say that x^2 is just another explanatory variable—if you called it w instead, it would be clear that this model is an example of multivariate linear regression. On the other hand, $Y \sim \text{Normal}(ax^b, \sigma^2)$ is nonlinear: it is linear with respect to a (the second derivative of ax^b with respect to a is zero) but nonlinear with respect to b ($d^2(ax^b)/db^2 = b \cdot (b-1) \cdot ax^{b-2} \neq 0$).

9.2.1 Simple Linear Regression

Simple, or ordinary, linear regression predicts y as a function of a single continuous covariate x. The model is

$$Y \sim \text{Normal}(a + bx, \sigma^2). \tag{9.2.2}$$

The equivalent R code is

```
> lm.reg = lm(y ~ x)
```

The intercept term a is implicit in the R model. If you want to force the intercept to be equal to zero, fitting the model $y \sim \text{Normal}(bx, \sigma^2)$, use `lm(y~x-1)`.

Typing `lm.reg` by itself prints only the formula and the estimates of the coefficients; `summary(lm.reg)` also gives summary statistics (range and quartiles) of the residuals, standard errors and p-values for the coefficients, and R^2 and F statistics for the full model; `coef(lm.reg)` gives the coefficients alone, and `coef(summary (lm.reg))` pulls out the table of estimates, standard errors, t statistics, and p-values. `confint(lm.reg)` calculates confidence intervals. The function `plot(lm.reg)` displays various graphical diagnostics that show how well the assumptions of the model fit and whether particular points have a strong effect on the results; see `?plot.lm` for details. `anova(lm.reg)` prints an ANOVA table for the model.* If you need to extract numeric values of, e.g., R^2 values or F statistics for further analysis, wade through the output of `str(summary(lm.reg))` to find the pieces you need (e.g., `summary (lm.reg)$r.squared`).

To do linear regression by brute force with `mle2`, you could write this negative log-likelihood function:

```
> linregfun = function(a, b, sigma) {
+       Y.pred = a + b * x
+       -sum(dnorm(Y, mean = Y.pred, sd = sigma, log = TRUE))
+ }
```

or use the formula interface:

```
> mle2(Y ~ dnorm(mean = a + b * x, sd = sigma), start = ...)
```

When using `mle2` you must explicitly fit a standard deviation term σ, which is implicit in the `lm` approach.

9.2.2 Multiple Linear Regression

It's easy to extend the simple linear regression model to multiple continuous predictor variables (covariates). If the extra covariates are powers of the original variable $(x^2, x^3, \ldots)$, the model is called *polynomial* regression (*quadratic* if just the x^2 term is added):

$$Y \sim \text{Normal}(a + b_1 x + b_2 x^2, \sigma^2). \tag{9.2.3}$$

Or you can use completely separate variables $(x_1, x_2, \ldots)$:

$$Y \sim \text{Normal}(a + b_1 x_1 + b_2 x_2 + b_3 x_3, \sigma^2) \tag{9.2.4}$$

As with simple regression, the intercept a and the coefficients of the different covariates (b_1, b_2) are implicit in the R formula:

```
> lm.poly = lm(y ~ x + I(x^2))
```

---

* anova gives so-called *sequential sums of squares*, which SAS calls "type I" sums of squares. If you need SAS-style "type III" sums of squares, you can use the Anova function in the car package. However, be aware that type III sums of squares are problematic, and indeed controversial (Venables, 1998).

(surround x^2 and other powers of x with I(), meaning "as is") or

```
> lm.mreg = lm(y ~ x1 + x2 + x3)
```

You can add interactions among covariates, testing whether the slope with respect to one covariate changes linearly as a function of another covariate—e.g., $Y \sim \text{Normal}(a + b_1 x_1 + b_2 x_2 + b_{12} x_1 x_2, \sigma^2)$; in R, lm.intreg = lm(y~x1*x2).

Use the anova function with test="Chisq" to perform Likelihood Ratio tests on a nested series of multivariate linear regression models (e.g., anova(lm1,lm2, lm3,test="Chisq")). If you wonder why anova is a test for regression models, remember that regression and analyses of variance are just different subsets of the general linear model.

While multivariate regression is conceptually simple, models with many terms (e.g., models with many covariates or with multiway interactions) can be difficult to interpret. Blind fitting of models with many covariates can get you in trouble (Whittingham et al., 2006). If you absolutely must go on this kind of fishing expedition, you can use step, or stepAIC in the MASS package, to do stepwise modeling, or regsubsets in the leaps package to search for the best model.

9.2.3 One-Way Analysis of Variance (ANOVA)

If the predictor variables are discrete (factors) rather than continuous (covariate), the general linear model becomes an analysis of variance. The basic model is

$$Y_i \sim \text{Normal}(\alpha_i, \sigma^2); \tag{9.2.5}$$

in R it is

```
> lm.1way = lm(y ~ f)
```

where f is a factor. If your original data set has names for the factor levels (e.g., {N,S,E,W} or {high,low}), then R will automatically transform the treatment variable into a factor when it reads in the data. However, if the factor levels look like numbers to R (e.g., you have site designations 101, 227, and 359, or experiments numbered 1 to 5), R will interpret them as continuous rather than discrete predictors and will fit a linear regression rather than doing an ANOVA—not what you want. Use v=factor(v) to turn a numeric variable v into a factor, and then fit the linear model.

Executing anova(lm.1way) produces a basic ANOVA table; summary(lm.1way) gives a different view of the model, testing the significance of each parameter against the null hypothesis that it equals 0.

When fitting regression models, the parameters of the model are easy to interpret—they're just the intercept and the slopes with respect to the covariates. When you have factors in the model, however—as in ANOVA—the parameterization becomes trickier. By default, R parameterizes the model in terms of the differences between the first group and subsequent groups (*treatment contrasts*) rather than in terms of the mean of each group, although you can tell it to fit the means of each group by putting a −1 in the formula (e.g., lm.1way = lm(y~f-1)).

9.2.4 Multiway ANOVA

Multiway ANOVA models Y as a function of two or more different categorical variables (factors). For example, the full model for two-way ANOVA with interactions is

$$Y_{ij} \sim \text{Normal}(\alpha_i + \beta_j + \gamma_{ij}, \sigma^2) \tag{9.2.6}$$

where i is the level of the first treatment/group, and j is the level of the second. The R code using lm is

```
> lm.2way = lm(Y ~ f1 * f2)
```

(f1 and f2 are factors). As before, summary(lm.2way) gives more information, testing whether the parameters differ significantly from zero; confint(lm.2way) computes confidence intervals; anova(lm.2way) generates a standard ANOVA table; plot(lm.2way) shows diagnostic plots. If you want to fit just the main effects without the interactions, use lm(Y~f1+f2); use f1:f2 to specify an interaction between f1 and f2.

A negative log-likelihood function for mle2 could look like this:

```
> aov2fun = function(m11, m12, m21, m22, sigma) {
+     intval = interaction(f1, f2)
+     Y.pred = c(m11, m12, m21, m22)[intval]
+     -sum(dnorm(Y, mean = Y.pred, sd = sigma, log = TRUE))
+ }
```

(interaction(f1,f2) defines a factor representing the interaction of f1 and f2 with levels in the order (1.1, 2.1, 1.2, 2.2)). Using the formula interface:

```
> mle2(Y ~ dnorm(mean = m, sd = sigma),
+     parameters = list(m ~ f1 * f2))
```

For a multiway model, R's parameters are again defined in terms of contrasts. If you construct a two-way ANOVA with factors f1 (with levels A and B) and f2 (with levels I and II), the first ("intercept") parameter will be the mean of individuals in level A of the first factor and level I of the second (m11); the second parameter is the difference between A,II and A,I (m12-m11); the third is the difference between B,I and A,I (m21-m11); and the fourth, the interaction term, is the difference between the mean of B,II and its expectation if the effects of the two factors were additive (m22-(m11+(m12-m11)+(m21-m11)) = m22-m12-m21+m11).

9.2.5 Analysis of Covariance (ANCOVA)

Analysis of covariance defines a statistical model that allows for different intercepts and slopes with respect to a covariate x in different groups:

$$Y_i \sim \text{Normal}(\alpha_i + \beta_i x, \sigma^2). \tag{9.2.7}$$

In R:

```
> lm(Y ~ f * x)
```

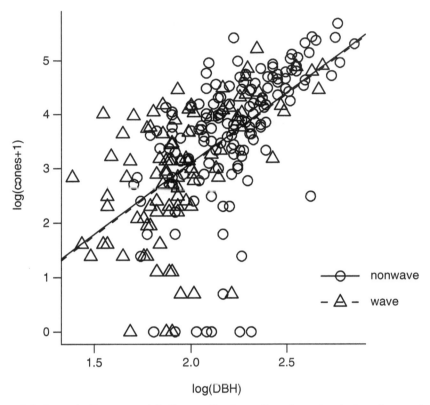

Figure 9.3 General linear model fit to fir fecundity data (analysis of covariance): `lm(log(TOTCONES+1)~log(DBH)+WAVE_NON,data=firdata)`. (Lines are practically indistinguishable between groups.)

where f is a factor and x is a covariate (the formula `Y~f+x` would specify parallel slopes, `Y~f` would specify zero slopes but different intercepts, `Y~x` would specify a single line). Figure 9.3 shows the fit of the model `lm(log(TOTCONES+1)~ log(DBH)+WAVE_NON)` to the fir data. As suggested by the figure, there is a strong effect of DBH but no significant effect of population (wave vs. nonwave).

As with other models, use `summary`, `confint`, `plot`, and `anova` to analyze the model. The parameters are now the intercept of the first factor level; the slope with respect to x for the first factor level; the differences in the intercepts for each factor level other than the first; and the differences in the slopes for each factor level other than the first.

A negative log-likelihood function for ANCOVA:

```
> ancovafun = function(i1, i2, slope1, slope2, sigma) {
+       int = c(i1, i2)[f]
+       slope = c(slope1, slope2)[f]
+       Y.pred = int + slope * x
+       -sum(dnorm(Y, mean = Y.pred, sd = sigma, log = TRUE))
+ }
```

9.2.6 More Complex General Linear Models

You can add factors (grouping variables) and interactions between factors in different ways to make multiway ANOVA, covariates (continuous independent variables) to make multiple linear regression, and combinations to make different kinds of analysis of covariance. R will automatically interpret formulas based on whether variables are factors or numeric variables.

9.3 Nonlinearity: Nonlinear Least Squares

Nonlinear least-squares models relax the requirement of linearity but keep the requirements of independence and normal errors. Two common examples are the power-law model with normal errors

$$Y \sim \text{Normal}(ax^b, \sigma^2) \qquad (9.3.1)$$

and the Ricker model with normal errors

$$Y \sim \text{Normal}(axe^{-rx}, \sigma^2). \qquad (9.3.2)$$

Before computers were ubiquitous, the only practical way to solve these problems was to *linearize* them by finding a transformation of the parameters (e.g., log-transforming x and y to do power-law regression). A lot of ingenuity went into developing transformation methods to linearize common functions. However, transforming variables changes the distribution of the error as well as the shape of the dependence of y on x. Ideally we'd like to find a transformation that simultaneously produces a linear relationship and makes the errors normally distributed with constant variance, but these goals are often incompatible. If the errors are normal with constant variance, they won't be after you transform the data to linearize $f(x)$.

The modern way to solve these problems without distorting the error structure, or to solve other models that cannot be linearized by transforming them, is to minimize the sums of squares (equivalent to minimizing the negative log-likelihood) computationally, using quasi-Newton methods similar to those built into `optim`. Restricting the variance model to normally distributed errors with constant variance allows the use of specific numeric methods that are more powerful and stable than the generalized algorithms that `optim` uses.

In R, use the `nls` command, specifying a nonlinear formula and the starting values (as a list); e.g., for the power model

```
> n1 = nls(y ~ a * x^b, start = list(a = 1, b = 1))
```

As usual, `summary(n1)` shows values of parameters and standard errors; `anova (n1,...)` does likelihood ratio tests for nested sequences of nonlinear fits; and `confint(n1)` computes profile confidence limits which are more accurate than the confidence limits suggested by `summary(n1)`. (Unfortunately, `plot(n1)` does nothing.) Figure 9.4 shows the fit of a nonlinear least-squares model (`nls(TOTCONES~ a*DBH^b)`) to the fir fecundity data set, along with the log-log fit (equivalent to a power-law fit with lognormal errors) calculated above. The power-lognormal model

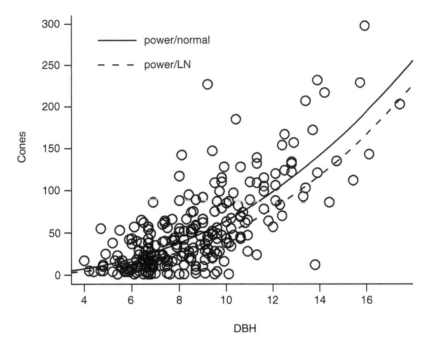

Figure 9.4 A nonlinear least-squares fit to the fir fecundity data (nls(TOTCONES~a*DBH^b,...)); the linear model fit to the log-log data (equivalent to a power-law fit with lognormal errors) is also shown.

is better from a biological point of view, since the normal distribution allows negative values, but both models are reasonable.

Fitting models with both nonlinear covariates and categorical variables (the non-linear analogue of ANCOVA—e.g., fitting different *a* and *b* parameters for wave and nonwave populations) is more difficult, but two functions from the nlme package, nlsList and gnls (generalized nonlinear least squares), can handle such models. nlsList does completely separate fits for separate groups—for example,

```
> nlsList(TOTCONES ~ a * DBH^b | WAVE_NON, data = firdata,
+     start = list(a = 0.1, b = 2.7))
```

would fit separate *a* and *b* parameters for wave and nonwave populations—all parameters will vary among groups. The gnls command can fit models with only a subset of the parameters differing among groups—for example,

```
> gnls(TOTCONES ~ a * DBH^b, data = firdata, start = c(0.1,
+     2.7, 2.7), params = list(a ~ 1, b ~ WAVE_NON))
```

will fit different *b* parameters but the same *a* parameter for wave and nonwave populations.

The numerical methods that nls uses are similar to mle2's in that (1) you must specify starting values and (2) if the starting values are unrealistic, or if the problem is otherwise difficult, the numerical optimization may get stuck. Errors such as

```
step factor [] reduced below 'minFactor' of ...
```

```
number of iterations exceeded maximum of ...
```

or

```
Missing value or an infinity produced when evaluating the model
```

indicate numerical problems. To solve these problems try to find better starting conditions, reparameterize your model, or adjust the control options of `nls` (see `?nls.control`).

As with ML models, you can often use simpler, more robust approaches like linear models to get a first estimate for the parameters (e.g., estimate the initial slope of a Michaelis-Menten function from the first 10% of the data and the asymptote from the last 10%, or estimate the parameters by linear regression based on a linearizing transformation). R includes some "self-starting" functions that do these steps automatically. The functions `SSlogis` and `SSmicmen`, for example, provide self-starting logistic and Michaelis-Menten functions. To fit a self-starting Michaelis-Menten model to the tadpole data with asymptote a and half-maximum b:

```
> data(ReedfrogFuncresp)
> nls(Killed ~ SSmicmen(Initial, a, b),
+       data = ReedfrogFuncresp)
```

Use `apropos("SS",ignore.case=FALSE)` to see a more complete list of self-starting models. The names are cryptic, so check the help system for information about each model.

Further reading: Bates and Watts (1988), Pinheiro and Bates (2000).

9.4 Nonnormal Errors: Generalized Linear Models

Generalized linear models (not to be confused with general linear models) allow you to analyze models that have a particular kind of nonlinearity and particular kinds of nonnormally distributed (but still independent) errors.

Generalized linear models can fit any nonlinear relationship that has a *linearizing transformation*. That is, if $y = f(x)$, there must be some function F such that $F(f(x))$ is a linear function of x. The procedure for fitting generalized linear models uses the function F to fit the data on the linearized scale ($F(y) = F(f(x))$) while calculating the expected variance on the untransformed scale in order to correct for the distortions that linearization would otherwise induce. In generalized-linear-model jargon F is called the *link* function. For example, when f is the logistic curve ($y = f(x) = e^x/(1 + e^x)$), the link function F is a the logit function ($F(y) = \log(y/(1-y)) = x$; see p. 83 for the proof that the logit is really the inverse of the logistic). R knows about a variety of link functions including the log ($x = \log(y)$, which linearizes $y = e^x$); square root ($x = \sqrt{y}$, which linearizes $y = x^2$); and inverse ($x = 1/y$, which linearizes $y = 1/x$): see `?family` for more possibilities.

The class of nonnormal errors that generalized linear models can handle is called the *exponential family*. It includes Poisson, binomial, Gamma and normal distributions, but not negative binomial or beta-binomial distributions. Each distribution has a standard link function: the log link is standard for a Poisson, a logit link is standard for a binomial distribution, etc. The standard link functions make sense for typical applications. For example, the logit transformation converts unconstrained values into values between 0 and 1, which are appropriate as probabilities in a binomial model. However, R does allow you some flexibility to change these associations for specific problems.

GLMs are fit by a process called *iteratively reweighted least squares*, which overcomes the basic problem that transforming the data to make them linear also changes the variance. The key is that given an estimate of the regression parameters, and knowing the relationship between the variance and the mean for a particular distribution, one can calculate the variance associated with each point. With this variance estimate, one reestimates the regression parameters weighting each data point by the inverse of its variance; the new estimate gives new estimates of the variance; and so on. This procedure quickly and reliably fits the models, without the user needing to specify starting points.

Generalized linear models combine a range of nonnormal error distributions with the ability to work with some reasonable nonlinear functions. They also use the same simple model specification framework as lm, allowing us to explore combinations of factors, covariates, and interactions among variables. GLMs include logistic and binomial regression and log-linear models. They use terminology that should now be familiar to you; they estimate log-likelihoods and test the differences between models using the LRT.

The glm function implements generalized linear models in R. By far the two most common GLMs are Poisson regression, for count data, and logistic regression, for survival/failure data.

- Poisson regression: log link, Poisson error ($Y \sim \text{Poisson}(ae^{bx})$):

```
> glm1 = glm(y ~ x, family = "poisson")
```

The equivalent likelihood function is

```
> poisregfun = function(a, b) {
+       Y.pred = exp(a + b * x)
+       -sum(dpois(y, lambda = Y.pred, log = TRUE))
+ }
```

- Logistic regression: logit link, binomial error ($Y \sim \text{Binom}(p = \exp(a + bx)/(1 + \exp(a + bx)), N)$):

```
> glm2 = glm(cbind(y, N - y) ~ x, family = "binomial")
```

or

```
> logistregfun = function(a, b) {
+       p.pred = exp(a + b * x)/(1 + exp(a + b * x))
+       -sum(dbinom(y, size = N, prob = p.pred, log =
+       TRUE))
+ }
```

(You could also say p.pred=plogis(a+b*x) in the first line of logistregfun.)

GLMs can also fit models of exponentially decreasing survival, $Y \sim \text{Binom}(p = \exp(a + bx), N)$. Strong et al. (1999) modeled the survival probability of ghost

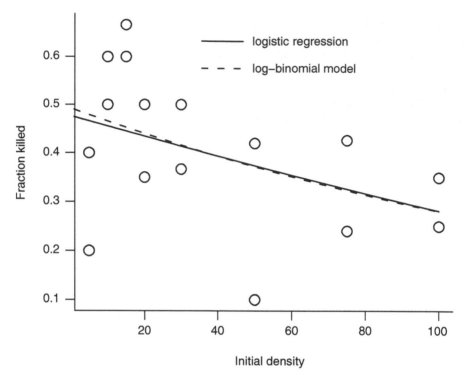

Figure 9.5 Logistic (binomial) regression and log-binomial regression of fraction of tadpoles killed as a function of tadpole density. Logistic regression:
```
glm(cbind(Killed,Initial-Killed)~Initial, family="binomial",
data=ReedfrogFuncresp)
```
Log-binomial regression: glm(...,family=binomial(link="log"),...)

moth caterpillars as a decreasing function of density (and as a function of the presence or absence of entomopathogenic nematodes); Tiwari et al. (2006) modeled the probability that nesting sea turtles would *not* dig up an existing nest as a decreasing function of nest density. You can fit such a model this way:

```
> glm3 = glm(cbind(y, N - y) ~ x, family = binomial(link =
+       "log"))
```

Use `family=binomial(link="log")` instead of `family="binomial"` to specify the log instead of the logit link function. The equivalent negative log-likelihood function is

```
> logregfun = function(a, b) {
+       p.pred = exp(a + b * x)
+       -sum(dbinom(y, size = N, prob = p.pred, log = TRUE))
+ }
```

You can use either a logistic or a log-binomial model to fit Vonesh's tadpole mortality data (Figure 9.5), but the fact that expected survival decreases exponentially at high densities in both models causes problems of interpretation. If the probability of survival declines exponentially with density—which is true for the log-binomial model and approximately true at high densities for the logistic—then the expected *number* surviving is $p(x) \cdot x = e^{-(a+bx)}x = cxe^{-bx}$. This is a Ricker function, which decreases to zero at high density rather than reaching an asymptote. In predator-prey systems for example, rather than this *overcompensation* response to density, we usually expect *compensatory* behavior—predation rate reaching an asymptote—the standard type II functional response model uses $p(x) = A/(1 + Ahx)$, which has a weaker dependence on x, and which makes the limit of $p(x)x$ as x becomes large equal to $1/h$. The GLM, while convenient, may not be ecologically appropriate in this case.

After you fit a GLM, you can use the same generic set of modeling functions—summary, coef, confint, anova, and plot—to examine the parameters, test hypotheses, and plot residuals. anova(glm1,glm2,...) does an *analysis of deviance* (Likelihood Ratio tests) on a nested sequence of models. As with lm, the default parameters represent (1) the intercept (the baseline value of the first treatment), (2) differences in the intercept between the first and subsequent treatments, (3) the slope(s) with respect to the covariate(s) for the first group, or (4) differences in the slope between the first and subsequent treatments. However, all of the parameters are given on the scale of the link function (e.g., log scale for Poisson models, logit scale for binomial models). To interpret them, you need to transform them with the inverse link function (exponential for Poisson, logistic (=plogis) for binomial). For example, the coefficients of the logistic regression shown in Figure 9.5 are intercept $= -0.095$, slope $= -0.0084$. To find the probability of mortality at a tadpole density of 60, calculate $\exp(-0.095 + -0.0084 \cdot 60)/(1 + \exp(-0.095 + -0.0084 \cdot 60)) = 0.355$.

Further reading: McCullagh and Nelder (1989); Dobson (1990); Hastie and Pregibon (1991); Lindsay (1997). R-specific: Crawley (2002); Faraway (2006).

9.4.1 Models for Overdispersion

To go beyond the exponential family of distributions (normal, binomial, Poisson, Gamma) you may well need to roll your own ML estimator. R has two built-in possibilities for the very common case of discrete data with *overdispersion*, i.e., more variance than would be expected from the standard (Poisson and binomial) models for discrete data.

9.4.1.1 QUASI LIKELIHOOD

Quasi-likelihood models "inflate" the expected variance of models to account for overdispersion (McCullagh and Nelder, 1989). For example, the expected variance of a binomial distribution with N samples and probability p is $Np(1 - p)$. The *quasi-binomial* model adds another parameter, ϕ, which inflates the variance to $\phi Np(1 - p)$. The *overdispersion parameter* ϕ (Burnham and Anderson (2004) call it $\hat{c}$) is generally greater than 1—we usually find more variance than expected, rather than less. Quasi-Poisson models are defined similarly, with variance equal to $\phi\lambda$. This approach is called *quasi* likelihood because we don't specify a real likelihood model with a probability distribution for the data. We just specify the relationship between the mean and the variance. Nevertheless, the quasi-likelihood approach works well in practice. R uses the family function to specify quasi-likelihood models.

Because the quasi likelihood is not a true likelihood, we cannot use Likelihood Ratio tests or other likelihood-based methods for inference, but the parameter estimates and t statistics generated by summary should still work. However, various researchers have suggested that using an F test based on the ratio of deviances is appropriate: use anova(...,test="F") (Crawley, 2002; Venables and Ripley, 2002). Burnham and Anderson (2004) suggest using differences in "quasi-AIC" (qAIC) in this case, where the ΔqAIC uses the difference in deviance divided by the estimate of ϕ.

Since the log is the default link function for the quasipoisson family, you can fit a quasi-Poisson log-log model for fecundity as follows:

```
> glm(TOTCONES ~ log(DBH), data = firdata, family =
+     "quasipoisson")
```

9.4.1.2 NEGATIVE BINOMIAL MODELS

Although the exponential family does not strictly include the negative binomial distribution, negative binomial models can be fit by a small extension of the GLM approach, iteratively fitting the k (overdispersion) parameter and then fitting the rest of the model with a fixed k parameter. The glm.nb function in the MASS package fits linear negative binomial models, although they restrict the model to a single k parameter for all groups. (Use $theta to extract the estimate of the negative binomial k parameter from a negative binomial model.)

Because we can use a log link (which is glm.nb's default link), we can exactly replicate our original log-likelihood model (cones $\sim$ NegBin($a \cdot$ DBH^b, k)) with the following command:

```
> glm.nb(TOTCONES ~ log(DBH), data = firdata)
```

The only difference from our earlier model is that the estimated intercept parameter is $\log(a)$ rather than a.

9.5 R Supplement

Here's how to fit various linear models to the log-transformed fir data. Since the data (TOTCONES) contain some zero values, taking logarithms would give us negative infinite values. We need either to drop these values (subset=TOTCONES>0) or to add an offset of 1, in order to avoid infinities. However, since there are few zeros in the data (sum(firdata$TOTCONES==0) is 10 out of a total of 242 data points) and the mean number of cones is large, this adjustment shouldn't affect the results much. If zeros are frequent so that such an adjustment would affect your results significantly, or if the results vary depending on how large an offset you add, consider a different model (Section 9.4).

```
> logcones = log(firdata$TOTCONES + 1)
> lm.0 = lm(logcones ~ 1, data = firdata)
> lm.d = lm(logcones ~ log(DBH), data = firdata)
> lm.w = lm(logcones ~ WAVE_NON, data = firdata)
> lm.dw = lm(logcones ~ log(DBH) + WAVE_NON, data = firdata)
> lm.dwi = lm(logcones ~ log(DBH) * WAVE_NON,
+      data = firdata)
```

Since `log(DBH)` is a covariate and `WAVE_NON` is a factor, `lm.d` is a regression; `lm.w` is a one-way ANOVA; and `lm.dw` and `lm.dwi` are ANCOVA models with parallel and nonparallel slopes, respectively.

A few different ways to analyze the data:

```
> anova(lm.0, lm.d, lm.dw, lm.dwi)
```

Analysis of Variance Table

```
Model 1: logcones ~ 1
Model 2: logcones ~ log(DBH)
Model 3: logcones ~ log(DBH) + WAVE_NON
Model 4: logcones ~ log(DBH) * WAVE_NON
```

	Res.Df	RSS	Df	Sum of Sq	F	Pr(>F)	
1	241	384.53					
2	240	250.33	1	134.20	127.7512	<2e-16	***
3	239	250.29	1	0.04	0.0393	0.8431	
4	238	250.02	1	0.27	0.2535	0.6151	

```
---
Signif. codes:  0 '***' 0.001 '**' 0.01 '*' 0.05 '.' 0.1 ' ' 1
```

```
> AIC(lm.0, lm.d, lm.w, lm.dw, lm.dwi)
```

	df	AIC
lm.0	2	802.8349
lm.d	3	700.9556
lm.w	3	786.5281
lm.dw	4	702.9157
lm.dwi	5	704.6580

(I left `lm.w` out of the anova statement because it and `lm.d` cannot be nested.) anova compares the models sequentially, while AIC compares them simultaneously. AICtab in the emdbook package offers several more options such as sorting the table in order of increasing AIC or computing AIC weights. Try coef, summary, and confint on these models as well.

The full ANCOVA model fit via mle2:

```
> ancovafun = function(i1, i2, slope1, slope2, sigma) {
+     int = c(i1, i2)[WAVE_NON]
+     slope = c(slope1, slope2)[WAVE_NON]
+     Y.pred = int + slope * log(DBH)
+     -sum(dnorm(logcones, mean = Y.pred, sd = sigma,
```

```
+          log = TRUE))
+ }
> m1 = mle2(ancovafun, start = list(i1 = -2, i2 = -2,
+      slope1 = 2.5, slope2 = 2.5, sigma = 1),
+      data = firdata)
> AIC(m1)
```

[1] 704.658

The maximum likelihood fit gives the same AIC as the lm fit. You can't always take this equality for granted, since different models that are formally equivalent may include different constants in the likelihood, and different functions may count the number of parameters differently. This is especially true when comparing results from different statistics packages.

As pointed out in the text, the models are parameterized differently:

```
> coef(lm.dwi)
```

(Intercept)	log(DBH)	WAVE_NONw	log(DBH):WAVE_NONw
-2.3871702	2.7303449	0.5162954	-0.2630837

```
> coef(m1)
```

i1	i2	slope1	slope2	sigma
-2.387134	-1.870762	2.730329	2.467205	1.016441

You can check that the answers are equivalent; for example, the slope of the wave population is slope2 = 2.467 = log(DBH) + log(DBH):WAVE_NONw.

To do the full model comparison with mle2, you have to construct a series of nested models (analogous to lm.dw, lm.d, lm.w, lm.0). This is a bit tedious—one reason for using built-in functions where possible. You may want to read about the model.matrix function, which can simplify model construction. model.matrix uses a user-specified formula to construct a *design matrix* that, when multiplied by a vector of parameters, gives the expected value of each data point. By default the design matrix uses parameters that represent baseline levels and differences among groups, as in lm and glm. mle2's formula interface uses model.matrix internally, so that (e.g.) you can easily fit the full ANCOVA model by specifying

```
> mle2(log(TOTCONES + 1) ~ dnorm(i + slope*log(DBH), sd),
> parameters = list(i ~ WAVE_NON, slope ~ WAVE_NON),
+      data = firdata, start = ...)
```

Congratulations

You have now finished the first part of the book, which covers all the important basic tools. You know everything you need to know to fit reasonably complex, realistic ecological models to you data.

 **Warning**

Models with multiple levels of variability and dynamical models, the subjects of the last two chapters, are much harder to create and fit from scratch. Powerful and specialized statistical methods that have been developed to handle these problems are beginning to make their way into ecology. The second part of the book will give a brief overview of these topics, but to use them in any serious way you will have to go to a specialized reference such as Gelman and Hill (2006) or Clark (2007) to learn more. The good news is that the concepts and terminology you have now learned should speed up the learning process considerably.

If your brain is full after the first part of the book, stop here. If you are eager for more, read on. If you are already swamped but desperately need to incorporate multiple levels of variability in your analysis, see Section 10.4.3 for ways of avoiding multilevel models. If you are swamped but must do something to estimate parameters for a dynamic model, see Section 11.4.

10   Modeling Variance

This chapter addresses models that incorporate more than one kind of variability, variously called *mixed, multilevel, multistratum,* or *hierarchical* models. It starts by considering data with (1) changing amounts of variability or (2) correlation among data points. These kinds of data can be modeled adequately with the tools introduced in previous chapters. The last part of the chapter considers data with two or more qualitatively different sources of variability. These kinds of data are much more challenging to model, but they can be fitted with analytical or numerical integration techniques or via MCMC. This chapter is more conceptual and less technical than previous chapters.

10.1 Introduction

Throughout this book we have partitioned ecological models into deterministic (Chapter 3) and stochastic (Chapter 4) submodels. For example, we might use a deterministic logistic function to describe changes in mean population density with increasing rainfall and a Gamma stochastic distribution to describe the natural variation in population density at a particular level of rainfall. We have focused most of our attention on constructing the deterministic model and testing for differences in parameters among groups or as a function of covariates. For the stochastic model, we have stuck to fitting single parameters such as the variance (for normally distributed data) or the overdispersion parameter (for negative binomially distributed data) to describe the variability around the deterministic expectation. We have assumed that the variance parameter is constant within groups, and that we can use a single distribution to describe the variability.

This chapter will explore more sophisticated stochastic models for ecological data. Section 10.2 is a warmup, presenting models where the variance may differ among groups or change as a function of a covariate. These models are easy to fit with our existing tools. Section 10.3 briefly reviews models for *correlation* among observations, useful for incorporating spatial and temporal structure.

Sections 10.4 and 10.5 tackle models that incorporate more than one type of variability, referred to as *mixed, multilevel, multistratum,* or *hierarchical* models.

Section 10.4 discusses how to use canned procedures in R to fit multilevel models—or to cheat and avoid fitting multilevel models at all—while Section 10.5 briefly describes strategies for tackling more general multilevel models.

The most common multilevel models are *block* models, which divide observations into discrete groups according to their spatial or temporal locations, genetic identity, or other characteristics. The model assigns each block a different random mean, typically drawn from a normal or log-normal distribution. Individual responses within blocks vary around the block mean, most frequently according to a normal distribution but sometimes according to a Poisson or binomial sampling distribution. In survival analysis, block models are called *shared frailty* models (Therneau et al., 2003).

Individual-level models allow for variation at the individual rather than the group level. When individuals are measured more than once, the resulting *repeated-measures* models can be analyzed in the same way as block models. They may also allow for temporal correlation among measurements, and more sophisticated versions can quantify random variation among individuals in the parameters of nonlinear models (Vigliola et al., 2007).

When individuals are measured only once, we lose most of our power to discriminate among-individual and within-individual variation. We can still make progress, however, if we assume that the observed variation in individual responses is a combination of among-individual variation in a mean response and a random sampling process (either in the ecological process itself, or in our measurement of it) that leads to variation around the mean. The combination of among-individual and sampling variation results in a *marginal* distribution (the distribution of observations, including both levels of variability) that is *overdispersed*, or more variable than expected from the sampling process alone.

Finally, when individuals or populations are measured over time we have to distinguish between *measurement* and *process error*, because these two sorts of variability act differently on ecological dynamics. Process error feeds back to affect the ecological system in the next time step, while measurement error doesn't. We defer analysis of such dynamical models to Chapter 11.

Building models that combine several *deterministic* processes is straightforward. For example, if we know the functions for mean plant biomass as a function of light availability (say $B(L) = aL/(b+L)$) and for mean fecundity as a function of biomass (say $F(B) = cB^d$), we can easily combine them mathematically in order to use the combined function in a maximum likelihood estimate ($F(L) = c(aL/(b+L))^d$).* Incorporating multiple levels of variability in a model is harder because we usually have to compute an integral in order to average over all the different ways that different sources might combine to produce a particular observation. For example, suppose that the probability that a plant establishes in an environment with light availability L is $P_L(L)\,dL$ (where P_L is the probability density) and the probability that it will grow to biomass B at light level L is $P_{BL}(B|L)\,dB$. Then the marginal probability density $P_B(B)$ that a randomly chosen plant will have biomass B is the combination of all the different probabilities of achieving biomass B at different light levels: $\int P_{BL}(B|L)P_L(L)\,dL$.

---

* Although we may not be able to estimate all the parameters separately. In this case, we can estimate ca^d but cannot estimate c and a^d separately; see p. 333.

Computing these integrals analytically is often impossible, but for some models the answer is known (Section 10.5.1). Otherwise, we can either use numerical brute force to compute them (Section 10.5.2) or use Markov chain Monte Carlo to compute a stochastic approximation to the integral (Section 10.5.3). These methods are challenging enough that, in contrast to the models of previous chapters, you will often be better off finding an existing procedure that matches the characteristics of your data rather than coding the model from scratch.

10.2 Changing Variance within Blocks

Once we've thought of it, it's simple to incorporate within-block changes in variance into an ecological model. All we have to do is define a sensible model that describes how a variance parameter changes as a function of predictor variables. For example, Figure 10.1 shows data on glacier lilies (*Erythronium glandiflorum*) that display the typical triangular, or "factor-ceiling," profile of many ecological data sets (Thomson et al., 1996). The triangular distribution is often caused by an environmental variable that sets an upper limit on an ecological response rather than determining its precise value. In this case, the density of adult flowers (or something associated with adult density) appears to set an upper limit on the density of seedlings, but the number of seedlings varies widely below the upper limit. I fitted the model

$$S \sim \text{NegBin}(\mu = a, k = ce^{d \cdot f}), \tag{10.2.1}$$

where S is the observed number of seedlings and f is the number of flowers. The mean μ is constant, but the overdispersion parameter k increases (and thus the variance decreases) as the number of flowers increases. The R negative log-likelihood function for this model is

```
> nlikfun = function(a, c, d) {
+     k = c * exp(d * flowers)
+     -sum(dnbinom(seedlings, mu = a, size = k, log = TRUE))
+ }
```

Alternatively, the `mle2` formula would be

```
> mle2(seedling ~ dnbinom(mu = a,
+     size = c * exp(d * flowers)), ...)
```

This function resembles our previous examples, except that the variance parameter rather than the mean parameter changes with the predictor variable (f or `flowers`). Figure 10.1 shows the estimated mean (which is constant) and the estimated upper 90%, 95%, and 97.5% quantiles, which look like stair-steps because the negative binomial distribution is discrete. It also shows a nonparametric density estimate for the mean and quantiles as a function of the number of flowers, as a cross-check of the appropriateness of the parametric negative binomial model. The patterns agree qualitatively—both models predict a roughly constant mean and decreasing upper quantiles. Testing combinations of models that allow mean, variance, both, or neither

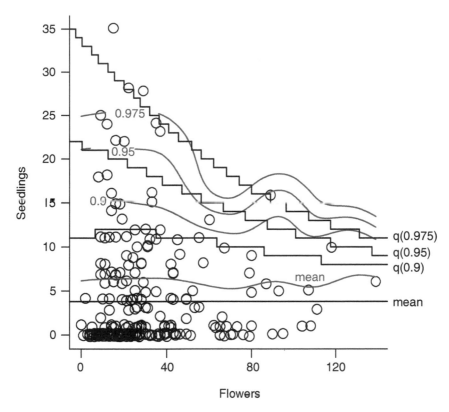

Figure 10.1 Lily data from Thomson et al. (1996); jittered numbers of seedlings as a function of number of flowers. Black lines show fit of a negative binomial model with constant mean and increasing k/decreasing variance; gray lines are from a nonparametric density estimate.

to vary with the numbers of flowers suggests that the best model is the constant model, but the model fitted here (constant mean, varying k) is the second best, better than allowing the mean to decrease while holding k constant. Thomson et al. (1996) suggested a pattern where the ceiling actually increases initially at small numbers of flowers, but this pattern is hard to establish definitively.

The same general strategy applies for the variance parameter of other distributions such as the variance of a normal distribution or log-normal distribution, the shape parameter of the Gamma distribution, or the overdispersion parameter of the beta-binomial distribution. Just as with deterministic models for the mean value, the variance might differ among different groups or treatment levels (represented as factors in R), might change as a function of a continuous covariate as in the example above, or might depend on the interactions of factors and covariates (i.e., different dependence of variance on the covariate in different groups). Just the variance, or both the mean and the variance, could differ among groups. Use your imagination and your biological intuition to decide on a set of candidate models, and then use the LRT or AIC values to choose among them.

Variance parameters are best fit on the log scale ($\log \sigma^2$, $\log k$, etc.) to make sure the variances are always positive. For fitting to continuous covariates, use nonnegative functions such as the exponential ($\sigma^2 = ae^{bx}$) or power ($\sigma^2 = ax^b$), for the same reason.

When using a normal distribution to model the variability is reasonable, you can use the built-in functions gls and gnls from the nlme package: gls fits linear and gnls fits nonlinear models. These models are called *generalized (non)linear least squares* models.* The weights argument allows a variety of relationships (exponential, power, etc.) between covariates and the variance. For example, using the fir data, suppose we wanted to fit a power-law model for the mean number of cones as a function of DBH and include a power-law model for the variance as a function of DBH:

$$\text{cones} \sim \text{Normal}(\mu = a\text{DBH}^b, \sigma^2 = c\text{DBH}^d). \qquad (10.2.2)$$

Fitting the model with gnls:

```
> data(FirDBHFec)
> firdata = na.omit(FirDBHFec)
> gnls(TOTCONES ~ a * DBH^b, data = firdata,
+      weights = varPower(form = ~DBH), start = list(a = 1,
+         b = 1))
```

The syntax of weights argument is a bit tricky: Pinheiro and Bates (2000) is essential background reading.

10.3 Correlations: Time-Series and Spatial Data

Up to now we have assumed that the observations in a data set are all independent. When this is true, the likelihood of the entire data set is the product of the likelihoods of each observation, and the negative log-likelihood is the sum of the negative log-likelihoods of each observation. Every negative log-likelihood function we have written has contained code like -sum(ddistrib(...,log=TRUE)), either explicitly or implicitly, to capitalize on this fact. With a bit more effort, however, we can write and numerically optimize likelihood functions that allow for correlations among observations.

It's best to avoid correlation entirely by designing your observations or experiments appropriately. Correlation among data points is a headache to model, and it always reduces the total amount of information in the data: correlation means that data points are more similar to each other than expected by chance, so the total amount of information in the data is smaller than if the data were independent. However, if you are stuck with correlated data—for example, because your samples come from a spatial array or a time series—all is not lost. Moreover, sometimes the correlation in the data is biologically interesting: for example, the range of spatial correlation might indicate the spatial scale over which populations interact.

---

* Not to be confused with general linear models or generalized linear models!

The standard approach to correlated data is to specify a likelihood of correlated data, usually using a multivariate normal distribution. The probability distribution of the multivariate normal distribution (dmvnorm in the emdbook package) is

$$\text{MVN}(\mathbf{x}, \boldsymbol{\mu}, \mathbf{V}) = \frac{1}{\sqrt{(2\pi)^n |\mathbf{V}|}} \exp\left(-\frac{1}{2}(\mathbf{x} - \boldsymbol{\mu})^T \mathbf{V}^{-1}(\mathbf{x} - \boldsymbol{\mu})\right), \qquad (10.3.1)$$

where $\mathbf{x}$ is a vector of data values, $\boldsymbol{\mu}$ is a vector of means, and $\mathbf{V}$ is the variance-covariance matrix ($|\mathbf{V}|$ is the determinant of $\mathbf{V}$—see below—and T stands for transposition). The formula looks scary, but like most matrix equations you can understand it by making analogies to the scalar (i.e., nonmatrix) equivalent, in this case the univariate normal distribution

$$N(x) = 1/\sqrt{2\pi\sigma^2} \exp\left(-(x - \mu)^2/(2\sigma^2)\right). \qquad (10.3.2)$$

The term $\exp\left(-\frac{1}{2}(\mathbf{x} - \boldsymbol{\mu})^T \mathbf{V}^{-1}(\mathbf{x} - \boldsymbol{\mu})\right)$ is the most important part of the formula, equivalent to the $\exp\left(-(x - \mu)^2/(2\sigma^2)\right)$ term in the univariate normal distribution. $(\mathbf{x} - \boldsymbol{\mu})$ is the deviation of the observations from their theoretical mean values. Multiplying by $\mathbf{V}^{-1}$ is equivalent to dividing by the variance, and multiplying by $(\mathbf{x} - \boldsymbol{\mu})^T$ is equivalent to squaring the deviations from the mean. The stuff in front of this term is the normalization constant. The $|\mathbf{V}|$ matches the σ^2 in the normalization constant of the univariate normal, and the $\sqrt{2\pi}$ term is raised to the nth power to normalize the n-dimensional probability distribution.

If all the points are actually independent and have identical variances, then the variance-covariance matrix is a diagonal matrix with σ^2 on the diagonal:

$$\mathbf{V} = \begin{pmatrix} \sigma^2 & 0 & \cdots & 0 \\ 0 & \sigma^2 & \cdots & 0 \\ \vdots & \vdots & \ddots & \vdots \\ 0 & 0 & \cdots & \sigma^2 \end{pmatrix}. \qquad (10.3.3)$$

Comparing (10.3.1) and (10.3.2) one term at a time shows that the multivariate normal reduces to a product of identical univariate normal probabilities in this case. For a diagonal matrix with σ^2 on the diagonal, the inverse is a diagonal matrix with $1/\sigma^2$ on the diagonal, so the matrix multiplication $(\mathbf{x} - \boldsymbol{\mu})^T \mathbf{V}^{-1}(\mathbf{x} - \boldsymbol{\mu})$ works out to $\sum_i (x_i - \mu_i)^2/\sigma^2$—the sum of squared deviations. Exponentiating this sum gives a product. For a diagonal matrix, the determinant $|\mathbf{V}|$ is the product of the diagonal elements, so for the identical-variance case $|\mathbf{V}| = (\sigma^2)^n$, which is proportional to the product of the normalizing constants for n independent normal distributions.

If the points are independent but each point has a different variance—like assigning different variances to different groups with one individual in each group—then

$$\mathbf{V} = \begin{pmatrix} \sigma_1^2 & 0 & \cdots & 0 \\ 0 & \sigma_2^2 & \cdots & 0 \\ \vdots & \vdots & \ddots & \vdots \\ 0 & 0 & \cdots & \sigma_n^2 \end{pmatrix}. \qquad (10.3.4)$$

Carrying through the exercise of the previous paragraph will show that the multivariate normal reduces to a product of univariate normals, each with its own variance.

The exponent term is now half the *weighted* sum of squares $\sum (x_i - \mu_i)^2/(\sigma_i^2)$, with the deviation for each point weighted by its own variance; data points with larger variances have less influence on the total.

Most generally,

$$
\mathbf{V} = \begin{pmatrix} \sigma_1^2 & \sigma_{12} & \cdots & \sigma_{1n} \\ \sigma_{21} & \sigma_2^2 & \cdots & \sigma_{2n} \\ \vdots & \vdots & \ddots & \vdots \\ \sigma_{n1} & \sigma_{n2} & \cdots & \sigma_n^2 \end{pmatrix}. \tag{10.3.5}
$$

The off-diagonal elements σ_{ij} quantify the covariance between points i and j. We could also specify this information in terms of the *correlation matrix* $\mathbf{C} = \{\rho_{ij}\} = \left\{\sigma_{ij}/\sqrt{\sigma_i^2 \sigma_j^2}\right\}$. The diagonal elements of the correlation matrix ρ_{ii} all equal 1, and the off-diagonal elements range between -1 (perfect anticorrelation) and 1 (perfect correlation). The variance-covariance matrix $\mathbf{V}$ must be symmetric (i.e., $\sigma_{ij} = \sigma_{ji}$). In this case there is no way to express the exponent as a sum of independent normals, but its meaning is the same as in the previous cases: it weights combinations of deviations by the appropriate variances and covariances.

To fit a correlated multivariate normal model, you would need to specify parameters for the variance-covariance matrix $\mathbf{V}$. In principle you could specify $n(n+1)/2$ different parameters for each of the distinct entries in the matrix (since the matrix is symmetric there are $n(n+1)/2$ rather than n^2 distinct entries), but there are two reasons not to. First, such a general parameterization takes lots of parameters. Unless we have lots of data, we probably can't afford to use up so many parameters specifying the variance-covariance matrix. Second, in addition to being symmetric, variance-covariance matrices must also be *positive definite*, which means essentially that the relationships among points must be consistent.* For example,

$$
\mathbf{V} = \begin{pmatrix} 1 & 0.9 & -0.9 \\ 0.9 & 1 & 0.9 \\ -0.9 & 0.9 & 1 \end{pmatrix} \tag{10.3.6}
$$

is not a valid correlation matrix, even though it is symmetric.[†] It states that site 1 is strongly positively correlated with site 2 ($\rho_{12} = 0.9$), and site 2 is strongly correlated with site 3 ($\sigma_{23} = 0.9$), but site 1 is strongly *negatively* correlated with site 3 ($\sigma_{13} = -0.9$), which is not possible.

For these two reasons, modelers usually select from established correlation models that (1) use a small number of parameters to construct a full variance-covariance (correlation) matrix and (2) ensure positive definiteness. For example,

$$
\mathbf{V} = \sigma^2 \mathbf{C} = \sigma^2 \begin{pmatrix} 1 & \rho & \rho^2 & \cdots & \rho^n \\ \rho & 1 & \rho & \cdots & \rho^{n-1} \\ \vdots & \vdots & \vdots & \ddots & \vdots \\ \rho^n & \rho^{n-1} & \rho^{n-2} & \cdots & 1 \end{pmatrix}, \tag{10.3.7}
$$

---

* Technically, positive definiteness means that the matrix must have all positive eigenvalues.
† Its eigenvalues are 1.9 (repeated twice) and -0.8.

with $|\rho| < 1$, specifies a correlation matrix **C** corresponding to sites arranged in a line—or data taken in a temporal sequence—where correlation falls off with the number of steps between sites (sampling times): nearest neighbors (sites 1 and 2, 2 and 3, etc.) have correlation ρ, next-nearest neighbors have correlation ρ^2, and so forth.* Other correlation models allow correlation ρ_{ij} to drop to zero at some threshold distance, or to be a more general function of the spatial distance between sites.

Three different areas of classical statistics use multivariate normal distributions to describe correlation among observations. *Repeated-measures ANOVA* is a form of analysis of variance that allows the errors to be nonindependent in some way, particularly by building in individual-level variation (see Section 10.4.1) but also by allowing for correlation between successive points in time. *Time-series* models (Chatfield, 1975; Diggle, 1990; Venables and Ripley, 2002) and *spatial models* (Ripley, 1981; Cressie, 1991; Kitanidis, 1997; Venables and Ripley, 2002; Haining, 2003) specify correlation structures that make sense in temporal and spatial contexts, respectively.

Generalized least-squares (g[n]ls) allows for correlation among observations, using the correlation argument. These functions include a variety of standard models for temporal and spatial autocorrelation; see ?corClasses (in the nlme package) and Pinheiro and Bates (2000) for more details. The lme and nlme functions, which fit repeated-measures models, also have a correlation argument. The ts package implements time-series models, which incorporate correlation, although the description and methods used are different from the more general models described here. The spatial package will fit trend lines and surfaces with spatial correlation between points.

To construct variance-covariance functions to use in your own custom-made likelihood functions, start with the matrix function (for general matrices) or diag (for diagonal matrices) and extend them. For example, diag(4) produces a 4×4 identity matrix (with 1 on the diagonal); diag(c(2,3,4)) produces a 3×3 diagonal matrix with variances of 2, 3, and 4 on the diagonal. The row and col functions, which return matrices encoding row and column numbers, are also useful. For example, d=abs(row(M)-col(M)) produces a absolute-value distance matrix $\{|i-j|\}$ with the same dimensions as M, and rho^d produces correlation matrix (10.3.7). The slightly tricky code ifelse(d==0,1,ifelse(d==1,rho,0)) produces a matrix with 1 on the diagonal, rho on the first off-diagonal, and 0 elsewhere, corresponding to correlation only among nearest-neighbor sites.

If your data are irregularly spaced or two-dimensional, you can start by computing a matrix of the distances between points, using d=as.matrix(dist(cbind(x,y))).† Then you can easily use the distance matrix to compute exponential (proportional to e^{-d}), Gaussian (proportional to e^{-d^2}), or other spatial correlation matrices. Consult Venables and Ripley (2002) or a spatial statistics reference for more details.

---

* This correlation matrix is sometimes referred to as AR(1), meaning "autoregressive order 1," meaning that each point is correlated directly with its first neighbor. The higher powers of ρ with distance arise because of a chain of correlation: next-nearest neighbors are correlated through their mutual neighbor, and so on.

† dist(cbind(x,y)) computes the distances between x and y but returns the answer as a dist object. It is more useful in this case to use as.matrix to convert it to a matrix.

Once you have constructed a correlation matrix, multiply it by the variance to get a covariance matrix suitable for use with the dmvnorm density function, for example, negative log-likelihood

```
> -dmvnorm(z, mu, Sigma = V, log = TRUE)
```

where z is a vector of data, mu is a mean vector, and V is one of the variance-covariance matrices defined above. You can use the mvrnorm command from the MASS package to generate random, correlated normal deviates. For example, mvrnorm(n=1,mu=rep(3,5),Sigma=V) produces a five-element vector with a mean of 3 for each element and variance-covariance matrix V. Asking mvrnorm for more than one random deviate (n > 1) will produce a five-column matrix where each row is a separate draw from the multivariate distribution.

Statisticians usually deal with correlated data that is nonnormal (e.g., Poisson or binomial) by combining a multivariate normal model for the underlying mean values with a nonnormal distribution based on these varying means. We usually exponentiate the MVN distribution to get a multivariate lognormal distribution so that the means are always positive. For example, to model correlated Poisson data we could assume that Λ, the vector of expected numbers of counts at each point, is the exponential of a multivariate normally distributed variable:

$$\mathbf{Y} \sim \text{Poisson}(\mathbf{\Lambda})$$
$$\mathbf{\Lambda} \sim \exp\left(\text{MVN}(\boldsymbol{\mu}, \mathbf{V})\right). \tag{10.3.8}$$

Here $\mathbf{Y}$ is a vector of counts at different locations; $\mathbf{\Lambda}$ (a random variable) is a vector of expected numbers of counts (intensities); $\boldsymbol{\mu}$ is a vector of the logs of the average intensities; and $\mathbf{V}$ describes the variance and correlation of intensities. If we had already used one of the recipes above to construct a variance-covariance matrix V, and had a model for the mean vector mu, we could simulate the values as follows:

```
> Lambda = exp(mvrnorm(1, mu = mu, Sigma = V))
> Y = rpois(length(mu), Lambda)
```

Unfortunately, even though we can easily simulate values from this distribution, writing down a likelihood for this model is difficult because there are two different levels of variation. The rest of the chapter discusses how to formulate and estimate the parameters for such multilevel models.

10.4 Multilevel Models: Special Cases

While correlation models assume that samples depend on each other as a function of spatial or temporal distance, with overlapping neighborhoods in space or time, traditional multilevel models usually break the population into discrete groups such as family, block, or site. Within groups, all samples are equally correlated with each other. If samples 1 and 2 are from one site and samples 3 and 4 are from another, each pair would be correlated and we would write down this variance-covariance matrix:

$$\mathbf{V} = \sigma^2 \mathbf{C} = \sigma^2 \begin{pmatrix} 1 & \rho & 0 & 0 \\ \rho & 1 & 0 & 0 \\ 0 & 0 & 1 & \rho \\ 0 & 0 & \rho & 1 \end{pmatrix}. \tag{10.4.1}$$

Equivalently, we could specify a *random-effects* model that gives each group its own random offset from the overall mean value. The correlation model (10.4.1) is formally identical to a random-effects model where the value of the *j*th individual in the *i*th group is

$$Y_{ij} = \epsilon_i + \epsilon_{ij}, \tag{10.4.2}$$

where $\epsilon_i \sim N(0, \sigma_b^2)$ is the level of the random effect in the *i*th block and $\epsilon_{ij} \sim N(0, \sigma_w^2)$ is the difference of the *j*th individual in the *i*th block from the block mean. The among-group variance σ_b^2 and within-group variance σ_w^2 correspond to the correlation parameters (ρ, σ^2):

$$\sigma_b^2 = \rho \sigma^2, \qquad \sigma_w^2 = (1 - \rho)\sigma^2. \tag{10.4.3}$$

Classical models usually describe variability in terms of random effects because constructing huge variance-covariance matrices is very inefficient.

10.4.1 Fitting (Normal) Mixed Models in R

Some special kinds of block models can be fitted with existing tools in R and in many other statistics packages. Models with two levels of normally distributed variation are called *mixed-effect models* (or just *mixed* models) because they contain a mixture of random (between-group) and fixed effects. Classical block ANOVA models such as split-plot and nested block models fall into this category (Quinn and Keogh, 2002; Gotelli and Ellison, 2004). If the variation in your model is normally distributed, your predictors are all categorical, and your design is balanced, you can use the aov function with an Error term to fit mixed models (Venables and Ripley, 2002). The nlme package, and the newer, more powerful (but poorly documented) lme4 package fit a far wider range of mixed models (Pinheiro and Bates, 2000; Gelman and Hill, 2006). These packages allow for unbalanced data sets as well as random effects on parameters (e.g., ANCOVA with randomly varying slopes among groups), and nonlinear mixed-effect models (e.g., an exponential, power-law, logistic, or other nonlinear curve with random variation in one or more of the parameters among groups).

10.4.2 Generalized Linear Mixed Models

Generalized linear mixed models, or GLMMs, are a cross between mixed models and generalized linear models (p. 308). GLMMs combine link functions and exponential-family variation with random effects. The random effects must be normally distributed on the scale of the linear predictor—meaning on the scale of the data as transformed by the link function. For example, for a Poisson model with a log link, the between-group variation would be log-normal. GLMMs are cutting edge, and the methods for solving them are evolving rapidly. The `glmmPQL` function in the `MASS` package can fit an approximation to GLMMs, but one that is sometimes inaccurate (Breslow and Clayton, 1993; Breslow, 2003; Jang and Lim, 2005). The `glmmML` and `lme4` packages offer more robust GLMM fitting algorithms; so does J. Lindsey's `repeated` package, documented in his book (Lindsey, 1999) and available on his Web page (linked under "Related Projects" on the R project page). If you want to be thorough, it may also be worth cross-checking your results with PROC GLIMMIX or NLMIXED in SAS.

R has built-in capabilities for incorporating random effects into a few other kinds of models. *Generalized additive mixed models* (GAMMs) (`gamm` in the `mgcv` package) allow random effects and exponential-family variation with models where spline curves make up the deterministic part of the model. *Frailty* models (`frailty` in the `survival` package) incorporate Gamma, *t*, or normally distributed variation among groups in a survival analysis. Some other variations such as nonnormal repeated-measures models are described by Lindsey (1999*a*, 2001, 2004).

10.4.3 Avoiding Mixed Models

Even when canned packages are available, fitting mixed models can be difficult. The algorithms do not always converge, especially when the number of groups is small. If your design is linear, balanced, and has only nested random effects, then `aov` should always work, but otherwise you may be at a loss.

If you want to avoid fitting mixed models altogether, one option is to fit fixed-effect models instead, estimating a parameter for each group rather than a random variable for the among-group variation (Clark et al., 2005). You will probably lose some power this way, so the results are likely to be conservative.* As a second option, Gotelli and Ellison (2004, p. 182) suggest that when you have a simple nested design (i.e., subsamples within blocks) you should often just collapse each group's data by computing its mean and do a single-level analysis. This will be disappointing if you were hoping to glean information about the within-group variance, but it is simple and in many cases will give the same *p*-value as the classical algorithm coded by `aov`. Finally, you can try to convince yourself (and your reviewers, readers, or supervisor) that between-group variation is unimportant by fitting the model ignoring blocks and then examining the variation of the residuals between blocks both graphically and statistically. To justify ignoring between-group variation in the model, you must show that the between-group variation in the residuals is *both* statistically and biologically irrelevant. Biologically relevant variation is an important warning sign even if it is not statistically significant.

---

* The distinction between fixed effects and random effects is murky in any case; see Crawley (2002, p. 670) for some rules of thumb, and Gelman (2005, p. 20) for more than you ever wanted to know about the level of debate even among statisticians about the meaning of these terms.

10.5 General Multilevel Models

Now suppose that your data do not allow analysis by classical tools and that you are both brave and committed to finding out what a multilevel model can tell you.

10.5.1 Analytical Models for Marginal Distributions

In some cases, it may be possible to solve the integrals that arise in multilevel problems analytically. In this case you can fit the marginal distribution directly to your data. The marginal distribution describes the combination of multiple stochastic processes but doesn't attempt to provide information about the individual processes—analogous to knowing the row and column sums (the *marginal totals*) of a table without knowing the distribution of values within the table.

If you're comfortable with math, you can read an advanced treatment such as Bailey (1964) or Pielou (1977) to learn how to solve these problems. Otherwise, you should do some research to see if someone has already found the answer for your model. The most common of these distributions—the negative binomial arising from a Poisson sampling process with underlying Gamma-distributed variability, the beta-binomial arising from binomial sampling with underlying Beta-distributed variability, and the Student t distribution arising from normally distributed variation with Gamma-distributed variability in the inverse variance—were already discussed on p. 140. Zero-inflated distributions (p. 139) are another example of a multilevel process—the combination of a presence/absence process and a discrete sampling process—for which it is fairly easy to figure out the marginal distribution.

If your multilevel process involves summing several values—for example, if the measured value is the sum of a normally distributed block mean and a normally distributed individual variation—then the marginal distribution is called a *convolution*. If X and Y are random variables with probability densities P_X and P_Y, then the probability density of $X + Y$ is

$$P_{X+Y}(z) = \int P_X(x) \cdot P_Y(z - x) \, dx. \tag{10.5.1}$$

The intuition behind this equation is that we are adding up all the possible ways we could have gotten z. If $X = x$, then the value of Y must be $z - x$ in order for $X + Y$ to equal z, so we can calculate the total probability by integrating over all values of x. The convolutions of distributions with themselves—i.e., the distribution of sums of like variables—can sometimes be solved analytically. The sum of two normal variables $N(\mu_1, \sigma_1^2)$ and $N(\mu_2, \sigma_2^2)$ is also normal ($N(\mu_1 + \mu_2, \sigma_1^2 + \sigma_2^2)$); the sum of two Poisson variables is also Poisson ($\text{Pois}(\lambda_1) + \text{Pois}(\lambda_2) \sim \text{Pois}(\lambda_1 + \lambda_2)$); and the sum of n exponential variables with the same mean is Gamma distributed with shape parameter n. These solutions are simple, but they also warn us that we may sometimes face an *identifiability* problem (p. 333) when we try to separate multiple levels of variability. If we know only that the sum of two normally distributed variables is $N(\mu, \sigma^2)$, then we can't recover any more information about the means and variance of the individual variables—except that the variances are between 0 and σ^2. (In the case of classical block models we also know which group any individual belongs to, so we can partition variability within and among groups.)

10.5.2 Numerical Integration

If you can't find anyone who has computed the marginal distribution for your problem analytically, you may still be able to compute it numerically.

A common problem in forest ecology is estimating the distribution of growth rates g_i of individual trees in a stand from size measurements S_i in successive censuses: $g_i = S_{i,2} - S_{i,1}$. Foresters commonly assume that adult trees can't shrink, or at least not much, but it's typical to observe a small proportion of individuals in a data set whose measured size in the second census is smaller than their initial size. If we really think that measurement error is negligible, then we're forced to conclude that the trees actually shrank. It's standard practice to go through the data set and throw out negative growth values, along with any that are unrealistically big. Can we do better?

Although it is sensible to throw out really extreme values, which may represent transcription errors (being careful to keep the original data set intact and document the rules for discarding outliers), we may be able to extract information from the data set both about the "true" distribution of growth rates and about the distribution of errors. The key is that the distributions of growth and error are assumed to be different. The error distribution is symmetric and narrowly distributed (we hope) around zero, while the growth distribution is positive and right-skewed. Thus the negative tail of the distribution tells us about error—negative values must contain at least some error.

Specifically, let's assume a Gamma distribution of growth (we could equally well use a lognormal) and a normal distribution of error. The growth distribution has parameters a (shape) and s (scale), while the error distribution has just a variance σ^2—we assume that errors are equally likely to be positive or negative, so the mean is zero. Then

$$Y_{\text{true}} \sim \text{Gamma}(s, a)$$
$$Y_{\text{obs}} \sim \text{Normal}(Y_{\text{true}}, \sigma^2). \tag{10.5.2}$$

For normally distributed errors, we can also express this as the sum of the true value and an error term:

$$Y_{\text{obs}} = Y_{\text{true}} + \epsilon, \epsilon \sim \text{Normal}(0, \sigma^2). \tag{10.5.3}$$

According to the convolution formula, the likelihood of a particular observed value is

$$P(Y_{\text{obs}}|a, s, \sigma^2) = P(Y_{\text{true}} + \epsilon = Y_{\text{obs}}|a, s, \sigma^2)$$

$$= \int P(Y_{\text{true}} = Y_{\text{obs}} - \epsilon|a, s) \cdot P(\epsilon|\sigma^2) \, d\epsilon. \tag{10.5.4}$$

The log-likelihood for the whole data set is

$$L = \sum \log \int P(Y_{\text{true}} = Y_{\text{obs}} - \epsilon|a, s) \cdot P(\epsilon|\sigma^2) \, d\epsilon. \tag{10.5.5}$$

TABLE 10.1
Results of convolution analysis of tree growth

	True	MLE	Quadratic	Profile
Shape (*a*)	3	2.98	2.86– 3.10	2.58– 3.49
Scale (*s*)	10	10.37	9.79–10.95	8.84–12.03
σ	10	9.06	8.75– 9.37	8.06–10.59

Unfortunately we can't interchange the logarithm and the integral, which would make everything much simpler.

We can easily simulate some fake "data" from this system with plausible parameters in order to test our approach:

```
> set.seed(1001)
> x.true = rgamma(1000, shape = 3, scale = 10)
> x.obs = rnorm(1000, mean = x.true, sd = 10)
```

In the R supplement (Section 10.8), I defined a getdist function that implements (10.5.2)–(10.5.5). I then used mle2 and confint to find the maximum likelihood values and profile confidence limits (Table 10.1). It was slow, though: it took about 3 minutes (for a total of 137 function evaluations) to find the MLE using Nelder-Mead and 164 minutes to calculate the profile confidence intervals. The estimates are encouragingly close to the true values, and the confidence limits are reasonable. The quadratic confidence intervals, while too narrow for σ, are close enough to the profile confidence intervals that the extra 2.5 hours of computation may have been unnecessary. Figure 10.2 plots the observed histogram along with the estimated and true distributions of actual growth rates and the estimated and true distributions of measurement error. Excitingly, we can actually recover accurate information about the growth rates and the measurement process in this example.

Numerical integration works well here, although it's slow if we insist on calculating profile confidence limits. Hand-coded numerical integration in R will always be slower, and less stable, than the special-case algorithms built into packages like lme4 or glmmML. A commercial package called AD Model Builder uses sophisticated integration techniques to solve some very difficult multilevel modeling problems, but I have not evaluated it (Kitakado et al., 2006; Skaug and Fournier, 2006). Gelman and Hill (2006) primarily use R and BUGS to fit multilevel models, but they provide an appendix that describes how to fit some multilevel models in SAS, AD Model Builder, and other packages. Brute-force numerical integration can work reasonably well as long as you have enough data (both enough individual data points and enough groups) and as long as your problem has only one, or at most two, random variables to integrate. The estimation process is fairly straightforward, and numerical failures are usually obvious—although you should do all the usual checks for convergence.

10.5.3 MCMC for Mixed Models

Most of the time, brute-force numerical integration as illustrated above is just too hard. Once you have to integrate over more than one or two random variables,

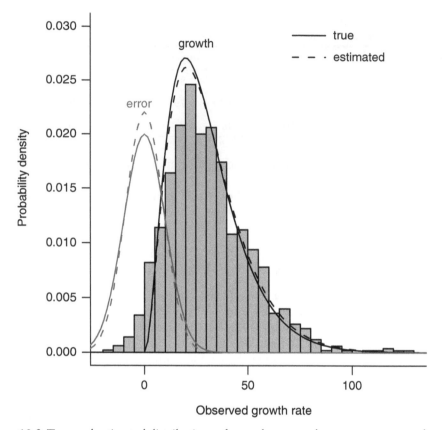

Figure 10.2 True and estimated distributions of growth rates and measurement error for simulated forest data. Histogram, observed data. Lines, true and estimated distributions of actual growth rate and measurement error.

computing the integrals becomes extremely slow. MCMC is an alternative way of doing these high-dimensional integrals, and it gets you confidence limits "for free." The disadvantages are that (1) it may be slower than sufficiently clever numerical integration approximations; (2) you have to deal with the Bayesian framework, including deciding on a set of reasonable priors (although Lele et al. (2007) recently suggested a simple method for using MCMC to do ML estimation for multilevel models); and (3) in badly determined cases where your model is poorly defined or where the data don't really contain enough information, BUGS may give you an answer that doesn't make sense instead of just crashing—which is *really* bad.

The BUGS input file for the Gamma-normal model is extremely simple:

```
model {
  for (i in 1:N) {
    x.true[i] ~ dgamma(sh,rate)
    x.obs[i] ~ dnorm(x.true[i],tau)
  }
  sh ~ dgamma(0.01,0.01)
  rate ~ dgamma(0.01,0.01)
  tau ~ dgamma(0.01,0.01)
}
```

The first half of the model statement is a direct translation of the model (10.5.2): for each value in the data set, the observed value is assumed to be drawn from a normal distribution centered on the true value, which is in turn drawn from a Gamma distribution.

The second half of the model statement specifies vague Gamma-distributed priors. As mentioned in Chapter 7, BUGS uses slightly different parameterizations from R for the normal and Gamma distributions. It specifies the normal by the mean and the *precision* τ, which is the reciprocal of the variance, and the Gamma by the shape parameter (just as in R) and the rate parameter, which is the reciprocal of the scale parameter. The mean of the gamma distribution is shape/rate and the variance is shape/rate^2; thus a standard weak Gamma prior uses equal shape and rate parameters (i.e., a mean of 1), with both the shape and the rate parameter small (i.e., a large variance). In this case I've chosen (0.01, 0.01) (but see Gelman (2006) for possible problems with this choice).

We'd like to run chains starting from overdispersed starting points, as recommended in Chapter 7, so that we can run a Gelman-Rubin diagnostic test to see if the chains have converged. However, it's still very important to start with sensible starting values (so that BUGS doesn't crash or take forever to burn in). A good strategy is to find a reasonable starting point and then change different parameters by up to an order of magnitude to get overdispersed starting points.

We can estimate a starting point for the Gamma parameters by fitting a Gamma distribution just to the nonzero observations:

```
> pos.obs = x.obs[x.obs > 0]
> f1 = fitdistr(pos.obs, "gamma")
```

We can estimate a starting value for the precision (τ) by selecting the negative observations, replicating them with the opposite sign, and calculating the reciprocal of the variance:

```
> neg.obs = x.obs[x.obs < 0]
> bineg.obs = c(neg.obs, -neg.obs)
> tau0 = 1/var(bineg.obs)
```

You're not required to specify starting values for all of the parameters—if you don't, BUGS will pick a random starting value from the prior distribution for that parameter. However, BUGS will often crash or converge slowly with random starting values, especially if you use very weak priors. We will start the chains for the true growth rates from 1 or from the observed growth rate for each individual, whichever is greater; this strategy avoids negative values for the true growth rate, which are impossible according to our model.

```
> tstart = pmax(x.obs, 1)
```

We specify a list of initial values for each chain; in this case I've decided to start one chain at the values of the crude estimates calculated above, and the other chains with perturbations of those estimates, with one or the other of the parameters halved or increased to 150%. I didn't bother to perturb the starting values for the rate, but you could in a more thorough analysis.

```
> clist = as.list(coef(f1))
> params = list(tau = tau0, sh = clist$shape,
```

```
+       rate = clist$rate, x.true = tstart)
> inits = rep(list(params), 5)
> inits[[2]]$tau = 0.5 * tau0
> inits[[3]]$tau = 1.5 * tau0
> inits[[4]]$sh = 0.5 * clist$shape
> inits[[5]]$sh = 1.5 * clist$shape
```

We need to specify the data for BUGS:

```
> N = length(x.obs)
> data = list("x.obs", "N")
```

Finally, we specify for which parameters of the model we want BUGS to save information. Of course we want to track the parameters of our model. In addition, it will be interesting to see what BUGS estimates for the true values, uncorrupted by measurement error, of both the minimum observed value in the data set (which obviously contains a lot of measurement error, since its value is −15.7) and one of the values closest to the median.*

```
> (minval = which.min(x.obs))
```

[1] 670

```
> (medval = which.min(abs(x.obs - median(x.obs))))
```

[1] 675

```
> parameters = c("sh", "rate", "tau", "x.true[670]",
+       "x.true[675]")
```

Running the model:

```
> gn1.bugs = bugs(data, inits, parameters.to.save =
+       parameters, model.file = "gammanorm.bug",
+       n.chains = length(inits))
```

BUGS took 264 minutes to run the models. The Gelman-Rubin statistics were all well below the rule-of-thumb cutoff of 1.2, suggesting that the chains had converged. Using as.mcmc to convert the BUGS result gn1.bugs to a coda object and running traceplot and densityplot confirmed graphically that the chains gave similar answers (Table 10.2), and that the posterior distributions are approximately normal: the Gelman-Rubin test makes this assumption, so it's good to check. While the estimated "true" value of the median point is close to its observed value of 28.2, the Bayesian model correctly adjusted the estimated true value for the minimum point from its observed value (−15.7) to a small but definitely positive value. The estimates of the values of particular data points have wide distributions—it would be too good to be true if we could use MCMC to magically get rid of observation error.

---

* Because the data set has an even number of points, the median value is halfway between the two values closest to the middle.

TABLE 10.2
Results of WinBUGS run on Gamma-normal model

	True	Mean	2.5%	97.5%
Shape	3.00	2.993	2.612	3.471
Rate	0.10	0.097	0.084	0.112
τ	0.01	0.012	0.009	0.017
Minimum value	10.70	7.556	1.554	16.780
Median value	24.20	27.053	11.971	42.445

10.6 Challenges

Multilevel models are hard to implement. General linear models like linear regression represent one extreme, where you can put in any kind of data and know that you will get a rapid, technically correct (if not necessarily sensible) answer. Multilevel models represent the other extreme: the methods for fitting them are computationally demanding and must be carefully tuned to a particular type of problem to ensure an efficient and correct solution. As with analytical solutions, if you are not ready for a technical challenge, your best bet is to find a prebuilt solution for your class of models. Another issue we have to face as we incorporate more levels of variability is *identifiability*. Unidentifiable parameters are those that are impossible to estimate separately from our data. For example, if we tried to estimate a linear model with two different intercepts ($Y = a_1 + a_2 + bX$), instead of the more sensible $Y = a + bX$, any answer where $a_1 + a_2 = a$ would fit equally well. We would have the same problem with the coefficients of two predictor variables that were perfectly correlated. Similarly, there is no way to identify catchability—the probability that you will observe an individual—from a single observational sample; you simply don't have the information to estimate how many animals or plants you failed to count. Most of these cases of "perfect" unidentifiability are easily detectable by common sense, although they can be obscured by complex models. Always remember that there is no free lunch: if a procedure for estimating parameters seems too good to be true, be suspicious.

Weak identifiability is more common than perfect unidentifiability. Weakly identifiable parameters are hard to estimate even with large, clean data sets—and practically impossible to estimate with moderate-size noisy data sets. Identifiability problems are not limited to multilevel models, but they are particularly common there. Weak identifiability is related to a lack of statistical power; it means that your model is structured in such a way that an enormous amount of data or very extreme kinds of data are needed in order to have the power to differentiate ecological processes.

The best outcome of analyzing an unidentifiable or weakly identifiable model is for your estimation procedure to crash: at least you will know something is wrong. The worst outcome is for your estimation procedure to give you a wrong answer, or one that depends in a sensitive way on the details of your numerical algorithm or your prior. Weak identifiability in MLE analysis leads either to numerical problems or to apparently well-defined, but misleadingly precise, answers; weak identifiability in MCMC analyses leads to poor convergence. In the worst case, chains may move through parameter space so slowly that it's hard to see that they are not converging. The best defenses against identifiability problems are (1) care and common sense in defining models (always remember the "no free lunch" principle); (2) examining model diagnostics—confidence intervals, convergence statistics; and the difference between prior and posterior distributions; and (3) running models with known inputs (i.e., simulations), with different amounts of data and different amounts of noise, to see when they are actually capable of getting the right answers.

Coming to grips with the difficulty of separating different types of variability, like coming to grips with limitations on statistical power (Chapter 5), can be sobering. Schnute (1994) says in a paper on state-space models (one kind of multilevel model) that

> the outcome of the analysis often depends critically on the values of [many variance] parameters, and it is generally impossible to estimate all of them . . . Statistically, the likelihood surface [may be] "flat", i.e. insensitive to large parameter changes. In cases like this, a large number of conflicting scenarios appear equally consistent with the known data, and the analyst has no objective means to choose among them.
>
> This apparent limitation can be turned into an advantage . . . State-space model design forces essential questions to be asked about underlying processes, observed data, and sources of variability. When these questions are answered honestly, the model may point to scenarios consistent with the data but in conflict with the prevailing view. If so, the modeling effort can help to delineate the limits of current knowledge and to establish rational priorities for future data collection.

Put another way, acknowledging the many different types of uncertainty that actually exist in our models may make us realize that we know less than we thought we did about the possible dynamics of our study system, and may drive us to make more observations, or more useful kinds of observations.

10.7 Conclusion

Why then are multilevel models worth so much effort?

They are clearly the wave of the future in ecological statistics. Books like Gelman and Hill (2006) and Clark (2007), along with a rapidly growing collection of papers in ecology and statistical ecology, demonstrate these models' potential for understanding ecological processes. We know that ecological systems are variable at every level, and multilevel models give us a framework for estimating this variability at multiple levels rather than lumping it into a single error term. Our statistical models should match our conceptual models as closely as possible. If we are interested in the differences among individuals or sites or genotypes, we would often be satisfied with knowing the amount of variability among groups rather than trying to estimate a value for every individual.

Furthermore, estimating the variability in this way may be more parsimonious than estimating fixed effects for every individual or group. While fixed effects require an additional parameter for every additional group in the model, estimating variability takes only a single parameter (the variance) no matter how many groups there are (Kenward and Roger, 1997).*

Although multilevel models are hard to implement and computationally challenging, things are getting better. Computers are faster every year, and infrastructure like grid computing is making it easier to access computing power beyond your desktop (Wang et al., 2005). Tools for Bayesian and maximum likelihood estimation are constantly improving, both in the power and breadth of algorithms available and in the convenience, interoperability, and robustness of software (Kerman and Gelman, 2006; Skaug and Fournier, 2006).

Multilevel models are data-hungry. If you are a typical ecological experimenter with a small, noisy data set, you may not be able to apply multilevel models to your data set (Clark et al., 2005), although Bayesian methods can help if you are willing to specify informative priors. On the other hand, if you have a *large*, but very noisy, data set, then multilevel models may be the perfect tool. Because of data collection tools such as data loggers, remote sensing, radiotelemetry, and climate proxies, and data synthesis tools such as Web databases, meta-analysis, and citizen science initiatives, more and more ecologists are finding themselves with data sets that are appropriate for multilevel modeling. Knowing how to use multilevel models, and knowing what dangers to avoid, will help you ask many more interesting questions about your data.

10.8 R Supplement

10.8.1 Numerical Integration

Here's a function that calculates the likelihood for a given value of the error (ϵ) and the parameters and the observed value (the integrand in (10.5.4)):

```
> tmpf = function(eps, shape, scale, sd, x) {
+     exp(dnorm(eps, mean = 0, sd = sd, log = TRUE) +
+         dgamma(x - eps, shape = shape, scale = scale,
+             log = TRUE))
+ }
```

Check that it gives a reasonable value (at least not an NA) for the first data point:

```
> tmpf(1, shape = 3, scale = 10, sd = 1, x = x.obs[1])
```

```
[1] 0.0002398506
```

---

* Statisticians are still deeply divided about the correct way to count the number of effective parameters associated with a random-effect term. The answer certainly lies between 1 (the variance) and $n - 1$ (the number of parameters required to estimate the differences of each group from the first), and the possible difference can be huge for models containing interaction terms. For a fixed-effect model, an interaction between two factors with m and n factor levels has $(m - 1)(n - 1)$ degrees of freedom, while the random-effect model might require as few as 1. Some researchers have suggested rules for calculating or estimating the number, but the question is still open (Kenward and Roger, 1997; Spiegelhalter et al., 2002; Burnham and Anderson, 2004, p. 315; Lee et al., 2006).

Integrate numerically, using `integrate`:

```
> i1 = integrate(f = tmpf, lower = -Inf, upper = Inf,
+       shape = 3, scale = 10, sd = 1, x = x.obs[1])
> i1$value
```

```
[1] 0.0009216708
```

Define a function to calculate this integral:

```
> tmpf2 = function(x, shape, scale, sd) {
+       integrate(f = tmpf, lower = -Inf, upper = Inf,
+           shape = shape, scale = scale, sd = sd, x = x)
+           $value
+ }
```

To calculate the integral for more than one data point at a time, we have to use `sapply`: if we give `tmpf2` a vector for x, R will do the wrong thing.

Define the negative log-likelihood function:

```
> getdist = function(shape, scale, sd, dat, debug = FALSE) {
+       v = -sum(log(sapply(dat, tmpf2, shape = shape,
+           scale = scale, sd = sd)))
+       if (debug)
+           cat(shape, scale, sd, v, "\n")
+       v
+ }
```

Try this function for one set of reasonable parameters:

```
> getdist(shape = 3, scale = 10, sd = 1, dat = x.obs)
```

```
[1] 5684.876
```

Now run `mle2` and `confint` to estimate the parameters and confidence intervals:

```
> m1 = mle2(minuslogl = getdist, start = list(shape = 3,
+       scale = 10, sd = 1), data = list(dat = x.obs),
+       method = "Nelder-Mead")
> m1.ci = confint(m2)
```

11 Dynamic Models

This chapter covers dynamic models, an important kind of multilevel model. It shows how to simulate dynamic models, discusses process and observation error, and illustrates methods for fitting models that assume only one or the other. For problems where we want to estimate process error when the magnitude of observation error is known, it introduces the SIMEX approach. Finally, it presents a brief introduction to fitting state-space models, which can estimate both process and observation error, via the Kalman filter or Markov chain Monte Carlo.

11.1 Introduction

This chapter covers concepts and techniques for fitting *dynamic* models—models that describe how ecological processes drive populations to change over time. Dynamic models are a special case of the multilevel models we introduced in Chapter 10. Dynamic models contain both *process error*, which feeds back on future states of the population, and *observation error*, which affects only the current observation.

  We introduce dynamic models by describing how to simulate them. Knowing how to simulate dynamic models is important because fitting dynamic models to data is so tricky that it's essential to fit models to simulated data to confirm that the methods work. (Most of the examples in this chapter use simulated "data.")

  The easiest way by far of dealing with observation and process error is to ignore one or the other (Section 11.4). If your data have little noise, you may be able to get away with this approach. When you can independently estimate the variance of the observation error, the more recently developed SIMEX (simulation-extrapolation) algorithm provides a way to get unbiased parameter estimates (Section 11.5).

  State-space models (Section 11.6) can in principle estimate both process and observation error from a single data set, subject to the very strong constraint that the data actually provide enough information to separate them reliably. The *Kalman filter* (Section 11.6.1) is a relatively simple algorithm for estimating the parameters of state-space models with normally distributed error. More generally, computationally intensive Bayesian (Millar and Meyer, 2000) and frequentist (de Valpine and Hastings, 2002; Thomas et al., 2005; Lele et al., 2007) methods can simultaneously estimate deterministic parameters, observation error, and process error in nonlinear, nonnormal ecological models (Section 11.6.2). The use of such methods has recently begun to explode in ecology (Solow, 1998; de Valpine and Hastings, 2002; Ellner et al., 2002; de Valpine, 2003; Jonsen et al., 2003, Buckland et al., 2004; Clark and Bjørnstad, 2004; Thompson et al., 2005). This chapter attempts to provide a basic and relatively painless introduction. If you want to explore this area further, you will have to dig into the literature (e.g., Calder et al., 2003).

11.2 Simulating Dynamic Models

Dynamic models describe the changes in the size and characteristics of a population over time. At each time step except the first, the size and characteristics of the population depend on the size and characteristics at the previous time step (or one or more times further in the past). Writing down the mathematical formula that describes the population size at time t is often much harder than describing how $N(t)$ depends on $N(t-1)$. The difference between observation and process error becomes vitally important in dynamic models, because they act differently. Process error affects future population dynamics, while observation error does not.

To simulate a dynamic model:

- Set aside space (a vector or matrix) to record the state of the population (numbers of organisms, possibly categorized by species/size/age).
- Set the starting conditions for all state variables.
- For each time step, apply R commands to simulate population dynamics over the course of one time step. Then apply R commands to simulate the observation process and record the current *observed* state of the population.
- Plot and analyze the results.

11.2.1 Examples

We can construct dynamic models corresponding to the two simple static models (linear/normal and hyperbolic/Poisson) introduced in Chapter 5.

Figure 11.1a shows a dynamic model analogous to the static model shown in Figure 5.1a (p. 149). The closest analogue of the static linear model, $Y \sim \text{Normal}(a + bx)$, is a dynamic model with observation error only:

$$N(1) = a$$

$$N(t+1) = N(t) + b \tag{11.2.1}$$

$$N_{\text{obs}}(t) \sim \text{Normal}\left(N(t), \sigma_{\text{obs}}^2\right).$$

The first line in (11.2.1) specifies the initial or starting condition (the value of N at time $t = 1$). The second line is the updating rule that determines the population size one time step in the future, which in this case is purely deterministic. The third line specifies the observation process, in this case that the observed value of the population size at time t, $N_{\text{obs}}(t)$, is normally distributed around the true value $N(t)$ with variance σ_{obs}^2.

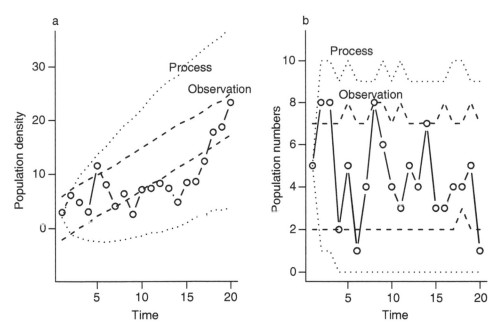

Figure 11.1 Dynamic models with process and observation error. (a) Linear, continuous (normal) model. (b) Nonlinear, discrete (hyperbolic/Poisson) model. In each case the envelopes (dotted and dashed lines) show the 95% confidence limits for equivalent models with pure process or pure observation error; the realizations shown are generated with a mixture of process and observation error.

The R code for this model would first specify nt, the number of time steps, and assign values for the parameters a, b, and sd.obs. Then:

```
> N = numeric(nt)
> Nobs = numeric(nt)
> N[1] = a
> for (t in 1:(nt - 1)) {
+       N[t + 1] = b + N[t]
+       Nobs[t] = rnorm(1, mean = N[t], sd = sd.obs)
+ }
> Nobs[nt] = rnorm(1, mean = N[nt], sd = sd.obs)
```

Since the for loop runs only from 1 to nt-1, we have to set the observed value for $t = $ nt at the end. If we ran the loop to nt, we would be predicting the state of the population at time nt+1, beyond the end of the vector we have set aside for the results. R would cooperate by extending the length of the vector, but the too-long vector might lead to confusion or errors in subsequent steps.

By contrast, a model with pure process error is defined as

$$N(1) = a$$

$$N(t+1) \sim \text{Normal}\left(N(t) + b, \sigma_{\text{proc}}^2\right) \qquad (11.2.2)$$

$$N_{\text{obs}}(t) = N(t).$$

The R code:

```
> N = numeric(nt)
> Nobs = numeric(nt)
> N[1] = a
> for (t in 1:(nt - 1)) {
+     N[t + 1] = rnorm(1, mean = b + N[t], sd = sd.proc)
+     Nobs[t] = N[t]
+ }
> Nobs[nt] = N[nt]
```

In this case, we assume that our observations are perfect ($N_{obs}(t) = N(t)$) but that the change in the population is noisy rather than deterministic.

The behavior of the mean in this dynamic model is exactly the same whether the variability in the model is caused by observation error or process error, and in fact it is identical to the deterministic part of a standard linear model $N = a + b(t-1)$. Furthermore, there is no way to separate process from observation error by simply looking at a single time series; the variation in the observed data will appear the same. (Figure 11.1 actually shows a single realization of a model with equal amounts of process and observation error; it falls outside the theoretical bounds of an observation-error-only model with slope $a = 1$, but only because we know the true slope. We couldn't tell the difference in a real data set.) The difference becomes apparent only when we simulate many realizations of the same process and look at how the variation *among realizations* changes over time (Figure 11.1a). With observation error only, the variance among realizations is constant over time; with process error only, there is initially no variance (we always start at the same density), but the variance among realizations increases over time.

Figure 11.1b shows a discrete-population model with process and observation error. In this case, the model is a rational function with the same form as the Beverton-Holt or Michaelis-Menten function. Suppose that per capita plant fecundity declines with population density according to the hyperbolic function $F(N) = a/(b+N)$. Then let the next year's expected population size $N(t+1)$ equal (population size) × (per capita fecundity) = $N(t)(a/(b+N(t)))$. The population grows asymptotically to a stable population size of $a - b$. (Convince yourself that when $N(t) = (a-b)$, $N(t+1) = N(t)$, and the simulated dynamics in Figure 11.1b are indeed nearly constant.)

For the observation error model, we assume that we have a probability of only p of counting each individual that is present in the population, which leads to a binomial distribution of observations:

$$N(1) = N_0$$
$$N(t+1) = aN(t)/(b+N(t)) \qquad (11.2.3)$$
$$N_{obs}(t) \sim \text{Binomial}(N(t), p).$$

The R code:

```
> N = numeric(nt)
> Nobs = numeric(nt)
> N[1] = N0
```

```
> for (t in 1:(nt - 1)) {
+     N[t + 1] = a * N[t]/(b + N[t])
+     Nobs[t] = rbinom(1, size = round(N[t]), prop = p)
+ }
> Nobs[nt] = rbinom(1, size = round(N[nt]), prop = p)
```

The only problem in this model is that $N(t+1)$ is usually not an integer, in which case the binomial doesn't make sense. I rounded the value in this case, although normally it would be more sensible to incorporate a more realistic process model with (discrete) process error.* Like the linear observation error model, the distribution of error stays constant over time—with a few random bumps on the upper confidence limit caused by sampling error (Figure 11.1b).

The process error model for the discrete population case is simpler:

$$N(1) = N_0$$

$$N(t+1) \sim \text{Poisson}(aN(t)/(b+N(t))) \qquad (11.2.4)$$

$$N_{\text{obs}}(t) = N(t).$$

The R code:

```
> N = numeric(nt)
> Nobs = numeric(nt)
> N[1] = N0
> for (t in 1:(nt - 1)) {
+     N[t + 1] = rpois(1, lambda = a * N[t]/(b + N[t]))
+     Nobs[t] = N[t]
+ }
> Nobs[nt] = N[nt]
```

The population size still converges to $a - b$ over time, but the distribution spreads out over the first few time steps. In fact, many of the simulated populations quickly go extinct. However, since this model has a stable equilibrium, the distribution of process error reaches its own equilibrium, rather than spreading out continuously like the linear model in Figure 11.1a.

11.2.1.1 CONTINUOUS-TIME MODELS

Many dynamic models in ecology are defined in continuous rather than discrete time. Typically these models are framed as ordinary differential equation (ODE) models. The rule $N(t+1) = f(N(t))$ is replaced by $dN/dt = f(N(t))$, which specifies the instantaneous population growth rate. Probably the best-known ODE model is the logistic, $dN/dt = rN(1 - N/K)$. Researchers use continuous-time models for a variety of reasons including realism (for populations with overlapping generations that can reproduce in any season), mathematical convenience (the dynamics of continuous-time models are often more stable than those of their discrete analogues), and consistency with theoretical models. Most dynamic models have no

---

*But Henson et al. (2001) describe some possible dynamic consequences of this kind of rounding, which they call "lattice effects," in ecological systems.

closed-form solution (we can't write down a simple equation for $N(t)$), so we often end up simulating them.

The simplest algorithm for simulating continuous-time models is *Euler's method*, which uses small time steps to approximate the continuous passage of time. Specifically, if we know the instantaneous growth rate $dN/dt = f(N(t))$, we can approximate the change in the population over a short time interval Δt by assuming that the population grows linearly at rate dN/dt, and thus that $\Delta N \approx dN/dt \cdot \Delta t$:

$$
\begin{aligned}
N(t + \Delta t) &= N(t) + \Delta N \\
&\approx N(t) + \frac{dN}{dt} \Delta t \\
&= N(t) + f(N(t)) \Delta t.
\end{aligned}
\tag{11.2.5}
$$

In order to find the population size at some arbitrary time t we make Δt "small enough" and work our way from the starting time to t, adding ΔN to the population at each time step Δt.

Euler's method is fine for small problems, but it tends to be both slow and unstable relative to more sophisticated approaches. If you are going to do serious work with continuous-time problems, you will need to solve them for thousands of different parameter values (which may in turn require experimenting with different values of Δt). The lsoda function in R's odesolve library, which implements an adaptive step size algorithm, will be much more efficient.

The central problem with comparing ODE models to data is that incorporating stochasticity in any other way than simply imposing normally distributed observation error is difficult. The mathematical framework that underlies stochastic differential equations is subtle (Roughgarden, 1997) and hard to apply to practical problems. For this reason, studies that attempt to estimate parameters of continuous-time models from data tend either to use simple least-squares criteria that correspond to normal observation error (Gani and Leach, 2001) or to revert to discrete-time models (Finkenstädt and Grenfell, 2000).

One can build dynamical models that are stochastic, are discrete-valued (and hence more sensible for populations), and run in continuous time, picking random numbers for the *waiting times* until the next event (birth, death, immigration, infection, etc.). The basic algorithm for simulating these models, called the *Gillespie algorithm* (Gillespie, 1977), is simple, but it and the advanced methods required to estimate parameters based on such models are beyond the scope of this chapter (Gibson and Renshaw, 1998; Gibson and Renshaw, 2001).

11.3 Observation and Process Error

In general, we describe dynamic data by setting up

- A deterministic function for the *expected* population dynamics—the relationship between the current density and the expected density at time $t+1$, $\bar{N}(t+1) = f(N(t))$—for example, the discrete logistic equation, $\bar{N}(t+1) = N(t) + rN(t)(1 - N(t)/K)$, with parameters r and K.

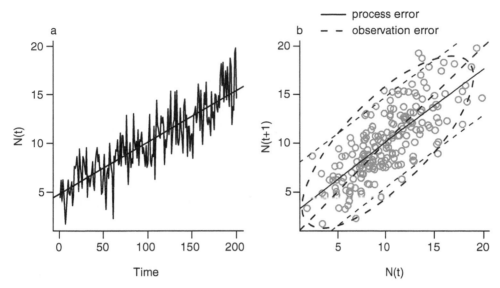

Figure 11.2 Time-series data: process or observation error?

- A model of process error—for example, $N(t)$ is negative binomially distributed with overdispersion parameter k, or $N(t) \sim \mathrm{NegBin}(\mu = \bar{N}(t), k)$.
- A model of observation error—for example, a binomial sample with capture probability p from $N(t)$, or $N_{\mathrm{obs}}(t) \sim \mathrm{Binom}(p, N(t))$.

To understand some of the basic issues of dynamic data, let's look at the simplest deterministic model for population growth—a constant increase in the population density per time step, $f(N(t)) = N(t) + b$, with normally distributed process and observation error. Formally:

$$N(t+1) \sim \mathrm{Normal}\left(N(t) + b, \sigma^2_{\mathrm{proc}}\right) \tag{11.3.1}$$

$$N_{\mathrm{obs}}(t) \sim \mathrm{Normal}\left(N(t), \sigma^2_{\mathrm{obs}}\right) \tag{11.3.2}$$

where σ^2_{proc} and σ^2_{obs} are the process and observation variances.

Suppose we recorded the data in Figure 11.2 and wanted to try to understand what was going on in the population. Depending on the combination of observation and process error that we assumed, we could draw very different conclusions about these data.

If we assumed there was only observation error, with no process error, then the simplest approach would be to solve the deterministic equation ($\bar{N}(t+1) = \bar{N}(t) + b$) as a function of time to get $\bar{N}(t) = \bar{N}(0) + bt$ and estimate b as the slope of an ordinary linear regression (`lm(N~time)`). We would interpret the population dynamics as a linear trend with time.

What if we instead wanted to use the plot of $N(t+1)$ against $N(t)$ (Figure 11.2b) to fit $f(N)$ ($f(N) = N + b$) directly? We would have to recognize that both $N_{\mathrm{obs}}(t)$ and $N_{\mathrm{obs}}(t+1)$ contain observation error, which doesn't fit the assumptions of ordinary linear regression. Instead we would minimize the *diagonal* deviations of points from a line $y = a + bx$, a procedure sometimes called *model II* regression.* Our model of the points $\{N(t), N(t+1)\}$ would be that they were bivariate normal, with the mean of $N(t+1)$ equal to $N(t) + b$; the ellipse in Figure 11.2b represents the confidence limits for the points in this model.

On the other hand, if we assumed process error only (with no observation error), then we should fit an ordinary linear regression to the plot of $N(t)$ vs. $N(t+1)$, because we assume that we know the x variable ($N(t)$) perfectly and the only uncertainty comes in the population growth from t to $t+1$. If we allow the full linear model $N(t+1) = aN(t) + b$, then we are fitting an *autoregressive model*: while the overall trend would be the same as the ordinary linear model (provided $a < 1$), the variance structure is different.[†] Figure 11.2b also shows that this assumption gives different answers from the model II regression, with a larger intercept (which corresponds to a larger population growth rate—remember this is the graph of $N(t)$ vs. $N(t+1)$, not the graph of $N(t)$ vs. t) and a flatter slope.

What if we can't reasonably assume either pure process error or pure observation error? Intermediate assumptions can lead to any answer between the two slopes shown in the figure, which might lead to a wide range of different biological conclusions! Unfortunately the data don't easily show us what assumption to make. The noisier our data, the more the results of the linear-trend and autoregressive models will diverge. In the extreme where we have almost no information, the linear-trend model will say that $N(t) = N(t+1)$ (a 45° regression line), while the autoregressive model will say that $N(t+1)$ is independent of $N(t)$ (a flat line). Since we have no information, our conclusions are entirely driven by the structure of our assumptions. This example is the first indication that in analyzing dynamic models we may sometimes be attempting to separate processes (process and observation variability) for which we have very little distinguishing information. We will return to this sobering theme at various points during the chapter.

11.4 Process and Observation Error

Now we will see how the extreme assumptions of only process error or only observation error play out if we want to fit a model with more interesting dynamics than simple linear increase or decrease with time. For problems with small amounts of error, or if you want to keep things simple, use one of these approaches, as suggested by Hilborn and Mangel (1997). For example, pure process error would be a reasonable model for small discrete populations that could be counted exactly or for experimental populations observed in the lab (Drury and Dwyer, 2005). Pure observation error seems less plausible, but you would still be in good company picking one or the other: many sensible analyses of dynamic data have used these crude but simple methods (Ives et al., 1999; Gani and Leach, 2001; van Veen et al., 2005).

---

* Model II regression is a big topic (Warton et al., 2006); in special cases like this one (dynamic data with only observation error) where we can assume that the variances in x and y are the same, we can use *reduced major axis* regression, which gives the slope as σ_y/σ_x, or equivalently as $\sqrt{b_{yx}/b_{xy}}$, where b_{yx} is the slope of the ordinary regression of y on x and b_{xy} is the slope of the ordinary regression of x on y.

[†] We can fit the restricted model $f(N) = b + N$, assuming the slope of $N(t+1)$ vs. $N(t)$ is exactly 1, with lm(y~offset(x)).

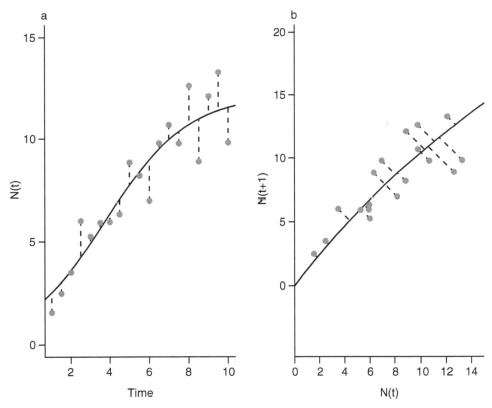

Figure 11.3 Logistic fit: shooting/trajectory matching (observation error only). True parameters $r = 1$, $K = 10$, $N(0) = 1$, $\sigma^2_{obs} = \sigma^2_{proc} = 1$. Estimated parameters $r = 0.48$, $K = 12.14$, $N(0) = 2.53$, $\sigma^2_{obs} = 1.41$. (a) Time dynamics, showing vertical residuals of observations from the fitted line. (b) Next vs. current observation, showing diagonal residuals from the fitted line.

11.4.1 Observation Error Only: Shooting or Trajectory Matching

If we assume observation error only we can start with the initial conditions of the system (e.g., the starting population sizes: we either assume we know these or take the starting values as additional parameters of the model) and "shoot" through the whole period, without correcting the model as we go along; this procedure is also called *trajectory matching* (Figure 11.3). If the deterministic dynamics are particularly simple (e.g., linear, exponential, or logistic), we may be able to derive a formula for $N(t)$ as a function of the starting conditions and calculate the predicted values in a single step (N=a+b*time or N=a*exp(b*time)), but much more often we will be able to compute the expected values only by using a loop to go from the value at each time step to the value at the next time step. (With a continuous-time model, you can use the odesolve package to solve numerically for each set of parameter values.) One way or the other, we compute the predicted values at all observation times, ignoring the variability in the actual data, and then compare the overall fit of the predicted curve to the data.

Since we assume there is no uncertainty in the predicted values for each time step given the starting conditions and the parameters, the only error is between the predicted values and the observed values. We can then do what we've been doing all along: assume independent observations and add up the log-likelihoods of observation error for every data point based on our model of observation error.

Trajectory matching is widely used because it is simple and requires no consideration of process variability. If one assumes normally distributed observation error with constant variance, it simplifies still further to least-squares fitting of the deterministic trajectory (e.g., Gani and Leach, 2001; van Veen et al., 2005). Trajectory matching also works with missing data or unobserved variables (Wood, 2001), although Ellner et al. (2002) warn that trajectory matching can be seriously misleading in cases where process variability qualitatively changes the dynamics of the population (e.g., Ellner et al., 1998).

11.4.2 Process Error Only: One-Step-Ahead Fitting

Alternatively, we can assume there is no observation error. Then the only uncertainty is in the relationship between N_t and N_{t+1}. If we plot the expected value of each N_{t+1} as a function of the (perfectly known) N_t, we have errors only in the Y variable. Instead of starting with the initial conditions and "shooting" (forecasting) through the whole observation time period, we take the observation from each time step and predict just the next time step (Figure 11.4). This way we need not worry about how process errors compound from step to step. (This procedure is more difficult with missing time points, because we then have to somehow figure out the expected relationship, including the process error, between (e.g.) $N(t)$ and $N(t+2)$ (Clark and Bjørnstad, 2004).) This procedure is called *one-step-ahead prediction*. For population dynamics modeled in continuous rather than discrete time, a slightly more sophisticated analogue is called *gradient matching* (Ellner et al., 2002).

Shooting and one-step-ahead prediction are approximations, but they are simple and usually worth trying before you do anything more sophisticated. If the answers are not (biologically) significantly different, the fancier techniques may not be worth the effort. Furthermore, if you find in the end that the distinction between process and observation variability is unidentifiable, stating the results of process-error-only and observation-error-only analyses and saying that the true value is likely to be somewhere between those answers may be the best you can do.

11.5 SIMEX

In our one-step-ahead example, ignoring observation error led to a high estimate of r (1.22 vs. true value 1) and a low estimate of K (9.88 vs. true value 10). It's impossible to infer from a single example, but in fact ignoring observation error will generally give upward biased answers for r because observation error suggests that the population is changing faster than it really is. In this example K is biased downward as well. It's hard to figure out in general what direction of bias to expect—it depends in detail on the nonlinearities in the model—but estimates of nonlinear

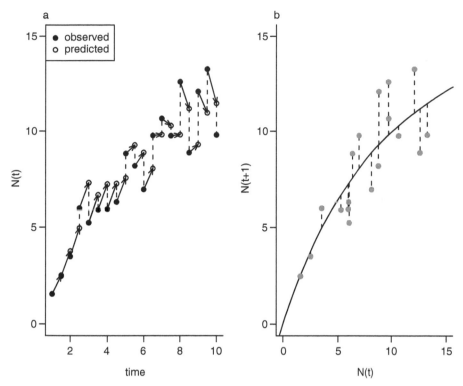

Figure 11.4 Logistic fit: one-step-ahead (process error only). True parameters $r = 1$, $K = 10$, $N(0) = 1$, $\sigma^2_{obs} = \sigma^2_{proc} = 1$. Estimated parameters $r = 1.22$, $K = 9.88$, $\sigma^2_{obs} = 2.66$. (a) Time dynamics and predictions. (b) Current vs. next observations, showing vertical residuals from the fitted line.

model parameters that ignore observation error are very likely to be biased one way or the other.

However, if you do have an estimate of the magnitude of the observation error, you can use the SIMEX (simulation-extrapolation) algorithm to correct for the bias caused by neglecting observation error. SIMEX works by inflating the observation error—adding additional noise to the data set—and reestimating the parameters (Cook and Stefanski, 1994; Carroll et al., 1995; Carroll et al., 1999; Stefanski and Cook, 1995). After estimating how increasing levels of observation error change the parameter estimates, you can then extrapolate to estimate the parameter values you *would* get with zero observation error. (Yes, this seems like black magic, but it works.)

More specifically, the procedure for SIMEX is as follows:

- Based on your estimate of observation error, pick a range of increased error values; tripling the existing observation variance in four to eight steps is a reasonable rule of thumb. (For example, if the estimate of observation error is σ^2_{obs}, pick observation variances of $\{1.5\sigma^2_{obs}, 2\sigma^2_{obs}, 2.5\sigma^2_{obs}, 3\sigma^2_{obs}\}$.)
- For each error magnitude in your range, generate a data set with that increased error. The procedure is more stable if you pick a single set of normally distributed random values and then multiply them by increasing factors for each

simulation. (If y_i are your values and ϵ_i is a set of normal deviates with variance σ^2_{obs}, the first simulated data set with the inflation factors above would be $y_i + \sqrt{0.5}\epsilon_i$; the variance of this data set is $\sigma^2_{obs} + 0.5\sigma^2_{obs} = 1.5\sigma^2_{obs}$. The second data set with $y_i + \epsilon_i$ would have variance $2\sigma^2_{obs}$.)

- For each simulated data set, estimate the values of the parameters using one-step-ahead prediction and save them.
- Estimate a relationship between the total variance and the values of the parameters (a separate regression for each parameter, typically a linear or quadratic regression: `lm1 = lm(param~measerr+I(measerr^2))`).
- Find the SIMEX bias-corrected estimates of the parameters by extrapolating the regressions to zero variance (for a linear or quadratic regression, the first coefficient is the intercept: `coef(lm1)[1]`).

11.6 State-Space Models

The final, most sophisticated and most general but most challenging category of statistical estimation procedures for dynamic data are so-called *state-space models*. In principle state-space models can allow you to estimate parameters of the deterministic process, observation error, *and* process error from a single observed time series—always subject to the constraints of identifiability. Trying to fit state-space models to time series that are too short, vary too little, or otherwise contain insufficient information to identify the parameters will lead to numerical problems and wide confidence intervals if you're lucky (and skilled), and misleading answers if you're unlucky. Schnute's warning about identifiability (p. 334) refers specifically to state-space models. With that warning in mind, here we go.

In general, we know that the amount of observation error will be the same for each observation (or at least will depend only on the true value, and not on when it is measured), while the amount of process error will tend to increase over time. The longer we wait between observations, the more random variation will decrease our certainty about the state of the system.

The key insight of state-space models is that every observation we make does several things:

- It provides information that shrinks the cloud of uncertainty around the true but unknown current state of the system.
- It provides *indirect* information about the likelihood of the next state. For example, a higher-than-expected population count in 2000 increases the expectation for the 2001 count.
- It also provides indirect information about the *previous* state of the system. For example, a higher-than-expected population count in 2000 also makes us think that the true population size in 1999 might have been higher than we previously thought.

If this discussion sounds Bayesian to you (updating our expectations of the probability of the state of the system based on prior observations), you're right. Lots of state-space modeling has a Bayesian flavor, although it can also be done in a frequentist framework (de Valpine and Hastings, 2002; Ionides et al., 2006; Lele et al., 2007). At

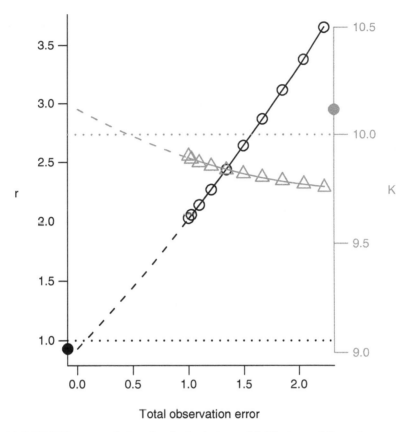

Figure 11.5 SIMEX extrapolation for the logistic model. Horizontal lines show true values: black circles show estimates for r and gray triangles show estimates for K, both extrapolated quadratically back to $\sigma^2 = 0$.

this cutting-edge level, there's much more interplay between Bayesian and frequentist approaches than at more basic levels.

Estimation algorithms for state-space models are essentially systems that carry out the complicated bookkeeping required to keep track of the current estimates of the true state of the system at a particular time. For each new choice of parameters, the algorithm works through the data set one observation at a time, updating estimates of the true value and variance at that time based on the parameters and the current estimate of the previous time step (and in some systems, of the next time step as well). Once this is done for the whole data set, you can use the estimates and variances to calculate the likelihood for the new set of parameters and decide how to pick the next one, using a standard algorithm such as Nelder-Mead or MCMC.

11.6.1 Kalman Filter

The *Kalman filter* is an algorithm for calculating the expected means and covariances of the observed values for a whole time series in the presence of observation and process error. In its original form it works only for linear population models (i.e.,

$N(t + 1)$ is a linear function of $N(t)$) with multivariate normal error; the *extended Kalman filter* uses an approximation that works for nonlinear population dynamics. The Kalman filter's great strengths are its relative simplicity and speed.

The Kalman filter works by stepping through the data set one observation at a time, updating what we know about the mean and variance of the true state variables at time t. It is an inductive procedure, giving the rules for figuring out the mean and variance at time t if we already know the mean and variance at time $t - 1$. Clearly, then, if we can figure out starting values for the mean and variance at time 1, we can work through the whole data set this way.

I'll illustrate this with a very simple example (keeping in mind that we can add many realistic complications), with a single population growing linearly at rate a per year, with an autoregressive term b that means that N_{t-1} and N_t have a correlation coefficient of b (over and above the general linear trend with time). I assume there is both process (σ_{proc}^2) and observation (σ_{obs}^2) error, both normally distributed. So our model is

$$N_t \sim \text{Normal}(a + bN_{t-1}, \sigma_{proc}^2) \tag{11.6.1}$$

$$N_{obs,t} \sim \text{Normal}(N_t, \sigma_{obs}^2). \tag{11.6.2}$$

If $b < 1$, then the population is stable, because random deviations in N shrink by a factor b every year; if $b > 1$, then the population is unstable and random deviations grow over time.

Suppose, based on all the observations up through time $t - 1$, we believe that the mean of the true population size at time $t - 1$, N_{t-1}, is μ_0, and its variance is σ_0^2. We can calculate based only on the population parameters a and b what we expect the mean and variance to be at the next time step. The change in the mean is a direct reflection of the population model; the variance term is a combination of multiplying the previous variance by b^2 since we have multiplied the population size by b, and adding the new variability introduced by process error between $t - 1$ and t. So

$$\text{mean}(N_t | N_{obs,t-1}) = \mu_1 = a + b\mu_0 \tag{11.6.3}$$

$$\text{Var}(N_t | N_{obs,t-1}) = \sigma_1^2 = b^2\sigma_0^2 + \sigma_{proc}^2. \tag{11.6.4}$$

More stable populations, indicated by low values of b, imply lower variance. As b gets very small, no variance carries over from one time step to the next and the standing variance of the population becomes just σ_{proc}^2.

The mean of the observation at time t equals the mean of the true value (we assume variance, but no bias, in the observation process). The variance equals the current variance of the true population size plus the observation variance:

$$\text{mean}(N_{obs,t} | N_{obs,t-1}) = \mu_2 = \mu_1 \tag{11.6.5}$$

$$\text{Var}(N_t | N_{obs,t-1}) = \sigma_2^2 = \sigma_1^2 + \sigma_{obs}^2. \tag{11.6.6}$$

The last step of the Kalman filter, taking the information about the current observation into account, is the hardest. The current observation, changes our estimate of the mean of the true population state. How much it changes it depends on how far the current observation is from where it was expected to be based on the previous information ($N_{\text{obs},t} - \mu_2$), as well as the ratio of the variances of the true value and of the observation. If there is no observation error, then the variance of the observation is the same as the variance of the true state of the population, and (as shown by the formula below) we simply set the mean of the population equal to the current observation. If there is lots of observation error, then the current observation doesn't tell us very much and we don't let an unexpected observation change our value of the mean.

$$\text{mean}(N_t | N_{\text{obs},t}) = \mu_3 = \mu_1 + \frac{\sigma_1^2}{\sigma_2^2}(N_{\text{obs},t} - \mu_2). \qquad (11.6.7)$$

The Bayesian approach suggests another interpretation of this equation: our best estimate of the current population size is a weighted average of our prior—what we think the population size is based on previous time steps (μ_1)—and the current observational data ($N_{\text{obs},t}$).

Finally, we need to update the variance based on the current observation. Here we actually reduce the current variance of the true value, again based on the ratio of the variance of the true value to the variance of the observation.

$$\text{Var}(N_t | N_{\text{obs},t}) = \sigma_3^2 = \sigma_1^2 \left(1 - \frac{\sigma_1^2}{\sigma_2^2}\right). \qquad (11.6.8)$$

If there is no observation error, then $\sigma_1^2 = \sigma_2^2$ and the variance of the true value becomes zero. Unlike the mean, the variance is independent of the observed data.

Now that we've figured out the mean and variance of N and N_{obs} based on all the observations up to time t, we can repeat the procedure to calculate the values at time $t + 1$. Once we have worked through the whole data set, we know the expected mean and variance at each time step, and we can calculate the standard normal log-likelihood for the observed values.

The concepts are the same but the formulas are considerably more complicated in the general case described by Schnute (1994). The one extension I will describe here is how to estimate a nonlinear population growth function $f(N)$, called the *extended Kalman filter*.

All we have to do is replace (11.6.3) and (11.6.4) with appropriate generalizations. For example, let's replace the linear equation in (11.6.1) with the discrete logistic equation:

$$N_t \sim \text{Normal}(N_{t-1} + rN_{t-1}\left(1 - \frac{N_{t-1}}{K}\right), \sigma_{\text{proc}}^2). \qquad (11.6.9)$$

Then substitute this equation for (11.6.3):

$$\text{mean}(N_t | N_{\text{obs},t-1}) = \mu_1 = \mu_0 + r\mu_0 \left(1 - \frac{\mu_0}{K}\right). \qquad (11.6.10)$$

For the variance, we need to find the equivalent of b, the per capita growth rate, to substitute into (11.6.4). A reasonable approximation for the current per capita growth rate is the derivative of the population growth rate with respect to the population size:

$$\frac{\partial f}{\partial N} = \frac{\partial(N + rN(1 - N/K))}{\partial N} = 1 + r - 2rN/K. \tag{11.6.11}$$

Since this equation is based on a first-order Taylor expansion, it is good only for relatively small noise or short time steps.

Evaluating the derivative at the current mean value of the population size ($N = \mu_1$) gives

$$\text{Var}(N_t | N_{\text{obs},t-1}) = \sigma_1^2 = (1 + r - 2r\mu_1/K)^2 \sigma_0^2 + \sigma_{\text{proc}}^2. \tag{11.6.12}$$

If the population is currently growing ($\frac{\partial f}{\partial N} > 1$), the variance is inflated. If it is shrinking, the variance is deflated.

So how do we implement this in R?

The R supplement defines a function `nlkfpred` that calculates the nonlinear Kalman filter predictions for a set of time-series data and a `nlkflik` that uses those predictions to compute the negative log-likelihood for a set of parameters. We fit all the parameters on the log scale to avoid the possibility of negative parameter values.

We need to pick starting values for the estimation. I'm going to cheat here since I know the true values, but it would be easy enough to do a one-step-ahead or trajectory-matching fit to the data, or even eyeball, to estimate reasonable starting values for r, K, and the variances.

```
> startvec = list(logr = log(0.25), logK = log(10),
+     logprocvar = log(0.5), logobsvar = log(0.5),
+     logM.n.start = log(3), logVar.n.start = -2)
```

Maximum-likelihood estimation of the parameters:

```
> m4 = mle2(minuslogl = nlkflik, start = startvec,
+     data = list(obs.data = y.procobs2),
+     method = "Nelder-Mead", control = list(maxit = 2000))
```

The fitted parameters are reasonable (although K appears slightly biased upward) and the confidence intervals bracket the true values, as seen in Figure 11.6 and Table 11.1. It's not surprising that the confidence intervals are narrow for K and slightly wider for r (the population spends more time around its carrying capacity than in the growth phase), or that the confidence intervals for the variances are larger than the confidence intervals for the deterministic parameters.

The Kalman filter has been widely used in fisheries modeling, where the need to squeeze information out of rare data is so strong that researchers are always looking for the next powerful technique. Early on, researchers applied the technique to abundant but noisy catch-per-unit-effort data. More recent applications have used the Kalman filter as a way to estimate the locations of animals from noisy telemetry data, allowing the observed position at a previous time to help constrain the expected location at the current time (Jonsen et al., 2003). The KF has more recently begun to make its way into mainstream terrestrial ecology, as a way of estimating parameters for the growth of species of conservation concern in the presence of both observation and observation error (Lindley, 2003). While the assumption of linear population

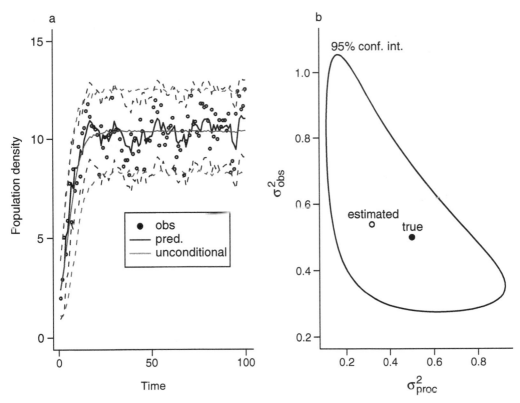

Figure 11.6 Results of Kalman filter: (a) observed, predicted, and results of unconditional simulations; (b) MLE, true value, and approximate 95% bivariate confidence interval.

dynamics in the standard Kalman filter might seem constraining, the autoregressive equation $N_t = a + bN_{t-1}$ does allow a range of population dynamics, from fluctuation around a stable equilibrium if $a > 0$ and $b < 1$, to exponential dynamics if $a = 0$ (declining if $b < 1$, increasing if $b > 1$), to a pure random walk if $a = 0$ and $b = 1$. Much of conservation biology is built on linear models, which will often apply when species are rare and thus intraspecific competition is low (Caswell, 2000). And if you do need nonlinearity, you can use the extended Kalman filter.

TABLE 11.1
Results of Kalman filter estimation

	True	Fitted	2.5%	97.5%
r	0.25	0.30	0.18	0.48
K	10.00	10.43	10.01	10.87
σ^2_{proc}	0.50	0.32	0.13	0.74
σ^2_{obs}	0.50	0.54	0.31	0.92

There are ways to incorporate many other biological complexities in the Kalman filter such as multiple species, time lags, bias and imperfect catchability in observations, correlated observations, time-varying control parameters, and covariates measured with error; see Schnute (1994) for details. Don't be scared by the notation. If you follow through it carefully, you can match up the special case here with all the details in that paper.

11.6.2 Markov Chain Monte Carlo Approaches (WinBUGS et al.)

The Kalman filter has limitations. In particular, it assumes normal, or lognormal, distributions. More subtly, the Kalman filter is a *prospective* algorithm (Schnute, 1994). It uses only the information up to time t to predict the mean and variance of the population size, even though the observation at time $t + 1$ also gives us information about the population size at time t—*retrospective* information that we might be able to use to improve estimation.

Retrospective bookkeeping can be done several ways. Schnute discusses a frequentist approach called the *errors-in-variables* method, and de Valpine (de Valpine and Hastings, 2002; de Valpine, 2003) has also developed such a frequentist method. Here I'm going to present Bayesian methods (e.g., Millar and Meyer, 2000), which are rapidly growing in popularity because BUGS makes it simple to develop and estimate the parameters of relatively complex population dynamic models (Lele et al. (2007) suggest a way to use BUGS to calculate maximum likelihood estimates for complex models). The basic idea carries over from the Kalman filter; if you assume you know all the observations and the true values at every *other* time step, you can use them to estimate the population size *now*. The Markov chain Monte Carlo approach alternates between picking new random values for each true population size, one at a time (at each time step pretending you know the population sizes at all the other time steps), and picking new random values for the parameters that are consistent with the current assumed population sizes. Figure 11.7 shows the dependency graph for the first four steps of a logistic process. Each observed value depends on the true value at that time step and the observation error; each true value depends on the parameters and determines the observed value and the value at the next time step. In this kind of graph, though, you can also follow arrows *backward* to see that as well as depending on the value at time 1, our information about the true value at time 2 is also influenced by the observed value at time 2 and by the true value at time 3—and hence indirectly by the observation at time 3, just as I suggested above.

To use BUGS to analyze dynamic data you must first decide on a model. Here we'll again use the discrete logistic equation with normally distributed observation and process error for comparison, although BUGS would allow us to be much more flexible. You also need to set priors for the parameters. Translating (11.6.3) and (11.6.4) into BUGS syntax produces the following model:

```
model {
  t[1] <- n0
  o[1] ~ dnorm(t[1],tau.obs)
  for (i in 2:N) {
    v[i] <- t[i-1]+r*t[i-1]*(1-t[i-1]/K)
```

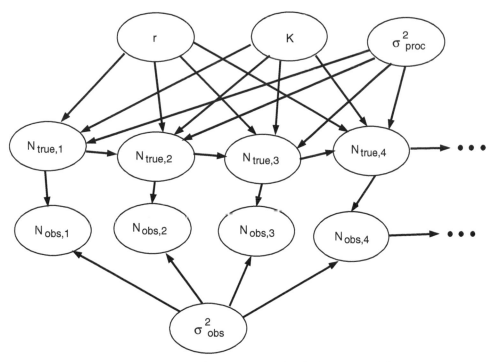

Figure 11.7 Dependency structure for the logistic model.

```
    t[i] ~ dnorm(v[i],tau.proc)
    o[i] ~ dnorm(t[i],tau.obs)
}
```

The first two lines define the initial conditions. The rest of the model steps through the data set, calculating the deterministic expectation (v[i]) and then defining the distribution of the true values (t[i]) and observed values (o[i]).

The rest of the model file defines the priors:

```
r ~ dunif(0.1,maxr)
K ~ dgamma(0.005,0.005)
tau.obs ~ dgamma(0.005,0.005)
tau.proc ~ dgamma(0.005,0.005)
n0 ~ dgamma(1,n0rate)
}
```

The prior growth rate r is uniformly distributed between 0.1 and a maximum value, which I made a parameter so I could vary it within R without changing my BUGS input file. The priors for the carrying capacity K and the precisions (inverse variances) τ_{obs} and τ_{proc} are Gamma distributions with rate and shape parameters of 0.005, giving them a mean of 1 and a large variance ($0.005/0.005^2 = 200$: remember that BUGS uses a shape + rate parameterization rather than R's shape + scale parameterization). The initial density n_0 has a prior distribution that is Gamma with shape parameter 1 and a rate parameter equal to the reciprocal of the first observed value—again defined as a parameter to be calculated in R—which gives it an exponential distribution with mean equal to the first observed value. Though weak, this prior is stronger than the prior distributions for the carrying capacity and precisions.

A bit of R code to define the upper limit of the r prior and the parameter of the initial-state prior:

```
> maxr = 2
> n0rate = 1/y.procobs2[1]
```

We will set up the model, using the same data series y.procobs2 as before and defining five different chains, using the perturb.params function from the emdbook package to change the values of r and the precisions (τ_{obs}, τ_{proc}). We should probably vary the starting values of the precisions a bit more systematically, although BUGS tends to crash if the starting values are too extreme.

```
> o = y.procobs2
> N = length(y.procobs2)
> statespace.data = list("N", "o", "maxr", "n0rate")
> inits = perturb.params(list(n0 = y.procobs2[1], r = 0.2,
+     K = 10, tau.obs = 1, tau.proc = 1), alt = list(r =
+     c(0.1, 0.4), tau.obs = 3, tau.proc = 3))
```

We next define the parameters we want to keep track of; we could also track the estimated true values at each time step.

```
> parameters = c("r", "K", "tau.obs", "tau.proc", "n0")
```

After running WinBUGS from within R, we convert the output to a coda object—the coda format has slightly different uses from the format returned by R2WinBUGS.

```
> statespace.sim = bugs(data = statespace.data, inits,
+     param = parameters, model = "statespace.bug",
+     n.chains = length(inits), n.iter = 15000)
> s1 = as.mcmc.bugs(statespace.sim)
```

R2WinBUGS's defaults for running an MCMC analysis are to take the total number of iterations (the default is 2,000); set aside half of them as "burn-in"; divide the other half equally among all the chains specified by the user (the default is 3); and "thin" the results to save a total of 1000 iterations across all chains. In this case I chose to run 15,000 iterations with five chains, so each chain ran for 3000 steps; the first 1500 were discarded; and then 13% of the remaining iterates were kept for a total of 1000.

Checking convergence:

```
> gelman.diag(s1)
```

```
Potential scale reduction factors:

          Point est.    97.5% quantile
r              1.00          1.02
K              1.01          1.02
tau.obs        1.02          1.05
```

```
tau.proc      1.05          1.09
n0            1.01          1.02
deviance      1.02          1.05
```

Multivariate psrf

1.02

Based on the G-R rule of thumb that a scale reduction factor < 1.2 for all variables means adequate convergence, the G-R diagnostic suggests that the chains did in fact run long enough to mix with each other.

The summary of a coda object provides the quantiles of the chains; these results are practically identical to those from the Kalman filter (Table 11.2). I have inverted the precisions ($\tau_{proc} = 1/\sigma^2_{proc}$, $\tau_{obs} = 1/\sigma^2_{obs}$) to make it easier to compare directly with the KF results; the median is not identical to the mode (which in turn is close to the maximum likelihood estimate if the priors are weak), but it's close.

Figure 11.8 shows the results of the R2WinBUGS run for σ^2_{obs} and σ^2_{proc}. The values from the Kalman filter (Figure 11.6) are shown in gray. The 95% credible interval matches the approximate 95% confidence interval reasonably well, especially considering that the confidence interval is an approximation based on the local curvature. The mode of the posterior density, as expected, is very close to the MLE—with a weak prior probability distribution, the likelihood surface and the posterior probability distribution are close to the same shape. The mean is slightly larger than the mode—there is some skew toward large values of the process variance—while the median, not shown, falls between the mean and the mode. All four summary values (posterior mean, mode, and median, and the MLE) and the true value all fall within the 50% credible interval; as is often the case, all of these estimates give us *approximately* the same answer.

Finally, Figure 11.9 shows density plots for the R2WinBUGS analysis. The densities are all reasonably symmetric and bracket the known true values (the density of tau.obs extends to very high values; this is the result of a single freakish excursion in one of the chains to a very high value). Each chain's density is drawn with a different line type; they all fall on top of each other, reassuring us that the chains have converged and are all telling the same story.

TABLE 11.2
Results from WinBUGS analysis of logistic growth

	2.5%	Median	97.5%
r	0.17	0.30	0.48
K	9.99	10.45	10.98
σ^2_{proc}	0.14	0.36	0.82
σ^2_{obs}	0.22	0.52	0.90

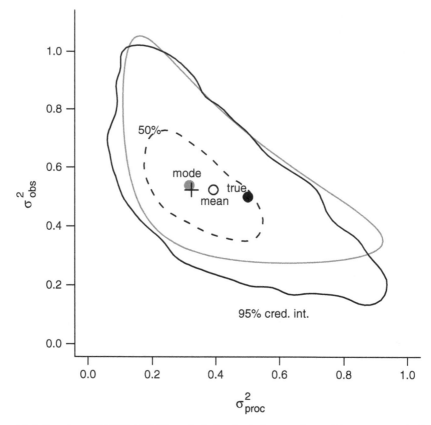

Figure 11.8 Results of R2WinBUGS analysis for logistic equation with process and observation error. Gray point and ellipse represent MLE and approximate (information-based) 95% confidence interval for Kalman filter fit (Section 11.6.1). Solid line is 95% credible interval based on BUGS chain; dashed line is 50% credible interval. Points show mean and mode of the posterior density and the true (known) value.

11.7 Conclusions

This chapter covered a variety of methods for estimating the parameters of dynamic models, ranging from crude (assuming either process error or observation error, but not both) to sophisticated (state-space models).

We largely skipped over the question of how to decide which dynamic models to fit. The logistic and a few simple discrete stochastic models were mentioned here, and the theta-logistic was noted in Chapter 3, but most of the book has focused on static models. Understanding dynamic models is a huge topic, mostly focused on deterministic models—Ellner and Guckenheimer (2006) or the other references listed in the ecological modeling section in Chapter 1 make a good starting point on the dynamics side, Clark (2007) includes a review of dynamic modeling in his presentation of Bayesian methods, and Bjørnstad and Grenfell (2001) give an overview of more recent advances in the field.

On the other hand, it's easy to incorporate some additional ecology in the single-species logistic model. For example, with either the Kalman filter or MCMC you

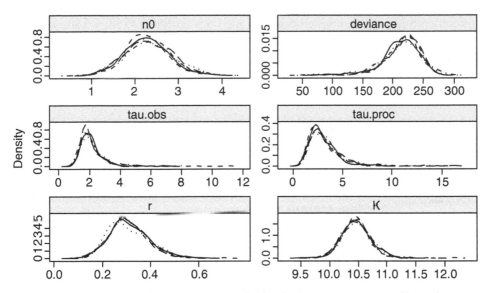

Figure 11.9 Density plots of R2WinBUGS results for the logistic equation. Different line types show results from different chains.

can incorporate the effects of covariates on the growth rate. To incorporate a linear effect of rainfall on the growth rate, you could just change the appropriate line of the BUGS model file to

```
v[i] <- t[i-1]+(r0+r1*rain[i-1])*t[i-1]*(1-t[i-1]/K)
```

and change the parameters and data values accordingly in the R code.

All too often, you can observe only one facet of a complicated ecological inter- action. For example, we might be able to sample just hare populations in a complex Canadian ecosystem consisting of lynx, hare, vegetation, and birds of prey. While trying to reconstruct an entire ecosystem from observations of a single species is hopeless, we could in principle include additional unobserved variables in a state- space model—remember that the "true" population sizes are also unobserved. Be very careful not to incorporate more complexity in your model than the data can support: try your model out with some optimistic, but plausible, simulation data. Such reconstruction has been shown to work, for example, in simple epidemic mod- els, where the number of new cases is observed but the number of possibly susceptible individuals left in the population is not. A formal process of "susceptible reconstruc- tion" provides a time series of susceptibles to go along with the time series of infected individuals, which then allows estimation of a transmission parameter (Finkenstädt and Grenfell, 2000; Lekone and Finkenstädt, 2006).

This chapter has presented only analyses of discrete-time models, where the methods are much better developed. It is unfortunate that continuous-time meth- ods for dynamical data are so sparse, since most theoretical models of ecological systems are defined in continuous time. Analysis is feasible if you assume only obser- vation error (Gani and Leach, 2001; van Veen et al., 2005) or know the amount of observation error and use SIMEX to correct bias (Ellner et al., 2002; Melbourne and Chesson, 2006), but the Kalman filter and MCMC approaches have been used almost exclusively in discrete time (although Fujiwara et al. (2005) provide a recent counterexample). Gibson developed such methods (Gibson, 1997; Gibson and Renshaw, 1998; Gibson and Renshaw, 2001; Streftaris and Gibson, 2004), but they have yet to be widely used or made practical.

Right now Bayesian analyses of dynamic models are easier than frequentist analyses, but the frequentists are catching up fast. *Particle filtering* and *sequential importance sampling* are powerful frequentist alternatives to the Bayesian MCMC methods presented here (Doucet et al., 2001; Buckland et al., 2004; Thomas et al., 2005; Harrison et al., 2006; Ionides et al., 2006). Particle filtering starts with a large number of random samples ("particles," e.g., 250,000 in Thomas et al. (2005)) from a prior (or pseudo-prior) distribution, including the distribution of the initial values of the state variables. Each sample is projected forward (simulated) one step, and a likelihood based on the first observation is calculated for each sample. The same number of particles is then resampled, but with weights proportional to their likelihoods. After simulating one more step, the likelihoods based on the next observation are calculated and the particles are resampled again (thus taking the observations at both t and $t+1$ into account). This process is iterated for the whole time series of observations, with various algorithms used to prevent all of the resamples from coming from a very small number of particles.

Estimating parameters of dynamic ecological models is still clearly an exercise on the cutting edge of science. Most of the papers that have appeared to date are technical and methods-oriented rather than applications to particular ecological questions. As time goes on the tools will improve and more examples will appear, giving potential users a better idea how much data (at least within an order of magnitude) is needed to apply these methods successfully. In the meantime, always check your answers against the results of simulations and against one-step-ahead (process error only) and trajectory-matching (observation error only) fits.

11.8 R Supplement

11.8.1 Kalman Filter

Here's a function that computes the Kalman filter predictions. Nobs is the data set; r and K are the population dynamic parameters; procvar and obsvar are the process and observation variances; and M.n.start and Var.n.start are the starting values of the mean and variance. This code sets aside numeric vectors for the results on the mean and variance of the observed population size at each time step. When estimating parameters we don't have to save the mean and variance of the true population size since we don't have anything to compare them with. The function then sets the starting values and works through the data set one time step at a time, applying the Kalman filter equations:

```
> nlkfpred = function(r, K, procvar, obsvar, M.n.start,
+       Var.n.start, Nobs) {
+       nt = length(Nobs)
+       M.nobs = numeric(nt)
+       Var.nobs = numeric(nt)
+       M.n = M.n.start
```

```
+      Var.n = Var.n.start
+      M.nobs[1] = M.n.start
+      Var.nobs[1] = Var.n.start + obsvar
+      for (t in 2:nt) {
+          M.ni = M.n + r * M.n * (1 - M.n/K)
+          b = 1 + r - 2 * r * M.n/K
+          Var.ni = b^2 * Var.n + procvar
+          M.nobs[t] = M.ni
+          Var.nobs[t] = Var.ni + obsvar
+          M.n = M.ni + Var.ni/Var.nobs[t] * (Nobs[t] -
+              M.nobs[t])
+          Var.n = Var.ni * (1 - Var.ni/Var.nobs[t])
+      }
+      list(mean = M.nobs, var = Var.nobs)
+ }
```

Our likelihood function takes a set of parameters (all fitted on the log scale so we don't run into trouble with negative values of the parameters), runs the Kalman filter to predict the values of the means and variances, and then plugs these values into a normal likelihood comparison with a set of observed values (taking the square root of the estimated variance since dnorm uses the standard deviation, not the variance, as a parameter):

```
> nlkflik = function(logr, logK, logprocvar, logobsvar,
+      logM.n.start, logVar.n.start, obs.data) {
+      pred = nlkfpred(r = exp(logr), K = exp(logK),
+          procvar = exp(logprocvar), obsvar =
+          exp(logobsvar), M.n.start = exp(logM.n.start),
+          Var.n.start = exp(logVar.n.start),
+          Nobs = y.procobs2)
+      -sum(dnorm(obs.data, mean = pred$mean,
+          sd = sqrt(pred$var), log = TRUE))
+ }
```

12 Afterword

So you read the whole thing . . . it was surely challenging and frustrating at times, but I hope the way was lightened by moments of clarity. Welcome to the cutting edge—you now have all the basic tools you need to pose, and answer, ecological questions in a quantitative way. There is more to learn, of course, but at this point you should be capable of picking your way through the primary literature to find new tools. New statistical ideas and new applications of statistics are appearing monthly in journals like *Ecology* and *Ecological Applications*, where they are generally phrased in "ecologist-friendly" terms, but you may also find yourself making your way to the pages of journals such as *Biometrika* and *Journal of the American Statistical Association* in search of new ideas. More important, however, you are now empowered to *make stuff up*—within the limits of common sense and the statistical tools you have learned, you can design and build your own models. Check them with simulations and ask statistically savvy colleagues to confirm that your methods are reasonable. You will be pleasantly surprised (I know I was the first time I brought a new statistical model to a statistician) when they say "gee, nobody's done that before, but it seems to make sense."

> . . . many places you would like to see are just off the map and many things you want to know are just out of sight or a little beyond your reach. But someday you'll reach them all, for what you learn today, for no reason at all, will help you discover all the wonderful secrets of tomorrow.
>
> —Norton Juster
> *The Phantom Tollbooth*

Appendix   Algebra and Calculus Basics

A.1 Exponentials and Logarithms

Exponentials are written as e^x or $\exp(x)$, where $e = 2.718\ldots$ By definition $\exp(-\infty) = 0$, $\exp(0) = 1$, $\exp(1) = e$, and $\exp(\infty) = \infty$. In R, e^x is `exp(x)`; if you want the value of e, use `exp(1)`. Logarithms are the solutions to exponential or power equations like $y = e^x$ or $y = 10^x$. *Natural* logs, ln or $\log_e$, are logarithms base e; *common* logs, $\log_{10}$, are typically logarithms base 10. When you see just "log" it's usually in a context where the difference doesn't matter (although in R $\log_{10}$ is `log10` and $\log_e$ is `log`).

1. $\log(1) = 0$. If $x > 1$, then $\log(x) > 0$, and vice versa. $\log(0) = -\infty$; logarithms are undefined for $x < 0$.
2. Logarithms convert products to sums: $\log(ab) = \log(a) + \log(b)$.
3. Logarithms convert powers to multiplication: $\log(a^n) = n\log(a)$.
4. You can't do anything with $\log(a + b)$.
5. Converting bases: $\log_x(a) = \log_y(a)/\log_y(x)$. In particular, $\log_{10}(a) = \log_e(a)/\log_e(10) \approx \log_e(a)/2.3$ and $\log_e(a) = \log_{10}(a)/\log_{10}(e) \approx \log_{10}(a)/0.434$. This means that converting between log bases just means multiplying or dividing by a constant. Here's the proof:

$$y = \log_{10}(x)$$

$$10^y = x$$

$$\log_e(10^y) = \log_e(x)$$

$$y\log_e(10) = \log_e(x)$$

$$y = \log_e(x)/\log_e(10)$$

    (compare the first and last lines).
6. The derivative of the logarithm, $d(\log x)/dx$, equals $1/x$. This is always positive for $x > 0$ (which are the only values for which the logarithm is defined anyway).
7. The fact that $d(\log x)/dx > 0$ means the function is *monotonic* (always either increasing or decreasing), which means that if $x > y$, then $\log(x) > \log(y)$ and if $x < y$, then $\log(x) < \log(y)$. This in turn means that if you find the maximum likelihood parameter, you've also found the maximum log-likelihood parameter (and the minimum negative log-likelihood parameter).

A.2 Differential Calculus

1. Notation: differentation of a function $f(x)$ with respect to x can be written, depending on the context, as $\frac{df}{dx}$; f'; $\dot{f}$; or f_x.
2. Definition of the derivative:

$$\frac{df}{dx} = \lim_{\Delta x \to 0} \frac{f(x + \Delta x) - f(x)}{(x + \Delta x) - x} = \lim_{\Delta x \to 0} \frac{f(x + \Delta x) - f(x)}{\Delta x}. \tag{A.2.1}$$

In words, the derivative is the slope of the line tangent to a curve at a point, or the instantaneous slope of a curve. The second derivative, d^2f/dx^2, is the rate of change of the slope, or the curvature.
3. The derivative of a constant (which is a flat line if you think about it as a curve) is zero (slope = 0).
4. The derivative of a linear equation, $y = ax$, is the slope of the line, a. (The derivative of $y = ax + b$ is also a.)
5. Derivatives of polynomials: $\frac{d(x^n)}{dx} = nx^{n-1}$.
6. Derivatives of sums: $\frac{d(f+g)}{dx} = \frac{df}{dx} + \frac{dg}{dx}$ (and $d(\sum_i y_i)/dx = \sum_i (dy_i/dx)$).
7. Derivatives of products: $\frac{d(fg)}{dx} = f\frac{dg}{dx} + g\frac{df}{dx}$.
8. Derivatives of constant multiples: $\frac{d(cf)}{dx} = c\frac{df}{dx}$, if c is a constant (i.e., if $\frac{dc}{dx} = 0$).
9. Derivative of the exponential: $\frac{d(\exp(ax))}{dx} = a\exp(ax)$, if a is a constant. (If not, use the chain rule.)
10. Derivative of logarithms: $\frac{d(\log(x))}{dx} = \frac{1}{x}$.
11. Chain rule: $\frac{d(f(g(x)))}{dx} = \frac{df}{dg} \cdot \frac{dg}{dx}$ (thinking about this as "multiplying fractions" is a good mnemonic but don't take it too literally!) *Example:*

$$\frac{d(\exp(x^2))}{dx} = \frac{d(\exp(x^2))}{d(x^2)} \cdot \frac{dx^2}{dx} = \exp(x^2) \cdot 2x. \tag{A.2.2}$$

Another example: people sometimes express the proportional change in x, $(dx/dt)/x$, as $d(\log(x))/dt$. Can you see why?
12. *Critical points* (maxima, minima, and saddle points) of a curve f have $df/dx = 0$. The sign of the second derivative determines the type of a critical point (positive = minimum, negative = maximum, zero = saddle).

A.3 Partial Differentiation

1. Partial differentiation acts just like regular differentiation except that you hold all but one variable constant, and you use a curly d (∂) instead of a regular d. So, for example, $\partial(xy)/\partial(x) = y$. Geometrically, this is taking the slope of a surface in one particular direction. (Second partial derivatives are curvatures in a particular direction.)
2. You can do partial differentiation multiple times with respect to different variables; order doesn't matter, so $\frac{\partial^2 f}{\partial x \partial y} = \frac{\partial^2 f}{\partial y \partial x}$.

A.4 Integral Calculus

For the material in this book, I'm not asking you to remember very much about integration, but it would be useful to remember that

1. The (definite) integral of $f(x)$ from a to b, $\int_a^b f(x)\,dx$, represents the area under the curve between a and b. The integral is a limit of the sum: $\sum_{x_i=a}^b f(x_i)\Delta x$ as $\Delta x \to 0$.
2. You can take a constant out of an integral (or put one in): $\int af(x)\,dx = a\int f(x)\,dx$.
3. Integrals are additive: $\int (f(x)+g(x))\,dx = \int f(x)\,dx + \int g(x)\,dx$.

A.5 Factorials and the Gamma Function

A *factorial*, written with an exclamation point !, means $k! = k \times (k-1) \times \cdots \times 1$. For example, $2! = 2$, $3! = 6$, and $6! = 720$. In R a factorial is factorial—you can't use the shorthand ! notation, especially since != means "not equal to" in R. Factorials come up in probability calculations frequently, e.g., as the number of permutations with k elements. The *gamma function*, usually written as Γ (gamma in R) is a generalization of factorials. For integers, $\Gamma(x) = (x-1)!$. Factorials are defined for integers only, but for positive, noninteger x, $\Gamma(x)$ is still defined and it is still true that $\Gamma(x+1) = x \cdot \Gamma(x)$.

Factorials and gamma functions get very large, and you often have to compute ratios of factorials or gamma functions (e.g., the binomial coefficient, $N!/(k!(N-k)!)$. Numerically, it is more efficient and accurate to compute the logarithms of the factorials first, add and subtract them, and then exponentiate the result: $\exp(\log N! - \log k! - \log(N-k)!)$. R provides the log-factorial (lfactorial) and log-gamma (lgamma) functions for this purpose. (Actually, R also provides choose and lchoose for the binomial coefficient and the log-binomial coefficient, but the log-gamma is more generally useful.)

The main reason that the gamma function (as opposed to factorials) comes up in ecology is that it is part of the *normalizing constant* (see Chapter 4) for the Gamma *distribution*, which is usually written as Gamma (not Γ): $\text{Gamma}(x, a, s) = \frac{1}{s^a \Gamma(a)} x^{a-1} e^{-x/s}$.

A.6 Probability

Most of the probability rules you need are discussed in Chapter 4.

1. Probability distributions always add or integrate to 1 over all possible values.
2. Probabilities of independent events are multiplied: $p(A \text{ and } B) = p(A)p(B)$.
3. The *binomial coefficient*,

$$\binom{N}{k} = \frac{N!}{k!(N-k)!},\tag{A.6.1}$$

is the number of different ways of choosing k objects out of a set of N, without regard to order.

A.7 The Delta Method

The formula for the delta method of approximating variances is

$$\text{Var}(f(x,y)) \approx \left(\frac{\partial f}{\partial x}\right)^2 \bigg|_{\substack{x=\bar{x} \\ y=\bar{y}}} \text{Var}(x) + \left(\frac{\partial f}{\partial y}\right)^2 \bigg|_{\substack{x=\bar{x} \\ y=\bar{y}}} \text{Var}(y) + 2\left(\frac{\partial f}{\partial x}\frac{\partial f}{\partial y}\right) \bigg|_{\substack{x=\bar{x} \\ y=\bar{y}}} \text{Cov}(x,y).$$

$$(A.7.1)$$

Lyons (1991) describes the delta method very clearly; Oehlert (1992) provides a short technical description of the formal assumptions necessary for the delta method to apply.

This formula is exact in some simple cases:

- Multiplying by a constant: $\text{Var}(ax) = a^2\text{Var}(x)$.
- Sum or difference of independent variables: $\text{Var}(x \pm y) = \text{Var}(x) + \text{Var}(y)$.
- Product or ratio of independent variables:

$$\text{Var}(x \cdot y) = \bar{y}^2\text{Var}(x) + \bar{x}^2\text{Var}(y) = \bar{x}^2\bar{y}^2\left(\frac{\text{Var}(x)}{\bar{x}^2} + \frac{\text{Var}(y)}{\bar{y}^2}\right).$$

A similar form holds for the coefficient of variation (CV): $(\text{CV}(x \cdot y))^2 = (\text{CV}(x))^2 + (\text{CV}(y))^2$.

- The formula is exact for linear functions of normal or multivariate normal variables.

The formula can be extended to more than two variables. The `deltavar` function in the `emdbook` package will calculate delta-method-based variances for functions with any number of parameters.

A.8 Linear Algebra Basics

This section is more of a "cheater's guide" than a real introduction to linear algebra: Lynch and Walsh (1997) and Caswell (2000) both give useful bare-bones linear algebra reviews. All you need to know for this book is how to understand the general meaning of a matrix equation.

In mathematics a matrix is a rectangular table of numbers, while a vector is a list of numbers (specified as either a 1 row $\times n$ column *row vector* or an n row $\times 1$ column *column vector* in some contexts). Matrices are usually uppercase, often denoted by boldface ($\mathbf{V}$). Vectors are usually lowercase, either bold ($\mathbf{x}$) or topped with arrows ($\vec{x}$). The *transpose* of a matrix or vector, which exchanges the rows and columns of a matrix or switches between row and column vectors, is written as $\mathbf{V}^T$ or $\mathbf{V}'$.

Matrices and vectors can be added to or subtracted from any matrix or vector with the same number of rows and columns. If the number of rows of $\mathbf{A}$ is equal to the number of columns of $\mathbf{B}$, then $\mathbf{A}$ can be multiplied on the left by $\mathbf{B}$ (i.e., $\mathbf{BA}$ is well-defined). Matrix multiplication is *noncommutative* in general: $\mathbf{AB} \neq \mathbf{BA}$, although *diagonal* matrices (matrices with nonzero entries only on the diagonal) do commute.

Matrices can be multiplied on the right by column vectors ($\mathbf{Ax}$) or on the left by row vectors ($\mathbf{x}^T\mathbf{A}$) or anywhere by scalars (i.e., plain numbers: $c\mathbf{A} = \mathbf{A}c$). The *inverse* of a matrix, $\mathbf{A}^{-1}$, is the matrix such that $\mathbf{A}\mathbf{A}^{-1}$ equals the *identity matrix* (**1** or **I**)—a matrix with ones on the diagonal and zero everywhere else. Multiplying by the inverse of a matrix is like dividing by the matrix.

The *inner product* of two (column) vectors $\mathbf{x}$ and $\mathbf{y}$ with each other is $\mathbf{x}^T\mathbf{y}$. The inner product of a vector with itself, $\mathbf{x}^T\mathbf{x}$, is the sum of squares of its elements. The *quadratic form* of a matrix $\mathbf{A}$ and a vector $\mathbf{x}$ is $\mathbf{x}^T\mathbf{Ax}$. The quadratic form that appears in the multivariate normal distribution, $(\mathbf{x} - \boldsymbol{\mu})^T\mathbf{V}^{-1}(\mathbf{x} - \boldsymbol{\mu})$, where $\mathbf{x}$ is the data vector, $\boldsymbol{\mu}$ is the vector of means, and $\mathbf{V}$ is the variance-covariance matrix, is roughly analogous to $(x - \mu)^2/\sigma^2$ in the univariate normal distribution. We could write the univariate form as $(x - \mu)(\sigma^2)^{-1}(x - \mu)$ to make the two expressions look more similar.

The *determinant* of a matrix, $|\mathbf{A}|$ or $\det(\mathbf{A})$, is complicated in general, but for diagonal matrices it is equal to the product of the diagonal entries. Similarly, the *trace*, $\text{tr}(\mathbf{A})$, is the sum of the diagonal entries for a diagonal matrix.

The best way to figure out a matrix equation is to think about the equivalent scalar equation, or see how the equation would simplify if all the matrices were diagonal; see p. 321 for an example.

Bibliography

Adler, F. R. 2004. *Modeling the Dynamics of Life: Calculus and Probability for Life Scientists*, 2d ed. Brooks/Cole, Pacific Grove, CA.

Agrawal, A. A. and M. Fishbein. 2006. Plant defense syndromes. *Ecology* 87:S132–S149.

Ågren, G. L. and E. Bosatta. 1996. *Theoretical Ecosystem Ecology: Understanding Element Cycles*. Cambridge University Press, Cambridge, England.

Agresti, A. 2002. *Categorical Data Analysis*, 2d ed. Wiley, Hoboken, NJ.

Aitchison, J. 1986. *The Statistical Analysis of Compositional Data*. Chapman & Hall, New York.

Anderson, R. M. and R. M. May. 1991. *Infectious Diseases of Humans: Dynamics and Control*. Oxford Science Publications, Oxford, England.

Bacon, D. W. and D. G. Watts. 1971. Estimating the transition between two intersecting straight lines. *Biometrika* 58:525–534.

———. 1974. Using a hyperbola as a transition model to fit two-regime straight-line data. *Technometrics* 16:369–373.

Bailey, N. T. J. 1964. *The Elements of Stochastic Processes with Applications to the Natural Sciences*. Wiley, New York.

Barrowman, N. J. and R. A. Myers. 2000. Still more spawner-recruitment curves: The hockey stick and its generalizations. *Canadian Journal of Fisheries and Aquatic Science* 57:665–676.

Bates, D. M. and D. G. Watts. 1988. *Nonlinear Regression Analysis and Its Applications*. Wiley, New York.

Begon, M., J. L. Harper, and C. R. Townsend. 1996. *Ecology: Individuals, Populations and Communities*, 3d ed. Blackwell Science, Cambridge, MA.

Bellows, T. S. 1981. The descriptive properties of some models for density dependence. *Journal of Animal Ecology* 50:139–156.

Berger, J. and D. Berry. 1988. Analyzing data: Is objectivity possible? *American Scientist* 76:159–165.

Billheimer, D. P., P. Guttorp, and W. F. Fagan. 1998. Statistical analysis and interpretation of discrete compositional data. Technical Report 11. University of Washington. Seattle, Washington. http://www.nrcse.washington.edu/pdf/trs11_interp.pdf.

Bjørnstad, O. N. and B. T. Grenfell. 2001. Noisy clockwork: Time series analysis of population fluctuations in animals. *Science* 293:638–643.

Blanco, J. A., M. A. Zavala, J. B. Imbert, and F. J. Castillo. 2005. Sustainability of forest management practices: Evaluation through a simulation model of nutrient cycling. *Forest Ecology and Management* 213:209–228.

Bolker, B. M., S. W. Pacala, and C. Neuhauser. 2003. Spatial dynamics in model plant communities: What do we really know? *American Naturalist* 162:135–148.

Bradlow, E. T., B. G. S. Hardie, and P. S. Fader. 2002. Bayesian inference for the negative binomial distribution via polynomial expansions. *Journal of Computational & Graphical Statistics* 11:189–201.

Brännström, Å. and D. J. T. Sumpter. 2005. The role of competition and clustering in population dynamics. *Proceedings of the Royal Society B* 272:2065–2072.

Breslow, N. 2003. Whither PQL? UW Biostatistics Working Paper Series, no.192. http://www.bepress.com/uwbiostat/paper192.

Breslow, N. E. and D. G. Clayton. 1993. Approximate inference in generalized linear mixed models. *Journal of the American Statistical Association* 88:9–25.

Brown, D. H. and B. M. Bolker. 2004. The effects of disease dispersal and host clustering on the epidemic threshold in plants. *Bulletin of Mathematical Biology* 66:341–371.

Buckland, S. T., K. B. Newman, L. Thomas, and N. B. Kösters. 2004. State-space models for the dynamics of wild animal populations. *Ecological Modelling* 171:157–175.

Burnham, K. P. and D. R. Anderson. 1998. *Model Selection and Inference: A Practical Information-Theoretic Approach*. Springer, New York.

———. 2002. *Model Selection and Multimodel Inference*, 2d ed. Springer, New York.

———. 2004. Multimodel inference: understanding AIC and BIC in model selection. *Sociological Methods & Research* 33:261–304.

Butler, M. I. and C. W. Burns. 1993. Water mite predation on planktonic cladocera: Parallel curve analysis of functional responses. *Oikos* 66:5–16.

Calder, C., M. Lavine, P. Müller, and J. S. Clark. 2003. Incorporating multiple sources of stochasticity into dynamic population models. *Ecology* 84:1395–1402.

Canham, C. D. and M. Uriarte. 2006. Analysis of neighborhood dynamics of forest ecosystems using likelihood methods and modeling. *Ecological Applications* 16:62–73.

Carroll, R. J., J. D. Maca, and D. Ruppert. 1999. Nonparametric estimation in the presence of measurement errors. *Biometrika* 86:541–554.

Carroll, R. J., D. Ruppert, and L. W. Stefanski. 1995. *Measurement Error in Nonlinear Models*. Chapman and Hall, New York.

Case, T. J. 1999. *An Illustrated Guide to Theoretical Ecology*. Oxford University Press, New York.

Caswell, H. 2000. *Matrix Population Models: Construction, Analysis and Interpretation*. Sinauer, Sunderland, MA.

Celeux, G., F. Forbes, C. P. Robert, and D. M. Titterington. 2006. Deviance information criteria for missing data models. *Bayesian Analysis* 1:651–674.

Chambers, J. M. and T. Hastie, editors. 1992. *Statistical Models in S*. Wadsworth & Brooks/Cole, Pacific Grove, CA.

Chatfield, C. 1975. *The Analysis of Time Series: Theory and Practice*. Chapman and Hall, London.

Chen, Y. 2004. Multiple periodic solutions of delayed predator-prey systems with type IV functional responses. *Nonlinear Analysis: Real World Applications* 5:45–53.

Clark, C. J., J. R. Poulsen, B. M. Bolker, E. F. Connor, and V. T. Parker. 2005. Comparative seed shadows of bird-, monkey-, and wind-dispersed trees. *Ecology* 86:2684–2694.

Clark, J. S. 2007. *Models for Ecological Data: An Introduction*. Princeton University Press, Princeton, NJ.

Clark, J. S. and O. N. Bjørnstad. 2004. Population time series: Process variability, observation errors, missing values, lags, and hidden states. *Ecology* 85: 3140–3150.

Clark, J. S., M. Silman, R. Kern, E. Macklin, and J. HilleRisLambers. 1999. Seed dispersal near and far: Patterns across temperate and tropical forests. *Ecology* 80:1475–1494.

Clarke, S. C., M. K. McAllister, E. J. Milner-Gulland, G. P. Kirwood, C. G. J. Michielsens, D. J. Agnew, E. K. Pikitch, H. Nakano, and M. S. Shivji. 2006. Global estimates of shark catches using trade records from commercial markets. *Ecology Letters* 9:1115–1126.

Cleveland, W. 1993. *Visualizing Data*. Hobart Press, Summit, NJ.

Cohen, J. E., F. Briand, and C. M. Newman. 1990. Community food webs: Data and theory. Springer-Verlag, Berlin.

Collings, J. B. 1997. The effects of the functional response on the bifurcation behavior of a mite predator-prey interaction model. *Journal of Mathematical Biology* 36:149–168.

Congdon, P. 2003. *Applied Bayesian Modelling*. Wiley, Hoboken, NJ.

Cook, J. R. and L. A. Stefanski. 1994. Simulation-extrapolation estimation in parametric measurement error models. *Journal of the American Statistical Association* 89:1314–1328.

Corless, R. M., G. H. Gonnet, D. E. G. Hare, D. J. Jeffrey, and D. E. Knuth. 1996. On the Lambert *W* function. *Advances in Computational Mathematics* 5:329–359.

Crawley, M. J. 2002. *Statistical Computing: An Introduction to Data Analysis Using S-PLUS*. Wiley, Hoboken, NJ.

———. 2005. *Statistics: An Introduction Using R*. Wiley, Hoboken, NJ.

———. 2007. *The R Book*. Wiley, Hoboken, NJ.

Cressie, N. A. C. 1991. *Statistics for Spatial Data*. Wiley, New York.

Crome, F. H. J. 1997. Researching tropical forest fragmentation: Shall we keep on doing what we're doing? Pages 485–501 in W. F. Laurance and R. O. Bierregard, editors. *Tropical Forest Remnants: Ecology, Management, and Conservation of Fragmented Communities*. University of Chicago Press, Chicago.

Crome, F. H. J., M. R. Thomas, and L. A. Moore. 1996. A novel Bayesian approach to assessing impacts of rain forest logging. *Ecological Applications* 6:1104–1123.

Crowder, M. J. 1978. Beta-binomial Anova for proportions. *Applied Statistics* 27: 34–37.

Dalgaard, P. 2003. *Introductory statistics in R*. Springer, New York.

Dalling, J. W., H. C. Muller-Landau, S. J. Wright, and S. P. Hubbell. 2002. Role of dispersal in the recruitment limitation of neotropical pioneer species. *Journal of Ecology* 90:714–727.

Damgaard, C. 1999. A test of asymmetric competition in plant monocultures using the maximum likelihood function of a simple growth model. *Ecological Modelling* 116:285–292.

Damgaard, C., J. Weiner, and H. Nagashima. 2002. Modelling individual growth and competition in plant populations: Growth curves of *Chenopodium album* at two densities. *Journal of Ecology* 90:666–671.

de Valpine, P. 2003. Better inferences from population-dynamics experiments using Monte Carlo state-space likelihood methods. *Ecology* 84:3064–3077.

de Valpine, P. and A. Hastings. 2002. Fitting population models incorporating process noise and observation error. *Ecology* 72:57–76.

Dennis, B. 1996. Discussion: Should ecologists become Bayesians? *Ecological Applications* 6:1095–1103.

Diaconis, P., S. Holmes, and R. Montgomery. 2004. Dynamical bias in the coin toss. Technical report 2004-32, Stanford University. http://www-stat.stanford.edu/reports/abstracts/04-32.pdf.

Diggle, P. J. 1990. *Time Series: A Biostatistical Introduction*. Oxford University Press, New York.

Doak, D. F., K. Gross, and W. F. Morris. 2005. Understanding and predicting the effects of sparse data on demographic analyses. *Ecology* 86:1154–1163.

Dobson, A. J. 1990. *An Introduction to Generalized Linear Models*. Chapman and Hall, London.

Dodd, M. E. and J. Silvertown. 2000. Size-specific fecundity and the influence of lifetime size variation upon effective population size in *Abies balsamea*. *Heredity* 85:604–609.

Doucet, A., N. de Freitas, and N. J. Gordon. 2001. *Sequential Monte Carlo Methods in Practice*. Springer-Verlag, New York.

Drury, K. L. S. and G. Dwyer. 2005. Combining stochastic models with experiments to understand the dynamics of monarch butterfly colonization. *American Naturalist* 166:731–750.

Duncan, R. S. and V. E. Duncan. 2000. Forest succession and distance from forest edge in an Afro-tropical grassland. *Biotropica* 32:33–41.

Dwyer, G., S. Levin, and L. Buttel. 1990. A simulation model of the population dynamics and evolution of myxomatosis. *Ecological Monographs* 60:423–447.

Edwards, A. W. F. 1992. *Likelihood: Expanded Edition*. Johns Hopkins University Press, Baltimore, MD.

Edwards, D. 1996. Comment: The first data analysis should be journalistic. *Ecological Applications* 6:1090–1094.

Ellison, A. M. 1996. An introduction to Bayesian inference for ecological research and environmental decision-making. *Ecological Applications* 6:1036–1046.

Ellner, S. P., B. A. Bailey, G. V. Bobashev, A. Gallant, B. T. Grenfell, and D. W. Nychka. 1998. Noise and nonlinearity in measles epidemics: Combining mechanistic and statistical approaches to population modeling. *American Naturalist* 151: 425–440.

Ellner, S. P. and J. Guckenheimer. 2006. *Dynamic Models in Biology*. Princeton University Press, Princeton, NJ.

Ellner, S. P., Y. Seifu, and R. H. Smith. 2002. Fitting population dynamic models to time-series data by gradient matching. *Ecology* 83:2256–2270.

Elston, D. A., R. Moss, T. Boulinier, C. Arrowsmith, and X. Lambin. 2001. Analysis of aggregation, a worked example: Numbers of ticks on red grouse chicks. *Parasitology* 122: 563–569.

Emlen, D. J. 1996. Artificial selection on horn length-body size allometry in the horned beetle *Onthophagus acuminatus* (Coleoptera: Scarabaeidae). *Evolution* 50:1219–1230.

Essington, T. E., J. F. Kitchell, and C. J. Walters. 2001. The von Bertalanffy growth function, bioenergetics, and the consumption rates of fish. *Canadian Journal of Fisheries and Aquatic Science* 58:2129–2138.

Etienne, R. S., M. E. F. Apol, and H. Olff. 2006a. Demystifying the West, Brown & Enquist model of the allometry of metabolism. *Functional Ecology* 20:394–399.

Etienne, R. S., A. M. Latimer, J. A. Silander, and R. M. Cowling. 2006b. Comment on "Neutral ecological theory reveals isolation and rapid speciation in a biodiversity hot spot." *Science* 311:610b.

Evans, M., N. Hastings, and B. Peacock. 2000. *Statistical Distributions*, 3d ed. Wiley, New York.

Everitt, B. and T. Hothorn. 2006. *A Handbook of Statistical Analyses Using R*. Chapman & Hall/CRC, Boca Raton, FL.

Faraway, J. J. 2004. *Linear Models with R*. CRC Press, Boca Raton, FL.

———. 2006. *Extending Linear Models with R: Generalized Linear, Mixed Effects and Nonparametric Regression Models*. Chapman & Hall/CRC, Boca Raton, FL.

Fenner, F., M. F. Day, and G. M. Woodroffe. 1956. Epidemiological consequences of the mechanical transmission of myxomatosis by mosquitoes. *Journal of Hygiene* **54**:284–303.

Ferrari, J. B. and S. Sugita. 1996. A spatially explicit model of leaf litter fall in hemlock-hardwood forests. *Canadian Journal of Forest Research* **26**:1905–1913.

Finkenstädt, B. F. and B. T. Grenfell. 2000. Time series modelling of childhood diseases: A dynamical systems approach. *Journal of the Royal Statistical Society Series C: Applied Statistics* **49**:187–205.

Flather, C. H. 1996. Fitting species-accumulation functions and assessing regional land use impacts on avian diversity. *Journal of Biogeography* **23**:155–168.

Forsythe, G., M. Malcolm, and C. Moler. 1977. *Computer Methods for Mathematical Computations*. Prentice Hall, Englewood Cliffs, NJ.

Fox, J. 2002. *An R and S-PLUS Companion to Applied Regression*. Sage Press, Thousand Oaks, CA.

Framstad, E., N. C. Stenseth, O. N. Bjørnstad, and W. Falck. 1997. Limit cycles in Norwegian lemmings: Tensions between phase-dependence and density-dependence. *Proceedings: Biological Sciences* **264**:31–38.

Fujiwara, M., B. E. Kendall, R. M. Nisbet, and W. A. Bennett. 2005. Analysis of size trajectory data using an energetic-based growth model. *Ecology* **86**:1441–1451.

Gani, R. and S. Leach. 2001. Transmission potential of smallpox in contemporary populations. *Nature* **414**:748–751.

Gelman, A. 2005. Analysis of variance—Why it is more important than ever. *Annals of Statistics* **33**:1–53.

———. 2006. Prior distributions for variance parameters in hierarchical models. *Bayesian Analysis* **1**:515–533.

Gelman, A., J. Carlin, H. S. Stern, and D. B. Rubin. 1996. *Bayesian Data Analysis*. Chapman and Hall, New York.

Gelman, A. and J. Hill. 2006. *Data Analysis Using Regression and Multilevel/Hierarchical Models*. Cambridge University Press, Cambridge, England.

Gelman, A. and D. Nolan. 2002. You can load a die, but you can't bias a coin. *American Statistician* **56**:308–311.

Gelman, A. and F. Tuerlinckx. 2000. Type S error rates for classical and Bayesian single and multiple comparison procedures. *Computational Statistics* **15**: 373–390.

Gibson, G. J. 1997. Markov chain Monte Carlo methods for fitting spatiotemporal stochastic models in plant epidemiology. *Journal of the Royal Statistical Society Series C* **46**:215–233.

Gibson, G. J. and E. Renshaw. 1998. Estimating parameters in stochastic compartmental models using Markov chain methods. *IMA Journal of Mathematics Applied in Biology and Medicine* **15**:19–40.

———. 2001. Likelihood estimation for stochastic compartmental models using Markov chain methods. *Statistics and Computing* **11**:347–358.

Gillespie, D. T. 1977. Exact stochastic simulation of coupled chemical reactions. *Journal of Physical Chemistry* **81**:2340–2361.

Goldman, N. and S. Whelan. 2000. Statistical tests of gamma-distributed rate heterogeneity in models of sequence evolution in phylogenetics. *Molecular Biology and Evolution* **17**:975–978.

Gonick, L. and W. Smith. 1993. *The Cartoon Guide to Statistics*. Harper-Perennial, New York.

Gotelli, N. J. 2001. *A Primer of Ecology*, 3d ed. Sinauer, Sunderland, MA.

Gotelli, N. J. and A. M. Ellison. 2004. *A Primer of Ecological Statistics*. Sinauer, Sunderland, MA.

Grenfell, B. T., K. Wilson, B. Finkenstädt, T. N. Coulson, S. Murray, S. D. Albon, J. M. Pemberton, T. H. Clutton-Brock, and M. J. Crawley. 1998. Noise and determinism in synchronized sheep dynamics. *Nature* 394:674–677.

Gurney, W. S. C. and R. Nisbet. 1998. *Ecological Dynamics*. Oxford University Press, Oxford, England.

Guthery, F. S., L. A. Brennan, M. J. Peterson, and J. J. Lusk. 2005. Invited paper: Information theory in wildlife science: Critique and viewpoint. *Journal of Wildlife Management* 69:457–465.

Haefner, J. W. 1996. *Modeling Biological Systems: Principles and Applications*. Kluwer, Dordrecht, Netherlands.

Haining, R. 2003. Spatial data analysis: Theory and practice. Cambridge University Press, Cambridge, England.

Halley, J. M., S. Hartley, A. S. Kallimanis, W. E. Kunin, J. J. Lennon, and S. P. Sgardelis. 2004. Uses and abuses of fractal methodology in ecology. *Ecology Letters* 7:254–271.

Hanski, I. 1999. *Metapopulation Ecology*. Oxford University Press, Oxford, England.

Hargrove, W. W. and J. Pickering. 1992. Pseudoreplication: A *sine qua non* for regional ecology. *Landscape Ecology* 6:251–258.

Harrison, P. J., S. T. Buckland, L. Thomas, R. Harris, P. P. Pomeroy, and J. Harwood. 2006. Incorporating movement into models of grey seal population dynamics. *Journal of Animal Ecology* 75:634–645.

Hassell, M. P. 1975. Density-dependence in single-species populations. *Journal of Animal Ecology* 45:283–296.

Hastie, T. J. and D. Pregibon. 1991. Generalized linear models. Pages 377–420 in J. M. Chambers and T. J. Hastie, editors. *Statistical Models in S*. Wadsworth & Brooks/Cole, Pacific Grove, CA.

Hastings, A. 1997. *Population Biology: Concepts and Models*. Springer-Verlag, New York.

Hatfield, J. S., W. A. Link, D. K. Dawson, and E. L. Lindquist. 1996. Coexistence and community structure of tropical trees in a Hawaiian montane rain forest. *Biotropica* 28:746–758.

Heffner, R. A., M. J. Butler, and C. K. Reilly. 1996. Pseudoreplication revisited. *Ecology* 77:2558–2562.

Heiberger, R. M. and B. Holland. 2004. *Statistical Analysis and Data Display: An Intermediate Course with Examples in S-PLUS, R, and SAS*. Springer, New York.

Henson, S. M., R. F. Costantino, J. M. Cushing, R. A. Desharnais, B. Dennis, and A. A. King. 2001. Lattice effects observed in chaotic dynamics of experimental populations. *Science* 294:602–605.

Hilborn, R. and M. Mangel. 1997. *The Ecological Detective: Confronting Models with Data*. Princeton University Press, Princeton, NJ.

Hoaglin, D. C., F. Mosteller, and J. W. Tukey, editors. 2000. *Understanding Robust and Exploratory Data Analysis*. Wiley, New York.

———. 2006. *Exploring Data Tables, Trends, and Shapes*, rev. ed. Wiley, New York.

Holt, R. D. 1983. Optimal foraging and the form of the predator isocline. *American Naturalist* 122:521–541.

Horne, J. S. and E. O. Garton. 2006. Selecting the best home range model: An information-theoretical approach. *Ecology* 87:1146–1152.

Hurlbert, S. 1984. Pseudoreplication and the design of ecological field experiments. *Ecological Monographs* 54:187–211.

Ingber, L. 1996. Adaptive simulated annealing (ASA): Lessons learned. *Control and Cybernetics* 25:33–54. http://www.ingber.com/asa96_lessons.pdf.

Inouye, B. D. 1999. Estimating competition coefficients: Strong competition among three species of frugivorous flies. *Oecologia* 120:588–594.

———. 2005. The importance of the variance around the mean effect size of ecological processes: comment. *Ecology* 86:262–264.

Ionides, E. L., C. Bretó, and A. A. King. 2006. Inference for nonlinear dynamical systems. *Proceedings of the National Academy of Sciences of the USA* 103:18438–18443.

Ives, A. R., S. R. Carpenter, and B. Dennis. 1999. Community interaction webs and zooplankton responses to planktivory manipulations. *Ecology* 80:1405–1421.

Jang, W. and J. Lim. 2005. PQL estimation biases in generalized linear mixed models. Discussion paper 2005-21, Institute for Statistics and Decision Sciences, Duke University. http://ftp.stat.duke.edu/WorkingPapers/05-21.html.

Jeffreys, H. 1961. *The Theory of Probability*. Clarendon Press, Oxford, England.

Jeschke, J. M., M. Kopp, and R. Tollrain. Consumer–food systems: why type I functional responses are exclusive to filter feeders. *Biological Reviews* 79:337–349.

Johnson, D. J. 1999. The insignificance of statistical significance testing. *Journal of Wildlife Management* 63:763–772.

Johnson, J. B. and K. S. Omland. 2004. Model selection in ecology and evolution. *Trends in Ecology and Evolution* 19:101–108.

Jonsen, I. D., R. A. Myers, and J. M. Flemming. 2003. Meta-analysis of animal movement using state-space models. *Ecology* 84:3055–3063.

Juliano, S. A. 1993. Nonlinear curve fitting: predation and functional response curves. Pages 159–182 in S. M. Scheiner and J. Gurevitch, editors. *Design and Analysis of Ecological Experiments*. Chapman & Hall, New York.

Kass, R. E. and A. E. Raftery. 1995. Bayes factors and model uncertainty. *Journal of the American Statistical Association* 90:773–795.

Katz, R. W., G. S. Brush, and M. B. Parlange. 2005. Statistics of extremes: Modeling ecological disturbances. *Ecology* 86:112–1134.

Keeling, K. B. and R. J. Pavur. 2007. A comparative study of the reliability of nine statistical software packages. *Computational Statistics & Data Analysis* 51:3811–3831.

Keller, J. B. 1986. The probability of heads. *American Mathematical Monthly* 93:191–197.

Kendall, M. and A. Stuart. 1979. *The Advanced Theory of Statistics*. Vol. 2: *Inference and Relationship*, 4th ed. Griffin, London.

Kenward, M. G. and J. H. Roger. 1997. Small sample inference for fixed effects from restricted maximum likelihood. *Biometrics* 53:983–997.

Kerman, J. and A. Gelman. 2006. Tools for Bayesian data analysis in R. *Statistical Computing and Graphics* 17:9–13.

Kirkpatrick, S., C. Gelatt, and M. Vecchi. 1983. Optimization by simulated annealing. *Science* 220:671–680.

Kitakado, T., S. Kitada, H. Kishino, and H. J. Skaug. 2006. An integrated-likelihood method for estimating genetic differentiation between populations. *Genetics* 173:2073–2082.

Kitanidis, P. K. 1997. *Introduction to geostatistics: Applications in Hydrogeology*. Cambridge University Press, Cambridge, England.

Lande, R., S. Engen, and B.-E. Sæther. 2003. *Stochastic Population Dynamics in Ecology and Conservation*. Oxford University Press, Oxford, England.

Latimer, A. M., J. A. Silander, and R. M. Cowling. 2005. Neutral ecological theory reveals isolation and rapid speciation in a biodiversity hot spot. *Science* 309: 1722–1725.

Lee, Y., J. A. Nelder, and Y. Pawitan 2006. *Generalized Linear Models with Random Effects: Unified Analysis via H-Likelihood*. Chapman & Hall/CRC, Boca Raton, FL.

Lekone, P. E. and B. F. Finkenstädt. 2006. Statistical inference in a stochastic epidemic SEIR model with control intervention: Ebola as a case study. *Biometrics* 62:1170–1177.

Lele, S. R., B. Dennis, and F. Lutscher. 2007. Data cloning: easy maximum likelihood estimation for complex ecological models using Bayesian Markov chain Monte Carlo methods. *Ecology Letters* 10:551–563.

Levins, R. 1966. The strategy of model building in population biology. *American Scientist* 54:421–431.

———. 1993. A response to Orzack and Sober: Formal analysis and the fluidity of science. *Quarterly Review of Biology* 68:547–555.

Levins, R. and D. Culver. 1971. Regional coexistence of species and competition between rare species. *Proceedings of the National Academy of Sciences of the USA* 6:1246–1248.

Lindley, S. T. 2003. Estimation of population growth and extinction parameters from noisy data. *Ecological Applications* 13:806–813.

Lindsey, J. K. 1997. *Applying Generalized Linear Models*. Springer, New York.

———. 1999a. *Models for Repeated Measurements*, 2d ed. Oxford University Press, New York.

———. 1999b. Some statistical heresies. *The Statistician* 48:1–40.

———. 2001. *Nonlinear Models in Medical Statistics*. Oxford University Press, New York.

———. 2004. *Introduction to Applied Statistics: A Modelling Approach*. Oxford University Press, New York.

Link, W. A. and R. J. Barker. 2006. Model weights and the foundations of multimodel inference. *Ecology* 87:2626–2635.

Lloyd-Smith, J. O. 2007. Maximum likelihood estimation of the negative binomial dispersion parameter for highly overdispersed data, with applications to infectious diseases. *PLoS ONE* 2:e180.

Ludwig, D. 1996. Uncertainty and the assessment of extinction probabilities. *Ecological Applications* 6:1067–1076.

Lynch, M. and B. Walsh. 1997. *Genetics and Analysis of Quantitative Traits*. Sinauer, Sunderland, MA.

Lyons, L. 1991. *A Practical Guide to Data Analysis for Physical Science Students*. Cambridge University Press, Cambridge, England.

Lytle, D. A. 2002. Flash floods and aquatic insect life-history evolution: Evaluation of multiple models. *Ecology* 83:370–385.

Maindonald, J. and J. Braun. 2003. *Data Analysis and Graphics Using R: An Example-Based Approach*. Cambridge University Press, Cambridge, England.

Mangel, M. 2006. *The Theoretical Biologist's Toolbox: Quantitative Methods for Ecology and Evolutionary Biology*. Cambridge University Press, Cambridge, England.

Martin, T. G., B. A. Wintle, J. R. Rhodes, P. M. Kuhnert, S. A. Field, S. A. Low-Choy, A. J. Tyre, and H. H. Possingham. 2005. Zero-tolerance ecology: Improving ecological inference by modelling the source of zero observations. *Ecology Letters* 8:1235–1246.

May, R. M. 1973. *Stability and Complexity in Model Ecosystems*. Princeton University Press, Princeton, NJ.

———. 1978. Host-parasitoid systems in patchy environments: A phenomenological model. *Journal of Animal Ecology* 47:833–844.

Maynard-Smith, J. and M. Slatkin. 1973. The stability of predator-prey systems. *Ecology* 54:384–391.

Mazerolle, M. J. 2004. Mouvements et reproduction des amphibiens en tourbières perturbées. Ph.D. Diss. Université Laval, Quebec. http://www.theses.ulaval.ca/2004/21842.html.

McCarthy, M. A. and K. M. Parris. 2004. Clarifying the effect of toe clipping on frogs with Bayesian statistics. *Journal of Applied Ecology* 41:780–786.

McCullagh, P. and J. A. Nelder. 1989. *Generalized Linear Models.* Chapman and Hall, London.

McCullough, B. D. and B. Wilson. 2005. On the accuracy of statistical procedures in Microsoft Excel 2003. *Computational Statistics & Data Analysis* 49: 1244–1252.

McGill, B. J., B. A. Maurer, and M. D. Weiser. 2006. Empirical evaluation of neutral theory. *Ecology* 87:1411–1423.

McKay, B., D. Bar-Natan, M. Bar-Hillel, and G. Kalai. 1999. Solving the Bible Code puzzle. *Statistical Science* 14:150–173.

Melbourne, B. A. and P. Chesson. 2006. The scale transition: Scaling up population dynamics with field data. *Ecology* 87:1478–1488.

Millar, R. B. and R. Meyer. 2000. Bayesian state-space modeling of age-structured data: Fitting a model is just the beginning. *Canadian Journal of Fisheries and Aquatic Sciences* 57: 43–50.

Miller, P. S. and R. C. Lacy. 2005. *VORTEX. A Stochastic Simulation of the Simulation Process. Version 9.50 User's Manual.* Conservation Breeding Specialist Group (IUCN/SSC), Apple Valley, MN. http://www.vortex9.org/vortex.html.

Mitzenmacher, M. 2003. A brief history of generative models for power laws and lognormal distributions. *Internet Mathematics* 1:226–251.

Moorcroft, P. R., M. W. Lewis, and R. L. Crabtree. 2006. Mechanistic home range patterns capture spatial patterns and dynamics of coyote territories in Yellowstone. *Proceedings of the Royal Society B* 273:1651–1659.

Morales, J. M., D. T. Haydon, J. Frair, K. E. Holsinger, and J. M. Fryxell. 2004. Extracting more out of relocation data: Building movement models as mixtures of random walks. *Ecology* 85:2436–2445.

Morris, W. F. 1997. Disentangling effects of induced plant defenses and food quantity on herbivores by fitting nonlinear models. *American Naturalist* 150:299–327.

Morris, W. F. and D. F. Doak. 2002. *Quantitative Conservation Biology: Theory and Practice of Population Viability Analysis.* Sinauer, Sunderland, MA.

Mossel, E. and E. Vigoda. 2006. Phylogenetic MCMC algorithms are misleading on mixtures of trees. *Science* 309:2207–2209.

Muggeo, V. M. R. 2003. Estimating regression models with unknown breakpoints. *Statistics in Medicine* 22:3055–3072.

Nakagawa, S. 2004. A farewell to Bonferroni: The problems of low statistical power and publication bias. *Behavioral Ecology* 15:1044–1045.

Nelder, J. A. 1961. The fitting of a generalization of the logistic curve. *Biometrics* 17:89–110.

Ness, J. H., W. F. Morris, and J. L. Bronstein. 2006. Integrating quality and quantity of mutualistic service to contrast ant species protecting *Ferocactus wislizeni. Ecology* 87:912–921.

Neuhauser, C. 2003. *Calculus for Biology and Medicine,* 2d ed. Prentice Hall, Upper Saddle River, NJ.

Niklas, K. J. 1993. The allometry of plant reproductive biomass and stem diameter. *American Journal of Botany* 80:461–467.

Nisbet, R. and W. Gurney. 1982. *Modelling Fluctuating Populations.* Wiley, New York. Reprint Blackburn Press, Caldwell, NJ, 2003.

Oehlert, G. W. 1992. A note on the delta method. *American Statistician* 46:27–29.

Oksanen, L. 2001. Logic of experiments in ecology: Is pseudoreplication a pseudoissue? *Oikos* 94:27–38.

Okubo, A. 1980. *Diffusion and Ecological Problems: Mathematical Models*. Springer-Verlag, New York.

Okuyama, T. and B. M. Bolker. 2005. Combining genetic and ecological data to estimate sea turtle origins. *Ecological Applications* 15:315–325.

Orzack, S. H. and E. Sober. 1993. A critical assessment of Levins's *The Strategy of Model Building in Population Biology* (1966). *Quarterly Review of Biology* 68:533–546.

Osenberg, C. W., C. M. St. Mary, R. J. Schmitt, S. J. Holbrook, P. Chesson, and B. Byrne. 2002. Rethinking ecological inference: Density dependence in reef fishes. *Ecology Letters* 5:715–721.

Otto, S. P. and T. Day. 2007. *A Biologist's Guide to Mathematical Modeling in Ecology and Evolution*. Princeton University Press, Princeton, NJ.

Ovaskainen, O. 2004. Habitat-specific movement parameters estimated using mark-recapture data and a diffusion model. *Ecology* 85:242–257.

Pacala, S. and J. Silander Jr. 1990. Field tests of neighborhood population dynamic models of two annual weed species. *Ecological Monographs* 60:113–134.

Pacala, S. W. and J. A. Silander Jr. 1987. Neighborhood interference among velvet leaf, *Abutilon theophrasti*, and pigweed, *Amaranthus retroflexus*. *Oikos* 48:217–224.

Paradis, E. 2006. *Analysis of Phylogenetics and Evolution with* R. Springer, New York.

Parris, K. M. 2006. Urban amphibian assemblages as metacommunities. *Journal of Animal Ecology* 75:757–764.

Pascual, M. A. and P. Kareiva. 1996. Predicting the outcome of competition using experimental data: Maximum likelihood and Bayesian approaches. *Ecology* 77:337–349.

Persson, L., K. Leonardsson, A. M. de Roos, M. Gyllenberg, and B. Christensen. 1998. Ontogenetic scaling of foraging rates and the dynamics of a size-structured consumer-resource model. *Theoretical Population Biology* 54:270–293.

Petersen, J. E., J. C. Cornwell, and W. M. Kemp. 1999. Implicit scaling in the design of experimental aquatic ecosystems. *Oikos* 85:3–18.

Piegorsch, W. W. 1990. Maximum likelihood estimation for the negative binomial dispersion parameter. *Biometrics* 46:863–867.

Pielou, E. 1977. *Mathematical Ecology*, 2d ed. Wiley, New York.

Pieters, E. P., C. E. Gates, J. H. Matis, and W. L. Sterling. 1977. Small sample comparison of different estimators of negative binomial parameters. *Biometrics* 33:718–723.

Pinheiro, J. C. and D. M. Bates. 2000. *Mixed-Effects Models in S and S-PLUS*. Springer, New York.

Post, E., N. C. Stenseth, R. O. Peterson, J. A. Vucetich, and A. M. Ellisa. 2002. Phase dependence and population cycles in a large-mammal predator-prey system. *Ecology* 83: 2997–3002.

Poulin, R. 1996. Measuring parasite aggregation: Defending the index of discrepancy. *International Journal for Parasitology* 26:227–229.

Press, W. H., S. A. Teukolsky, W. T. Vetterling, and B. P. Flannery. 1994. *Numerical Recipes in C: The Art of Scientific Computing*. Cambridge University Press, Cambridge, England.

Quinn, G. P. and M. J. Keough. 2002. *Experimental Design and Data Analysis for Biologists*. Cambridge University Press, Cambridge, England.

Quinn, T. J. and R. B. Deriso. 1999. *Quantitative Fish Dynamics*. Oxford University Press, New York.

Raftery, A. E. and S. M. Lewis. 1996. Implementing MCMC. Pages 115–130 in W. R. Gilks, S. Richardson, and D. J. Spiegelhalter, editors. *Markov Chain Monte Carlo in Practice*. Chapman and Hall, London.

Rand, D. A. and H. B. Wilson. 1991. Chaotic stochasticity: A ubiquitous source of unpredictability in epidemics. *Proceedings of the Royal Society B* 246:6.

Reeve, J. D. and W. W. Murdoch. 1985. Aggregation by parasitoids in the successful control of the California red scale: A test of theory. *Journal of Animal Ecology* 54:797–816.

Renshaw, E. 1991. *Modelling Biological Populations in Space and Time*. Cambridge University Press, Cambridge, England.

Ribbens, E., J. A. Silander, and S. W. Pacala. 1994. Seedling recruitment in forests: Calibrating models to predict patterns of tree seedling dispersion. *Ecology* 75:1794–1806.

Richards, F. J. 1959. A flexible growth function for empirical use. *Journal of Experimental Botany* 10:290–300.

Richards, S. A. 2005. Testing ecological theory using the information-theoretic approach: Examples and cautionary results. *Ecology* 86:2805–2814.

Ricketts, T. H. 2001. The matrix matters: Effective isolation in fragmented landscapes. *American Naturalist* 158:87–99.

Ripley, B. D. 1981. *Spatial Statistics*. Wiley, New York.

———. 2004. Selecting amongst large classes of models. Pages 155–170 in N. Adams, M. Crowder, D. J. Haud, and D. Stephens, editors. *Methods and Models in Statistics: In Honour of Professor John Nelder, FRS*. Imperial College Press, London.

Rogers, D. J. 1972. Random search and insect population models. *Journal of Animal Ecology* 41:369–383.

Ronquist, F., B. Larget, J. P. Huelsenbeck, J. P. Kadane, D. Simon, and P. van der Mark. 2006. Comment on "Phylogenetic MCMC algorithms are misleading on mixtures of trees." *Science* 312:3767a.

Rosenheim, J. A. and D. Rosen. 1991. Foraging and oviposition decisions in the parasitoid *Aphytis lingnanensis*: Distinguishing the influences of egg load and experience. *Journal of Animal Ecology* 60:873–893.

Roughgarden, J. 1997. *Primer of Ecological Theory*. Prentice Hall, Upper Saddle River, NJ.

Royama, T. 1992. *Analytical Population Dynamics*. Chapman and Hall, New York.

Ruel, J. J. and M. P. Ayres. 1999. Jensen's inequality predicts effects of environmental variation. *Trends in Ecology and Evolution* 14:361–366.

Sack, L., P. J. Melcher, W. H. Liu, E. Middleton, and T. Pardee. 2006. How strong is intra-canopy leaf plasticity in temperate deciduous trees? *American Journal of Botany* 93: 829–839.

Saha, K. and S. Paul. 2005. Bias-corrected maximum likelihood estimator of the negative binomial dispersion parameter. *Biometrics* 61:179–185.

Sandin, S. A. and S. W. Pacala. 2005. Fish aggregation results in inversely density-dependent predation on continuous coral reefs. *Ecology* 86:1520–1530.

Scheiner, S. M. and J. Gurevitch, editors. 2001. *Design and Analysis of Ecological Experiments*, 2d ed. Chapman and Hall, New York.

Schmitt, R. J., S. J. Holbrook, and C. W. Osenberg. 1999. Quantifying the effects of multiple processes on local abundance: A cohort approach for open populations. *Ecology Letters* 2:294–303.

Schnute, J. 1981. A versatile growth model with statistically stable parameters. *Canadian Journal of Fisheries and Aquatic Sciences* 38:1128–1140.

Schnute, J. T. 1994. A general framework for developing sequential fisheries models. *Canadian Journal of Fisheries and Aquatic Sciences* 51:1676–1688.

Self, S. G. and K.-Y. Liang. 1987. Asymptotic properties of maximum likelihood estimators and likelihood ratio tests under nonstandard conditions. *Journal of the American Statistical Association* 82:605–610.

Shaw, D. J. and A. P. Dobson. 1995. Patterns of macroparasite abundance and aggregation in wildlife populations: A quantitative review. *Parasitology* 111 Suppl:S111–S127.

Shono, H. 2000. Efficiency of the finite correction of Akaike's Information Criteria. *Fisheries Science* 66:608–610.

Sibly, R. M., D. Barker, M. C. Denham, J. Hone, and M. Pagel. 2005. On the regulation of populations of mammals, birds, fish, and insects. *Science* 309:607–610.

Silvertown, J. and M. Dodd. 1999. Evolution of life history in balsam fir (*Abies balsamea*) in subalpine forests. *Proceedings of the Royal Society of London B* 266:729–733.

Skaug, H. and D. Fournier. 2006. Automatic approximation of the marginal likelihood in non-Gaussian hierarchical models. *Computational Statistics and Data Analysis* 51:699–709.

Skellam, J. G. 1951. Random dispersal in theoretical populations. *Biometrika* 38:196–218. Reprinted in Leslie A. Real and James H. Brown, editors. *Foundations of Ecology: Classic Papers with Commentaries*, University of Chicago Press, Chicago, 1991.

Sokal, R. R. and F. J. Rohlf. 1995. *Biometry*, 3d ed. W. H. Freeman, New York.

Solow, A. R. 1998. On fitting a population model in the presence of observation error. *Ecology* 79:1463–1466.

Spiegelhalter, D. J., N. Best, B. P. Carlin, and A. Van der Linde. 2002. Bayesian measures of model complexity and fit. *Journal of the Royal Statistical Society B* 64:583–640.

Stefanski, L. A. and J. R. Cook. 1995. Simulation-extrapolation: The measurement error jackknife. *Journal of the American Statistical Association* 90:1247–1256.

Stephens, P. A., S. W. Buskirk, G. D. Hayward, and C. Martinez del Rio. 2005. Information theory and hypothesis testing: A call for pluralism. *Journal of Applied Ecology* 42:4–12.

Stram, D. O. and J. W. Lee. 1994. Variance components testing in the longitudinal fixed effects model. *Biometrics* 50:1171–1177.

Streftaris, G. and G. Gibson. 2004. Bayesian analysis of experimental epidemic of foot-and-mouth disease. *Proceedings of the Royal Society of London B* 271:1111–1117.

Strong, D. R., A. V. Whipple, A. L. Child, and B. Dennis. 1999. Model selection for a subterranean trophic cascade: Root-feeding caterpillars and entomopathogenic nematodes. *Ecology* 80:2750–2761.

Swartz, C. 2003. *Back-of-the-Envelope Physics*. Johns Hopkins University Press, Baltimore, MD.

Szymura, J. M. and N. H. Barton. 1986. Genetic analysis of a hybrid zone between the fire-bellied toads, *Bombina bombina* and *B. variegata*, near Cracow in southern Poland. *Evolution* 40:1141–1159.

Teh, C. 2006. *Introduction to Mathematical Modeling of Crop Growth: How the Equations Are Derived and Assembled into a Computer Model*. BrownWalker Press, Boca Raton, FL.

Therneau, T. M., P. M. Grambsch, and V. S. Pankratz. 2003. Penalized survival models and frailty. *Journal of Computational and Graphical Statistics* 12:156–175.

Thomas, L., S. T. Buckland, K. B. Newman, and J. Harwood. 2005. A unified framework for modelling wildlife population dynamics. *Australian and New Zealand Journal of Statistics* 47:19–34.

Thomas, W. R., M. J. Pomerantz, and M. E. Gilpin. 1980. Chaos, asymmetric growth and group selection for dynamical stability. *Ecology* 61:1312–1320.

Thompson, D., M. Lonergan, and C. Duck. 2005. Population dynamics of harbour seals *Phoca vitulina* in England: Monitoring growth and catastrophic declines. *Journal of Applied Ecology* 42:638–648.

Thomson, J. D., G. Weiblen, B. A. Thomson, S. Alfaro, and P. Legendre. 1996. Untangling multiple factors in spatial distributions: Lilies, gophers, and rocks. *Ecology* 77: 1698–1715.

Thornley, J. H. 2002. Instantaneous canopy photosynthesis: Analytical expressions for sun and shade leaves based on exponential light decay down the canopy and an acclimated non-rectangular hyperbola for photosynthesis. *Annals of Botany* 89:451–458.

Tilman, D. 1994. Competition and biodiversity in spatially structured habitats. *Ecology* 75:2–16.

Tiwari, M., K. A. Bjorndal, A. B. Bolten, and B. M. Bolker. 2005. Intraspecific application of the mid-domain effect model: Spatial and temporal nest distributions of green turtles, *Chelonia mydas*, at Tortuguero, Costa Rica. *Ecology Letters* 8:918–924.

———. 2006. Evaluation of density-dependent processes and green turtle *Chelonia mydas* production at Tortuguero, Costa Rica. *Marine Ecological Progress Series* 326:283–293.

Tjørve, E. 2003. Shapes and functions of species-area curves: A review of possible models. *Journal of Biogeography* 30:827–835.

Toms, J. D. and M. L. Lesperance. 2003. Piecewise regression: A tool for identifying ecological thresholds. *Ecology* 84:2034–2041.

Tracey, J. A., J. Zhu, and K. Crooks. 2005. A set of nonlinear regression models for animal movement in response to a single landscape feature. *Journal of Agricultural, Biological, and Environmental Statistics* 10:1–18.

Tucci, M. P. 2002. A note on global optimization in adaptive control, econometrics and macroeconomics. *Journal of Economic Dynamics and Control* 26:1739–1764.

Tufte, E. 2001. *The Visual Display of Quantitative Information*, 2d ed. Graphics Press, Cheshire, CT.

Tukey, J. W. 1977. *Exploratory Data Analysis*. Addison-Wesley, Reading, MA.

Turchin, P. 2003. *Complex Population Dynamics: A Theoretical/Empirical Synthesis*. Princeton University Press, Princeton, NJ.

Tyre, A. J., B. Tenhumberg, S. A. Field, D. Niejalke, K. Parris, and H. P. Possingham. 2003. Improving precision and reducing bias in biological surveys: Estimating false-negative error rates. *Ecological Applications* 13:1790–1801.

Underwood, A. J. 1996. *Experiments in Ecology: Their Logical Design and Interpretation Using Analysis of Variance*. Cambridge University Press, Cambridge, England.

van Veen, F. J. F., P. D. van Holland, and H. C. J. Godfray. 2005. Stable coexistence in insect communities due to density- and trait-mediated indirect effects. *Ecology* 86:1382–1389.

Vandermeer, J. H. and D. E. Goldberg. 2004. *Population Ecology: First Principles*. Princeton University Press, Princeton, NJ.

Venables, W. N. 1998. Exegeses on linear models. 1998 International S-PLUS User Conference. Washington, DC. http://www.stats.ox.ac.uk/pub/MASS3/Exegeses.pdf.

Venables, W. N. and B. D. Ripley. 2002. *Modern Applied Statistics with S*, 4th ed. Springer, New York.

Verzani, J. 2005. *Using R for Introductory Statistics*. Chapman and Hall/CRC, Boca Raton, FL.

Vesk, P. A. 2006. Plant size and resprouting ability: Trading tolerance and avoidance of damage? *Journal of Ecology* 94:1027–1034.

Vigliola, L., P. J. Doherty, M. G. Meekan, D. M. Drown, M. E. Jones, and P. H. Barber. 2007. Genetic identity determines risk of post-settlement mortality of a marine fish. *Ecology* 88:1263–1277.

Vonesh, J. R. and B. M. Bolker. 2005. Compensatory larval responses shift tradeoffs associated with predator-induced hatching plasticity. *Ecology* 86:1580–1591.

Walters, C. J. and D. Ludwig. 1981. Effects of measurement errors on the assessment of stock-recruitment relationships. *Canadian Journal of Fisheries and Aquatic Sciences* 38:704–710.

Wang, D., E. A. Carr, M. R. Palmer, M. W. Berry, and L. J. Gross. 2005. A grid service module for natural-resource managers. *IEEE Internet Computing*, 9:35–41.

Warton, D. I., I. J. Wright, D. S. Falster, and M. Westoby. 2006. Bivariate line-fitting methods for allometry. *Biological Reviews* 81:259–291.

Whittingham, M. J., P. A. Stephens, R. B. Bradbury, and R. P. Freckleton. 2006. Why do we still use stepwise modelling in ecology and behaviour? *Journal of Animal Ecology* 75: 1182–1189.

Wilmshust, J. F., J. M. Fryxell, and P. E. Colucci. 1999. What constrains daily intake in Thomson's gazelles? *Ecology* 80:2338–2347.

Wilson, J. and C. W. Osenberg. 2002. Experimental and observational patterns of density-dependent settlement and survival in the marine fish *Gobiosoma*. *Oecologia* 130:1432–1439.

Wilson, J. A. 2004. Habitat Quality, Competition and Recruitment Processes in Two Marine Gobies. Ph.D. diss. University of Florida, Gainesville.

Wilson, W. 2000. *Simulating Ecological and Evolutionary Systems in* C. Cambridge University Press, Cambridge, England.

Wintle, B. A. and D. C. Bardos. 2006. Modeling species-habitat relationships with spatially autocorrelated observation data. *Ecological Applications* 16:1945–1958.

Wood, S. N. 2001. Partially specified ecological models. *Ecological Monographs* 71:1–25.

———. 2006. *Generalized Additive Models: An Introduction with* R. Chapman and Hall/CRC, Boca Raton, FL.

Yoccoz, N. G. 1991. Use, overuse and misuse of significance tests in evolutionary biology and ecology. *Bulletin of the Ecological Society of America* 72:106–111.

Zar, J. H. 1999. *Biostatistical Analysis*, 4th ed. Prentice Hall, Upper Saddle River, NJ.

Index of R Arguments, Functions, and Packages

Functions in packages that are not automatically loaded are denoted as function [package]. As a reminder, you can search for help on R functions by using? or help (for functions in currently loaded packages); help.search (for functions in locally installed packages); or RSiteSearch (or the Firefox R Site Search extension, http://addictedtor.free.fr/rsitesearch/) for packages on the web.

General Index

abbott, 277n
Abies balsamea. See fir data
accuracy, 158
AD Model Builder, 243, 329
AIC, 209–213, 219, 250, 266,
 268, 270, 287, 291–292,
 313–314; corrected (AICc),
 210–211, 282; quasi-
 (qAIC), 312; weights, 215
allometric growth, 96, 202
analysis of covariance
 (ANCOVA), 300, 304–305,
 313
analysis of variance (ANOVA),
 35, 300, 303, 313
arguments, function, 26, 74,
 183
asymptote, 79–81, 83,
 91–92, 98
asymptotically valid tests, 194
autoregressive model, 323, 344

balloon plot, 42
bar plot, 43, 49
base graphics, 40
base rate fallacy, 111
Bayes factor, 213–214
Bayes' Rule, 107–114,
 177–179
Beers' Law, 94
Bernoulli distribution, 122
beta-binomial distribution,
 120, 126–128, 137,
 140–141, 265–266,
 275, 284, 327
Beta distribution, 115, 120,
 133–134, 137, 140,
 177–178

Beverton-Holt function, 91,
 151–152, 166, 204–205.
 See also Michaelis-Menten
 function
BFGS (Broyden-Fletcher-
 Goldfarb-Shanno) optimi-
 zation, 173, 226, 229n,
 245–247, 293
bias, 158, 162, 164, 166–167;
 tradeoff with variance, 167,
 204, 212
Bible Code, 29
BIC (Bayesian Information
 Criterion), 210–211,
 213–214, 219, 282
bimodal distribution, 44, 134,
 138
binomial distribution, 1, 13,
 115, 120–122, 137, 152,
 156, 170–172, 177–178,
 182, 263, 309, 340, 343
biological significance, 274
block random effects, 317, 325
Brent's method, 229
bubble plot, 42
BUGS, 185–186, 214, 239, 257,
 329, 330–332, 354–355. *See
 also* in the R Index: R2WinBUGS
 package
burn-in, for MCMC chains,
 237, 356

candidate distribution, 232,
 234, 236
catalytic curve, 94
Cauchy distribution, 138, 140
censored data, 56; interval,
 251, 280